Springer Geology

The book series Springer Geology comprises a broad portfolio of scientific books, aiming at researchers, students, and everyone interested in geology. The series includes peer-reviewed monographs, edited volumes, textbooks, and conference proceedings. It covers the entire research area of geology including, but not limited to, economic geology, mineral resources, historical geology, quantitative geology, structural geology, geomorphology, paleontology, and sedimentology.

More information about this series at http://www.springer.com/series/10172

Vladimir Karev · Dmitry Klimov
Konstantin Pokazeev
Editors

Physical and Mathematical Modeling of Earth and Environment Processes

3rd International Scientific School for Young Scientists, Ishlinskii Institute for Problems in Mechanics of Russian Academy of Science

Springer

Editors
Vladimir Karev
Institute for Problems in Mechanics of the Russian Academy of Sciences (IPMech RAS)
Moscow
Russia

Dmitry Klimov
Institute for Problems in Mechanics of the Russian Academy of Sciences (IPMech RAS)
Moscow
Russia

Konstantin Pokazeev
Faculty of Physics
Lomonosov Moscow State University
Moscow
Russia

ISSN 2197-9545 ISSN 2197-9553 (electronic)
Springer Geology
ISBN 978-3-030-08540-7 ISBN 978-3-319-77788-7 (eBook)
https://doi.org/10.1007/978-3-319-77788-7

Softcover re-print of the Hardcover 1st edition 2018

Printed on acid-free paper

This Springer imprint is published by the registered company Springer International Publishing AG part of Springer Nature
The registered company address is: Gewerbestrasse 11, 6330 Cham, Switzerland

Preface

This book, entitled “Physical and Mathematical Modeling of Earth and Environment Processes. 3rd International Scientific School for Young Scientists, Ishlinskii Institute for Problems in Mechanics of Russian Academy of Sciences” is the result of a collaborative work in the frame of the youth scientific conference held at the Ishlinskii Institute for Problems in Mechanics of RAS on November 1–3, 2017. This forum is held on a regular basis and causes great interest in the scientific community. For the third year in a row, more than one hundred scientists have taken part in it, two-thirds of who are young researchers.

The 3rd School, as well as the previous two, promoted to the solution of fundamental scientific problems arising in the study of natural processes in different media, the impact of anthropogenic activities on the environment. Intensive development of research in these areas is due to several factors. The widespread introduction of computer technology has allowed beginning calculation of complex phenomena, previously unavailable for analysis. Creation and improvement of a new generation of geophysical instruments and remote observing systems based on the ship, aircraft, and satellite allowed to obtain a large amount of data to objectively reflect the picture of the processes. International activities including the youth scientific schools are certainly an effective tool for exchange of information and the organizing of interdisciplinary research of environment processes.

One of the central topics for the School is associated with the elaboration of scientific bases of oil and gas production technologies. The creation of new breakthrough approaches to the development of hydrocarbon fields is very important today and requires the involvement of young minds and strength.

During all three Schools, participant’s reports were traditionally accompanied by active discussions which lasted beyond the end of the program sessions. The most interesting and promising areas of research were recognized in the following: the development of geomechanical approach to solving the problems of oil and gas production, physical and mathematical modeling of deformation and fracture of solid media and study of their interaction on the seepage, creation effective mathematical models and experimental base for research of flows in complex heterogeneous liquids, environmental issues, the study of the anthropogenic contribution to the dynamics of natural systems. As a result of the work of the third School, it was decided to publish the most interesting reports as the book.

The book presents the results of theoretical and experimental research of processes in the atmosphere, oceans, the lithosphere and their interaction; environmental issues; problems of human impact on the environment; methods of geophysical research. The conference papers included in the book describe the studies on the dynamics of natural systems, the human contribution to naturally occurring processes, laboratory modeling of these processes, and testing of new developed physical and mathematical models.

A wide range of problems associated with the production of hydrocarbons is the central topic of the book. A special attention is paid to the geomechanical approach to solving these problems. An alternative to the use of hydrocarbons as a main source of energy on the planet in the coming decades is unlikely to be found. At the same time, the resource base of hydrocarbons is quickly depleted, and new non-traditional sources are required. Among them, there is shale oil and gas, Arctic hydrocarbon reserves gas hydrates, as well as oil and gas from deep horizons (more than 5.5 km). "Deep oil" may become the most promising source of expanding the resource base of hydrocarbons according to many experts. At the same time, environmental problems are becoming increasingly acute, and the need arises to create environmentally friendly technologies. The basis for the creation of such technologies can become a geomechanical approach based on the use of a huge reserve of elastic energy stored in a rock mass by controlling the stress–strain state of the formation. The book includes the new results of the experimental and theoretical modeling deformation, destruction and filtration processes in the rocks related to issues of creating scientific fundamentals for new hydrocarbon production technologies. The investigation of the dependence of well stability and permeability of rocks on the stress–strain state in conditions of high rock pressure is represented too.

Our book

- Enriches the understanding of the geophysical processes taking place in various environments (lithosphere, hydrosphere, atmosphere), including anthropogenic, and promotes to the intensification of their studying.
- Includes the results of theoretical and experimental studies on the development of the geomechanical approach to creating new technologies in the field of hydrocarbon production based on controlling the stress–strain state in the reservoir.
- Contains the results of recent research in the field of interactions of the lithosphere, atmosphere, and hydrosphere of various scales, energy exchange of the atmosphere and the ocean, including under anthropogenic influences.
- Expands the understanding problems of providing oil and gas wells stability in the process of their construction and operation, increasing the efficiency of hydrocarbon production including on the ocean shelf, thereby raising the efficiency of training specialists in oil reservoir physics.

- Presents modern methods of research and modeling of various processes in environment, means and ways of monitoring of natural systems, methods of research and forecasting of natural and man-made disasters, and ways to eliminate their consequences.
- Proposes new physical and mathematical models of processes occurring in the environment, both natural and anthropogenic, as well as elaboration of existing ones.
- Contains additional material for specialists working in the oil and gas industry to expand, improve, and disseminate new acknowledgments in this field.

Program Committee of the School, which included the leading scientists on the scientific directions of the School, has conducted peer review of the papers submitted to the book and produced a competitive selection. Ninety-eight works were submitted. As a result of double-blind peer review, 38 papers were selected.

Organization

Organizers of the School

Ishlinskii Institute for Problems in Mechanics of Russian Academy of Sciences
Faculty of Physics of Lomonosov Moscow State University

Organizing Committee

V. I. Karev (Chairman)	Ishlinskii Institute for Problems in Mechanics of RAS, Russia
K. V. Pokazeev (Dep. Chair)	Lomonosov Moscow State University, Russia
A. Yu. Volkova	Lomonosov Moscow State University, Russia
Yu. F. Kovalenko	Ishlinskii Institute for Problems in Mechanics of RAS, Russia
A. L. Levitin	Ishlinskii Institute for Problems in Mechanics of RAS, Russia
E. V. Stepanova	Ishlinskii Institute for Problems in Mechanics of RAS, Russia
K. B. Ustinov	Ishlinskii Institute for Problems in Mechanics of RAS, Russia
T. O. Chaplina (Sc. Secretary)	Lomonosov Moscow State University, Russia
N. I. Shevstov	Ishlinskii Institute for Problems in Mechanics of RAS, Russia

Program Committee

A. A. Babanin	Swinburne University of Technology, Australia
T. O. Chaplina (Sc. Secretary)	Lomonosov Moscow State University, Russia
V. V. Fadeev	Lomonosov Moscow State University, Russia
N. N. Filatov	Northern Water Problems Institute Karelian Research Centre, Russia

Sponsors

Russia Foundation for Basic Research (project № 17-31-10210)

Russian Academy of Science Presidium Program for Fundamental Research I.4P «Deposits of strategic resources in Russia: innovative approaches for their prediction, evaluation and production. Oil from the deep horizons of sedimentary basins as the source of replenishment of resource base of hydrocarbons: theoretical and applied aspects»

School Venue

Ishlinskii Institute for Problems in Mechanics of RAS
119526, Russia, Moscow, Vernadskogo Avenue, 101-1

[1]R. V. Goldstein died on 24/09/2017.

Contents

The Tyrrhenian Continent Ragmentation

Al. A. Schreider[1], A. A. Schreider[2(✉)], and A. E. Sazhneva[2]

[1] Research Institute of Economics and Management in Gas Industry, NIIgazekonomika Company Ltd., Moscow, Russia
[2] Institute of Oceanology, Russian Academy of Sciences, Moscow, Russia
aschr@ocean.ru

Abstract. In the geological past, there was a continental formation in the place of the Tyrrhenian-Ligurian basin, that included Sicily, Sardinia, Corsica, and the Apennines in the southwest of the Italian peninsula. The stretching processes led to rifting, passing into diffuse spreading with the fragmentation of this formation. Calculation of Eulerian poles and rotation angles in the context of complex geological and geophysical interpretation of bottom geomorphology allowed to restore the spatial position of the axes of the initial split of the continental formation and to describe the kinematics of the microplates of analyzed region.

Keywords: West mediterranean paleogeodynamics · Tyrrhenian continent Euler poles

1 Introduction

The tectonic development of the western Mediterranean is inseparably linked with the multi-scale migration and rotations of the different age microplates of the continental and oceanic lithosphere [4, 5, 9, 13, 14, 16, 17]. Among the most important tectonic elements of this region include the Tyrrhenian Sea surrounding are such continental blocks as Corsica, Sardinia, Sicily and Apennine peninsula (Fig. 1). It is important to note that the geological and geophysical data accumulated to date do not contradict with the most diverse, sometimes mutually exclusive, paleogeodynamic constructions.

2 Tectonic Setting Peculiarities

In work [12] it is asserted that in the interval 170-67 million years ago the continental block uniting the islands of Corsica and Sardinia (hereinafter referred as the block CS) adjoined the Iberian Peninsula to the south of the Pyrenees. At a considerable distance from CS Sicily was located, and still further to the east of Sicily there were fragments of the structure of the modern Apennine peninsula. In contrast to that study, in [15] the block of CS 140 million years ago adjoined the Iberian Peninsula to the north of the Pyrenees. Sicily was located east of Corsica and further south of it was situated the south-western fragment of the modern Apennine peninsula. According to the data of [9], the rotation of the CS to its present day meridional position occurs successively for the last 30 million years, and it was shown in [10] that this rotation takes place only in

V. Karev et al. (Eds.): PMMEEP 2017, SPRINGERGEOL, pp. 1–8, 2018.
https://doi.org/10.1007/978-3-319-77788-7_1

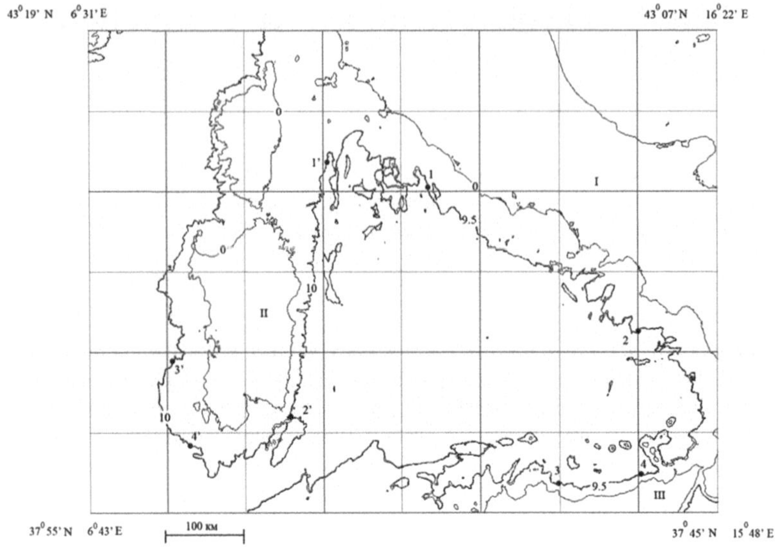

Fig. 1. The present day geographical position of the Apenninian peninsula (I), Corsica and Sardinia - block CS (II) and Sicily (III) along the periphery of the Tyrrhenian Sea. The position of the ends (points 1 and 2, 3 and 4) of the 0.95 km segments isobaths are shown, as well as the position of the conjugate sections of the 1 km isobaths ends (points 11 and 21, 31 and 41). Depths are in hundreds of meters according to [20].

the last 21.5 Ma in connection with the opening of the Ligur-Provencal basin. At the same time, according to [7, 18], this rotation was completed 13–16 million years ago. We add to above mentioned that in [11] the Sicily movement from the central Mediterranean in the northern direction to its current position is demonstrated in the interval 0–35 million years.

At the same time, it remains an open question how was created the Sicily north-eastern extremity which exists to the north-east of the southern tip of the Apennine peninsula at the present day.

The examples show that in the literature there is no consensus on an initial geographic position of Corsica, Sardinia, Sicily, Apennines etc. To restore their possible relative position in the present work a comprehensive paleogeodynamics analysis of the structure of the continental blocks surrounding the basin of the Tyrrhenian Sea is carried out. The basin located between Corsica and Sardinia in the West, Sicily in the South and the Apennine Peninsula in the East and has a triangular shape with a total elongation in longitudinal direction (Figs. 2 and 3). The bottom central part of the basin lies at depths greater than 3 km. The thickness of the continental crust at the periphery of the basin is 25–30 km and dramatically decreases toward its central part, where the continental crust is replaced by discovered by drilling of the ocean crust of 4–5 km [4].

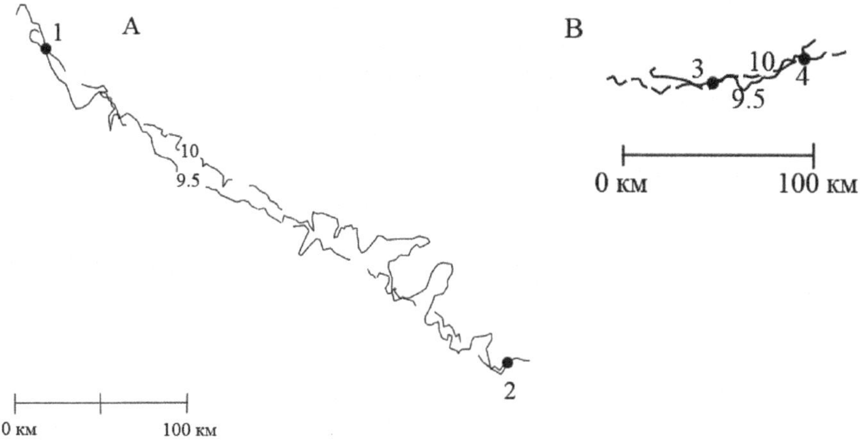

Fig. 2. Coinciding of isobaths 1 km on the slopes of the block CS and isobath 0.95 km on the slope of the Apenninian peninsula (A) and Sicily (B). The position of points 1–4 are shown in Fig. 1. Isobaths are in hundreds of meters by [20].

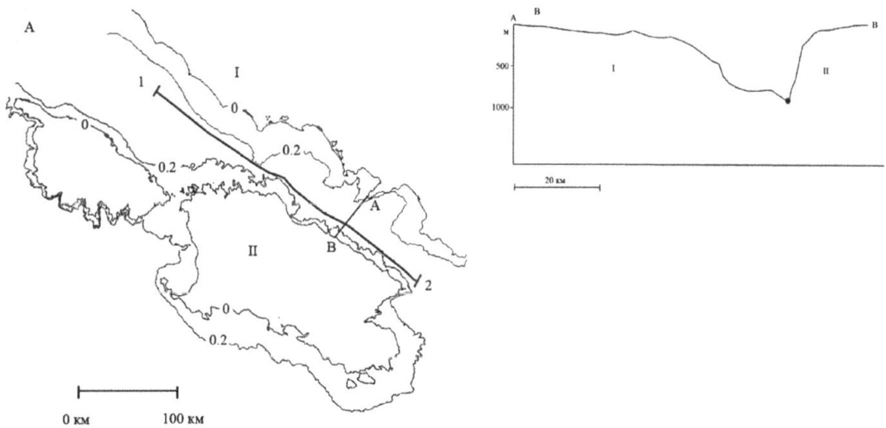

Fig. 3. A- Paleogeodynamic reconstruction of the clamping counter-slopes of the Apenninian Peninsula and the block CS. Bold line indicates the restored portion of the zone split axis. The position of the points 1 and 2 is shown in Fig. 1. B - Paleodepth profile before the split along the line A–B and the docking point of isobaths at the intersection of the line A–B and restored split axis along line 1–2. I and II are the same as in Fig. 1.

Here the consolidated crust directly overlaps by sediments of the upper Miocene – lower Pliocene (7-4 my) in the north-west part of the Vavilov basin, or by the sedimentary formations of the upper Pliocene (2-1, 8 my) in the south-east part of Marsili basin. Note that the North Tyrrhenian basin is adjacent to Ligurian basin with bottom depths of up to 2.8 km.

3 Calculation of Geodynamic Parameters

In the well-known paper [8] was firstly proposed computational method for the best combination of isobaths bounding the slopes of the continents at the edges of the Atlantic ocean. The combination was carried out by method of trials and errors, by minimizing the angular disagreement, measured along the Eulerian latitudes. The methodology is illustrated the principle that the best alignment can be performed for any of the circuits, as installed, or as expected, once constituted a single circuit. By implementing the principle of better alignment, it is possible to achieve reunification and rehabilitation of the primary continuity of all circuits, including isochronous, isobath, isohypse etc.

According to the electronic Bank on the bottom bathymetry [20] built the profiles in the direction perpendicular to the strike of the Tyrrhenian basin slopes with cross profiles distance of 5–10 miles. Analysis of the profiles indicates that almost all of them consist of three parts (Fig. 4). The upper (shallower than 1.5 km) and lower (deeper 2.9 km) parts of the slopes are no constant along the profile steepness. The Central part of the slopes, enclosed in the depth interval 0.7–1.9 km, is steep and has a relatively constant slope along each individual profile.

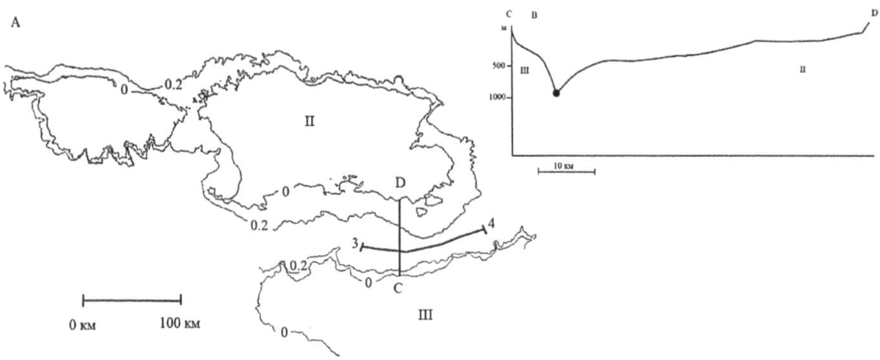

Fig. 4. A - Paleogeodynamics reconstruction of clamping of the counter slopes of Sicily and block CS. Bold line indicates the restored portion of the axis of the zone split. The position of points 3 and 4 is shown in Fig. 1. B - Profile bottom paleorelief before the split along the line C–D and the docking point of isobaths at the intersection of the C–D profile and line 3–4 restored split axis. II and III are the same that in Fig. 1.

Sedimentation in different areas of the Tyrrhenian basin has led to them filling in the sedimentary rocks, accompanied by decreasing angle of slopes due to sedimentation process. The sedimentation is uneven in time and space. Irregularity of it is associated with the distribution and redistribution of areas of drift, and with sediment slumping due to loss of stability. Instability precipitation occurs due to the accumulation of a critical mass, leading to slip along the interface or inside the precipitation along the surface of the foundation when varying the steepness of the slope. At small tilt angles and, ceteris paribus, the movement of sedimentary masses down the slope will be very low velocities.

Estimates show that in order to overcome the adhesion force between layers of sediments (in the Tyrrhenian basin is silt, clay, loam sand) and avalanche breakdown them down the slope with the development of considerable (up to close to 50 km/h) velocities of sliding, ceteris paribus, the slope of the sliding surface should be less than 3° [2]. Among the reasons for initiating the precipitation from sliding down the slope, is important impact on sediment mass exogenous (regular flow) and endogenous (e.g., earthquake) factors.

Based on the foregoing, in the present work, a modified method of E. Bullard was firstly applied for the case of combining the depth contours of the Tyrrhenian sea basin slopes. Numerous testing the connectivity of different areas different and the same isobaths showed that the most suitable for the purposes of paleogeodynamic analysis were portions of the isobaths in the range of 0.7–1.9 km.

In the depth interval 0.8–1.5 km the continental slope is steepest (the average angle of the slope surface greater than 100) and, using the information above about the nature of the sedimentary strata slipping, having the small thickness of sediments (or even completely devoid of them). On this basis, using a modified methodology E. Bullard joined us plots (Fig. 5) the 1 km isobaths on the slope of Sardinia and Corsica and 0.95 km, on the slopes of Sicily and the Apennine peninsula. Calculations of the Euler poles and rotation angles was carried out according to programs incorporated in the software environment, Global Mapper [6] and the principles which are set out in the works [3, 6]. In recent years, in the work [6] develops a methodology of automated selection conjugate plots of depth contours using statistical estimates of the extent of their connectivity.

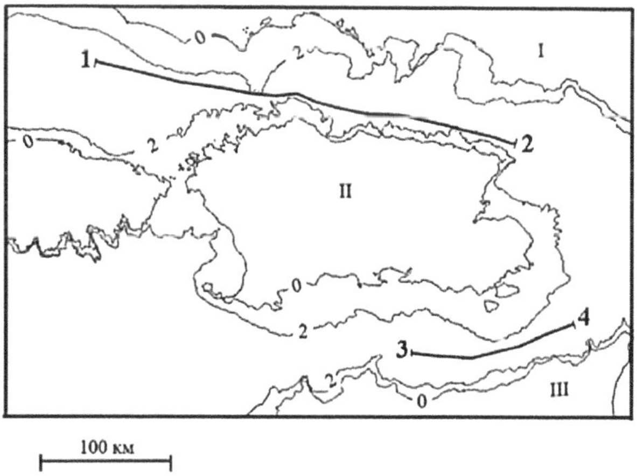

Fig. 5. Paleogeodynamic reconstruction of the Tyrrhenian continent. The lines in bold show the recovered split axis of the continent stretching areas. Designations I–III and the position of the points 1–4 are shown in Fig. 1.

4 Configuration of the Split Axis

According to the calculations at the position of the Euler poles of finite rotation coordinates 45.28° N 12.65° W manages North of 40° N for about 300 km to get a good combination isobath 0.95 km (section between points 1 and 2 in Fig. 1) the Western slope of the Apennine Peninsula and 1 km isobath (the area between the points 1^1 and 2^1 in Fig. 1) the Eastern slope of the block CS with an error in points of digitization ±8.7 km (16 points calculation). The module of the angle of rotation made 28.79° ± 0.7°. The results of the docking of the isobaths are given in Fig. 2A

At the position of the Euler poles of finite rotation coordinates 50.45° N 14.76° W manages the North 390 North latitude. over 100 km (the section between points 3 and 4 in Fig. 1) get a very good alignment of 0.95 km isobath of the Northern slope of Sicily and the 1 km isobath (the area between points 31 and 41 in Fig. 1) South slope block CS with an error in points of digitization ±5.4 km (8 points calculation). The module of the angle of rotation made 23.57° ± 0.4°. The results of docking the isobaths shown in Fig. 2.

Under the proposed approach, the combined plots of depth contours 1 km from the side of block KS and 0.95 km from the Apennine Peninsula (Fig. 3A) and from the side of Sicily (Fig. 4A) allow us to restore the planned configuration of the axes of the zones of continental rifting, which was the consequence of the processes of stretching. Docking of isobaths discussed above allows us to represent the profiles of the upper surface of the neck stretching of the continental crust immediately prior to rupture of the lithosphere and the initialization is split between the Apennine Peninsula and block CS (Fig. 3B), as well as Sicily and CS (Fig. 4B). These profiles show that the region of stretching in both cases had a width of over 150 km, and was more steep in the direction of the CS in the first case and in the direction of Sicily in the second.

The experimental data obtained are important for the recovery of process parameters of the rupture of the continental crust during rifting of the lithosphere in the Western Mediterranean. Splits the continental lithosphere in rifts could occur according to the scheme of Wernicke [19]. Listric quasilinear with sloping fault that separates the plates and reaches the surface, forming a separation zone "hanging" plate from the "underlying".

The result of the paleogeodynamic reconstruction is to restore the split axis of the chipping peripheral areas of the CS block from the Apenninian peninsula and from Sicily. An important factor of reconstruction is the difference of the abutting 0.05 km the latter fact probably reflects the circumstance of sliding along the lithospheric fault plane and downgliding in the process of stretching peripheral areas of the continental crust CS from the peripheral regions of Sicily and the Apennines in accordance with modification Schreider [6] scheme of Wernicke [19].

5 Paleogeodynamics of the Tyrrhenian Continent Split

Integrated geophysical analysis of available materials shows that paleogeodynamics evolution of the Western Mediterranean resulted in the formation of the Miocene continental array in place of the modern Tyrrhenian sea. Calculations of the Euler poles and rotation angles for the first time allowed to restore the position of the split axis of the stretched zone between the CS and the Apennines (Fig. 3) and the other split axis of the stretched zone between the CS and Sicily (Fig. 4). Joint analysis of two independent reconstructions allowed for the first time to make a general reconstruction of the Tyrrhenian continent (Fig. 5), which initially consisted of the Islands of Corsica, Sardinia, Sicily, parts of the Apennine peninsula. In the process of its destruction formed the basin of the Tyrrhenian sea - the youngest in the Western Mediterranean basins. Basin began to develop [5, 14, 17] in the late Miocene, about 9 million years ago when continental destruction has led to the emergence of the deep Vavilov basin. The pool continued to evolve until the turn of 4 million years ago, and after a break of 2 million years to the South-East it occurred the opening of the Marsili basin [4, 5], development of which is still ongoing. The sea floor spreading in the Tyrrhenian sea was diffuse [1, 5, 14, 17] and spread from North-West to South-East in several areas, which has created an extremely complex structure closed at present, precipitation of the consolidated crust in the form of horst and grabens, superimposed on the under crust.

Acknowledgement. The work was done in the framework of the State assignment, project 0149-2018-0015. While methodological issues of combining conjugate isobaths was supported by the Russian Foundation for Basic Research Project № 17-05-00075.

References

1. Verzjbitsky, E.V., Schreider, A.A., Stenshinsky, S.B.: Diffusive spreading rates in the Tyrrhenian sea. Izv. Ross. Akad. Nauk, Ser. Geol. **55**(8), 53–64 (1992). [in Russian]
2. Zhmur, V.V., Sapov, D.A., Nechaev, I.D., et al.: Intensive gravitational currents in the near bottom layer of the ocean. Izv. Ross. Akad. Nauk, Ser. Fiz. **66**(12), 1721–1726 (2002). [in Russian]
3. Zonenshtein, L.P., Lomize, M.G., Ryabukhin, A.G.: Practical Manual on Geotectonics. Moscow State University, Moscow (1990). [in Russian]
4. Khain, V.Y.: Tectonics of the land and ocean. Nauchnyi Mir, Moscow (2001)
5. Khain, V.Y., Limonov, A.F.: Regional geotectonics. Geos, Moscow (2004). [in Russian]
6. Schreider, Al.A.: Formation of the Black Sea deep basin. Nauchnyi Mir, Moscow (2011). [in Russian]
7. Argnani, A., Savelli, C.: Cenozoic volcanism and tectonics in the southern Tyrrhenian sea: space-time distribution and geodynamic significance. Geodynamics **27**, 409–432 (1999)
8. Bullard, E., Everett, J., Smith, A.: The fit of continents around Atlantic. Philos. Trans. Roy. Soc. Lond. **258A**, 41–51 (1965). Symposium on Continental Drift
9. Faccenna, C., Becker, W., Lucente, P., Joliet, L.: History of subduction and back arc extension in the central Mediterranean. Geophys. J. Int. **145**, 809–820 (2001)
10. Gattacceca, J., Deino, A., Rizzo, R., et al.: Miocene rotation of Sardinia: new paleomagnetic and geochronological constraints and geodynamic implications. Earth Planet. Sci. Lett. **258**, 359–377 (2007)
11. Goes, S., Giardini, D., Jenny, S., et al.: A recent tectonic reorganization in the south-central Mediterranean. Earth Planet. Sci. Lett. **226**, 335–345 (2004)
12. Handy, M., Schmid, S., Bouscquet, R., et al.: Reconciling plate-tectonic reconstructions of Alpine Tethys with the geological-geophysical record of spreading and subduction in the Alps. Earth Sci. Rev. **102**, 121–158 (2010)

13. Maffione, M., Speranza, F., Faccenna, C., et al.: A synchronous Alpine and Corsica-Sardinia rotation. J. Gophys. Res. **113**, 25–35 (2008)
14. Savelli, C., Schreider, A.A.: The opening processes in the deep Tyrrhenian basins of Marsili and Vavilov, as deduced from magnetic and chronological evidence of their igneous crust. Tectonophysics **190**, 119–131 (1991)
15. Scalera, G.: The Mediterranean as a slowly nascent ocean. Ann. Geophys. Suppl. **49**, 451–482 (2006)
16. Scheepers, P., Langereis, C., Zijderveld, J., Hilgen, F.: Paleomagnetic evidence for a Pleistocene clockwise rotation of the Calabro-Peloritan block (southern Italy). Tectonophysics **230**, 19–48 (1994)
17. Schreider, A.A., Yastrebov, V.S., Rimsky-Korsakov, N.A., Savelli, C.: Indagini e campionature di dettaglio di affioramenti rocciosi submarini dei Monti Baronie (Mar Tirreno): Primi risultati. Mem. Soc. Geol. It. **36**, 91–98 (1988)
18. Speranza, F., Villa, I., Sagnotti, L., et al.: Age of the Corsica-Sardinia rotation and Liguro-Provencal basin spreading: new paleomagnetic and Ar/Ar evidence. Tectonophysics **347**, 231–251 (2002)
19. Wernicke, B.: Low angle normal faults in the Basin and Range Province: nappe tectonics in an extending orogeny. Nature **291**, 645–648 (1981)
20. ftp://topex.ucsd.edu/pub/srtm15_plus/

Long Waves Influence on Polarization Ratio for Microwave Backscattering from the Sea Surface

Alexandr Zapevalov(✉)

Marine Hydrophysical Institute RAS, Sevastopol 299011, Russia
sevzepter@mail.ru

Abstract. The effect of slopes created by long waves on the resonance backscattering of microwave radio waves analyzed. The analysis is carried out within the framework of the Gaussian model of slopes distribution. The polarization ratio increases by approximately 10% as the wind speed increases up to 5 m/s if the sounding is performed along the direction of wind. If the sensing is accomplished across the direction of wind, as the wind speed tends to 5 m/s the polarization ratio increases to approximately 6%. The effect of the presence of long waves weakly depends on the incidence angle.

Keywords: Remote sensing · The microwave radiation · Resonance scattering Polarization ratio · Sea surface · Long waves

1 Introduction

The number of oceanographic spaceborne microwave sensors is continuously growing. Expands the range of parameters determined on the basis of data of remote sensing of the ocean. All this requires a detailed understanding of the characteristics of electromagnetic fields scattered by sea surface.

Models of the normalized radar cross-section of sea surface at incidence angles between 25° and 70° are usually treated as resonant (Bragg) scattering mechanism [1]. The resonance condition relates the wave number of radio wave and surface wave [2]

$$K_R = 2k \sin\theta, \tag{1}$$

where K_R is the wave number of surface resonance waves, k is the radar wave number, θ is the incidence angle.

If the resonant waves propagate along the flat surface, the normalized radar cross-section is proportional to the surface elevation spectrum at the resonance wave number [3, 4]

$$\sigma^0_{pp} = 8\pi k^4 \left|G_{pp}(\theta)\right|^2 \left[\Psi\left(\vec{K}_R\right) + \Psi\left(-\vec{K}_R\right)\right], \tag{2}$$

where pp is the polarization (the first index corresponds to emitted wave, the second one does to be received), $\left|G_{pp}(\theta)\right|^2$ is the polarization dependent reflection coefficient,

V. Karev et al. (Eds.): PMMEEP 2017, SPRINGERGEOL, pp. 9–16, 2018.
https://doi.org/10.1007/978-3-319-77788-7_2

$\Psi(\vec{K}_R)$ is the two-dimensional (Cartesian) wave-number spectrum of sea surface displacement. Resonance scattering of radio waves create surface waves traveling along the direction of sensing in forward or reverse direction. In this case, the polarization ratio

$$R^0(k,\theta) = \sigma^0_{HH}(k,\theta)/\sigma^0_{VV}(k,\theta) \tag{3}$$

is described by the expression

$$R^0(k,\theta) = |G_{HH}(\theta)|^2/|G_{VV}(\theta)|^2. \tag{4}$$

It should be noted that the polarization ratio (4) does not depend on the level of the sea surface roughness. This gives a fundamental possibility of remote determination of physical and chemical characteristics (temperature, salinity) of sea water. The presence on the sea surface waves longer than resonant waves modifies expression (2) [5]. Resonant waves propagate along the curved surface. The local incidence angle θ changes. As the result, the resonance condition (1) changes as well as the value of the spectrum $\Psi(\vec{K}_R)$ changes. Accordingly, the polarization ratio changes.

The main objective of this work is an analysis of the impact of long surface waves on polarization ratio. Long waves are waves with lengths much greater than the length of resonant waves.

2 Normalized Radar Cross-Section in the Presence of Long Waves

We assume that the size of radar spot on sea surface significantly exceeds the size of long waves. In this case, the effect of long waves on the backscattered signal can be taken into account by averaging expression (2) over the entire range of their slopes. Within the frame of the two-scale model, where resonant waves superposed on longer tilting waves in [3] obtained

$$\sigma^L_{pp} = 8\pi k^4 \int |G_{pp}(\theta,\delta)|^2 \left[\Psi(\vec{K}_R) + \Psi(-\vec{K}_R)\right] P(\xi_x,\xi_y)\, d\xi_x d\xi_y, \tag{5}$$

where δ is the out-of-plane tilt angle, ξ_x and ξ_y are orthogonal components of sea surface slope, $P(\xi_x,\xi_y)$ is the two-dimensional probability density function of slopes. Physically (5) is the averaging with the weight being proportional to $P(\xi_x,\xi_y)$ (the expected tilt of long waves).

The slopes ξ and their angles β are related by the nonlinear expression

$$\xi = \tan\beta. \tag{6}$$

Usually it was assumed that the slopes are small and one can use approximation $\xi \approx \beta$. An analysis of the errors associated with this approximation was carried out in [6].

It was shown that the variance slopes is of 8% higher than the variance of angles. The discrepancies for the higher statistical moments are higher. As a consequence, the probability density function of slopes, which is described using the Gram-Charlier series [7], differs significantly from the probability density function of angles.

We use a model in which the approximation $\xi \approx \beta$ is not used. This model has the form

$$\sigma_{pp}^{L} = \int \sigma_{pp}^{0}(k, \theta - \beta_{\uparrow}, \alpha) P(\beta_{\uparrow}) d\beta_{\uparrow}, \tag{7}$$

where $\beta_{\uparrow}$ is the angle of inclination of the sea surface in the direction of sounding, α is azimuth angle. To realize averaging (7), the spectrum of sea surface elevation represented as an explicit function of the incidence angle is required. The transition is carried out by using the normalization condition. According to this condition, the integral of the evaluation spectrum of elevations in all its variables is equal to the variance of elevations. In final form (6) received [5]

$$\sigma_{pp}^{L} = 2\pi k^2 \int \left|G_{pp}(\theta_\beta)\right|^2 \frac{S(2k \sin \theta_\beta)}{\sin \theta_\beta \cos \theta_\beta} \Theta(k2 \sin \theta_\beta, \alpha) P(\beta_{\uparrow}) d\beta_{\uparrow}, \tag{8}$$

where $\theta_\beta = \theta - \beta_{\uparrow}$, S is the one-dimensional wave-number spectrum of the sea surface displacement, Θ is the directional spreading function. The function Θ describes the angular distribution of wave energy. This function satisfies the normalization condition $\int_{-\pi}^{\pi} \Theta(\alpha) d\alpha = 1$.

In the general form, the coefficient $\left|G_{pp}(\theta)\right|^2$ depends on the relative dielectric constant of sea water. If the sounding is carried out in the centimeter range of the radio wave, we can use the simplified form proposed in the paper [8]. A simplified form is obtained under the assumption that the dielectric constant of sea water be equal to 81. For the vertical VV and horizontal HH polarizations, the coefficients $|G_{VV}(\theta)|^2$ and $|G_{HH}(\theta)|^2$ have the forms

$$|G_{VV}(\theta)|^2 = \frac{\cos^4 \theta (1 + \sin^2 \theta)^2}{(\cos \theta + 0.111)^4}, \tag{9}$$

$$|G_{HH}(\theta)|^2 = \frac{\cos^4 \theta}{(0.111 \cos \theta + 1)^4}. \tag{10}$$

We assume that long compared to resonance waves are longer than 20 cm. It is usually assumed that waves are long, if their length is 3–4 times greater than the length of the sensing radio waves. According to [9, 10], the variance of the slopes generated by these waves is approximately 1/3 of the total variance of slopes. Up/down wind $D(\xi_u)$ and crosswind $D(\xi_c)$ variance of slopes were determined from optical measurements [11]

$$D(\xi_u) = 0.000 + 0.00316W \pm 0.004, \tag{11}$$

$$D(\xi_c) = 0.003 + 0.00192W \pm 0.002, \tag{12}$$

where W is wind speed at a height 10 m. Variances of slopes and angles are related as $D(\beta) = 0.92D(\xi)$ [6]. Here and further, the indices "u" and "c" correspond to directions up/down and cross wind. If the expression applies to both up/down and cross wind components of the slopes, the subscript is missing.

The wind speed dependences for standard deviation angles β (rms(β)) obtained in this way are shown in Fig. 1. The same figure shows the similar dependence, obtained according to measurements of wind speed and direction from the NASA Scatterometer and ocean reflectance from the POLDER (POLarization and Directionality of the Earth Reflectances) multi-directional radiometer [12].

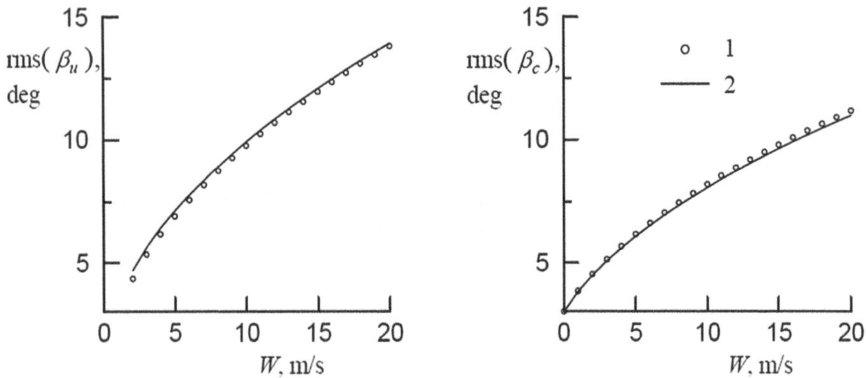

Fig. 1. The dependence for standard deviation angles rms(β) on wind speed W. Curve 1 is constructed according to [11], curve 2 – according to [12]

We introduce the non-dimensional parameter

$$\chi_{pp}(\theta, W) = \sigma^L_{pp}/\sigma^0_{pp} \tag{13}$$

which describes influence of long waves on normalized radar cross-section. We also assume that the angles β have a Gaussian distribution. The dependence of the parameter χ_{pp} on the incidence angle is shown in Fig. 2.

The dependencies $\chi_{pp} = \chi_{pp}(\theta)$ shown in Fig. 2 are constructed for three wind speeds. It can be seen that the effect of slopes produced by long waves is more pronounced on vertical polarization than on horizontal polarization.

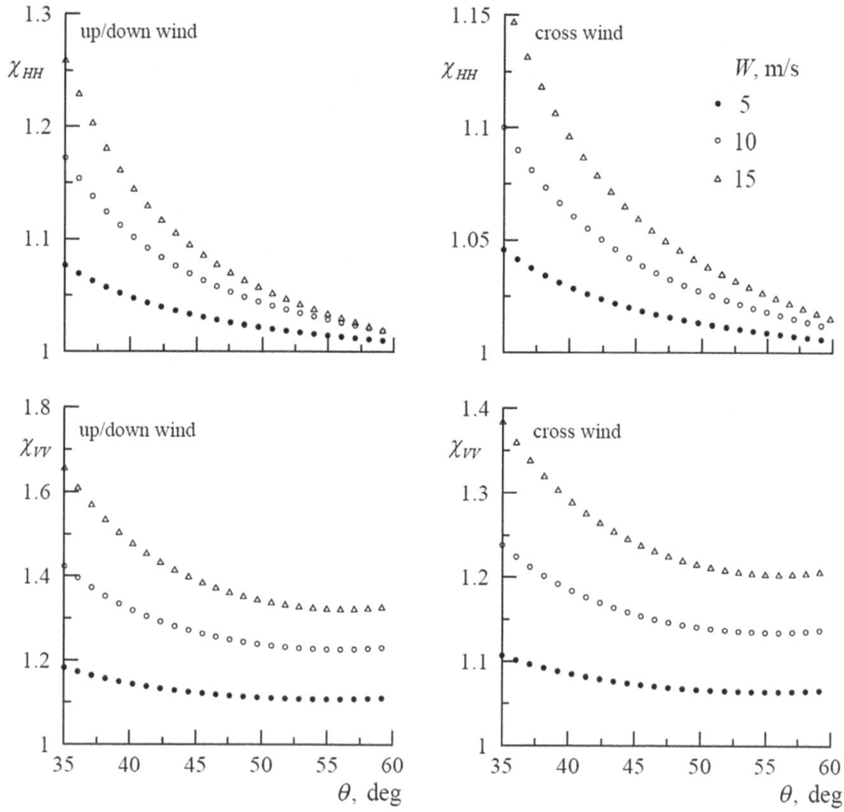

Fig. 2. The dependence of radar cross-section χ_{pp} on wind speed W

3 Polarization Ratio in the Presence of Long Waves

In the presence of long waves the polarization ratio can be represented in the form

$$R^L = \frac{\sigma^L_{HH}}{\sigma^L_{VV}} = \frac{\int \left|G_{HH}(\theta_\beta)\right|^2 \frac{S(2k \sin \theta_\beta)}{\sin \theta_\beta \cos \theta_\beta} \Theta(k2 \sin \theta_\beta, \alpha) P(\beta_\uparrow) d\beta_\uparrow}{\int \left|G_{VV}(\theta_\beta)\right|^2 \frac{S(2k \sin \theta_\beta)}{\sin \theta_\beta \cos \theta_\beta} \Theta(k2 \sin \theta_\beta, \alpha) P(\beta_\uparrow) d\beta_\uparrow}. \quad (14)$$

Polarization ratios (14) predicted by multiscale composite models based on the resonance theory scattering models that include the effects of long-wave tilt.

Directional spreading function is the narrowest on the scale of the dominant waves [13]. As the wave number increases, it expands. In the range of gravity-capillary waves the angular distribution of wave energy become isotropic. This suggests that the change in the length of the resonant waves do not lead to significant changes in the function of angular distribution and we can eliminate the function $\Theta(2k \sin \theta_\beta, \alpha)$ in (14).

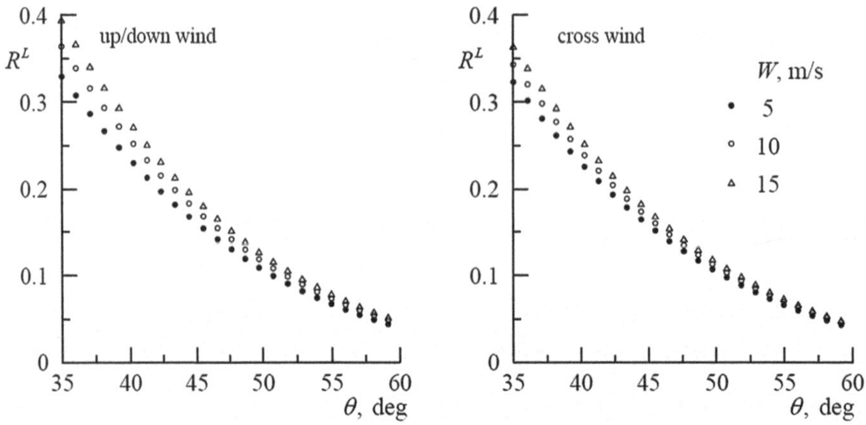

Fig. 3. The dependence of the polarization ratio R^L on the angle of incidence θ and the wind speed W

Graphs on Fig. 3 are obtained within assumption that waves of the gravitational-capillary range are resonant. According to [14], in this range the spectrum of surface waves is approximated by the dependence

$$S(K) \sim K^{-3}. \tag{15}$$

In order to quantify the change in the polarization ratio with the change in wind speed, we investigate the parameter

$$\psi(\theta, W) = R^L/R^0, \tag{16}$$

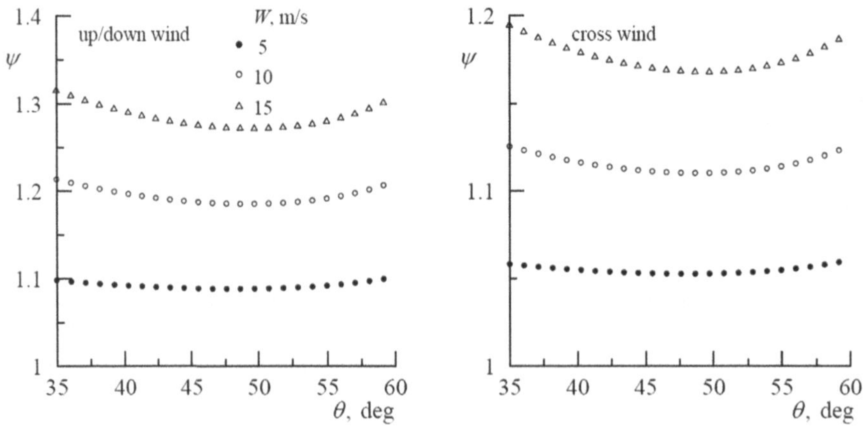

Fig. 4. The dependence of the parameter ψ on the angle of incidence θ and wind speed W

where $R^0 = \sigma^0_{HH}/\sigma^0_{VV}$. The parameter ψ describes the change in polarization relations compared to the situation when the resonance waves propagate along the flat surface. How the parameter ψ changes when the wind speed or the incidence angle change is shown in Fig. 4.

4 Conclusion

An analysis of the effect of long surface waves on the field of backscattered radio waves, when sounding the sea surface in the microwave range is made. The situation is considered when resonant scattering is the dominant mechanism of creating a scattered field. The calculations were performed for incidence angles from 35° to 60°. With wind speed increasing at 5 m/s, the polarization ratio increases by approximately 10% if the sounding is performed along the wind direction. If the sounding is performed across the wind direction, then with wind speed increase at 5 m/s, the polarization ratio is increased approximately on 6%. The effect of the presence of long waves weakly depends on the incidence angle.

Acknowledgements. This work was carried out in the context of the State project № 0827-2014-0011.

References

1. Valenzuela, G.: Theories for the interaction of electromagnetic and ocean waves - a review. Bound. Layer Meteorol. **13**(1–4), 61–85 (1978)
2. Bass, F.G., Fuchs, I.M.: Wave Scattering from Statistically Rough Surfaces. Pergamon Press, Oxford (1978)
3. Thompson, D., Elfouhaily, T., Chapron, B.: Polarization ratio for microwave backscattering from the ocean surface at low to moderate incidence angles. In: Proceedings of the 1998 IEEE International Geoscience and Remote Sensing Symposium, IGARSS 1998, pp. 1671–1673. IEEE, Seattle (1998)
4. Kudryavtsev, V., Hauser, D., Caudal, G., Chapron, B.: A semiem-pirical model of the normalized radar cross-section of the sea surface 1. Background model. J. Geophys. Res. **108** (C3), 8054 (2003). https://doi.org/10.1029/2001JCOO1003
5. Zapevalov, A.S.: Bragg scattering of centimeter electromagnetic radiation from the sea surface: the effect of waves longer than Bragg components. Izv. Atmos. Ocean Phys. **45**(2), 253–261 (2009)
6. Zapevalov, A.S., Lebedev, N.E.: Simulation of statistical characteristics of sea surface during remote optical sensing. Atmos. Ocean. Opt. **27**(6), 487–492 (2014)
7. Zapevalov, A.S., Pustovoitenko, V.V.: Modeling of the probability distribution function of sea surface slopes in problems of radio wave scattering. Radiophys. Quantum Electron. **53** (2), 100–110 (2010)
8. Plant, W.J.: A two-scale model of short wind-generated waves and scatterometry. J. Geophys. Res. **91**, 10735–10749 (1986)
9. Hollinger, J.P.: Passive microwave measurements of sea surface roughness. IEEE Trans. Geosci. Electron. (GE) **9**, 165–169 (1971)

10. Wilheit, T.T.: A model for the microwave emissivity of the ocean's surface as a function of wind speed. IEEE Trans. Geosci. Electron. (GE) **17**(4), 244–249 (1979)
11. Cox, C., Munk, W.: Measurements of the roughness of the sea surface from photographs of the sun glitter. J. Opt. Soc. Am. **44**(11), 838–850 (1954)
12. Bréon, F.M., Henriot, N.: Spaceborne observations of ocean glint reflectance and modeling of wave slope distributions. J. Geoph. Res. **111**(56), 1–10 (2006). C06005
13. Zapevalov, A.S.: On the estimation of the angular energy distribution function of dominant sea waves. Izv. Atmos. Ocean Phys. **31**(6), 802–808 (1996)
14. Monin, A.S., Krasitskii, V.P.: Phenomena on the Ocean Surface. Gidrometeoizdat, Leningrad (1985). [in Russian]

Elimination of Hydrocarbons Spills on Water Objects and Fluorescent Diagnostics of Water Pureness

T. O. Chaplina[1,2(✉)] and E. V. Stepanova[2]

[1] Faculty of Physics, Lomonosov Moscow State University, Moscow, Russia
chaplina_to@inbox.ru
[2] A. Yu. Ishlinsky Institute for Problems in Mechanics RAS, Moscow, Russia

Abstract. This paper presents a review of existing methods and sorbents for elimination of hydrocarbons spills from the water surface. The characteristics and operating principles of the existing devices are described. Experimental study of the sorption capacity of various substances and the usage of these substances for the elimination of hydrocarbon contamination of water surface are made. The measurements have shown that raw sheep's wool absorbs at least 6 times its weight in oil and other hydrocarbons. The sample of the device for liquidation of hydrocarbons spills on the surface of water is designed. The designed device can be used to eliminate surface spills of various hydrocarbons on water objects in two modes: from a board of surface craft and from waterfront.

Keywords: Sorption capacity · Sorption · Hydrocarbons · Sheep's wool Fluorescent diagnostics

1 Introduction

Currently, one of the greatest dangers to the ecological stability of water basins is posed by spills of oil and products of its processing. The volume of oil emergency discharges is intimidatingly high and reaches a critical value in the human-induced disasters (tanker wrecks, oil field accidents, pipelines ruptures) (see Table 1).

The oil spills in the coastline waters have the most significant destructive impact on ecosystems. The consequences of these disasters affect aquatic life, flora and fauna of the coastal zone. To restore ecological communities in such areas a long time (decades) and large financial injections are required [1, 2]. Only in Russia due to emergency situations and non-compliance with technological discipline, loss of oil and oil products annually reaches up to 5 million tons.

The study of methods for localization and liquidation of oil spills requires special attention and development of an additional complex of measures for collection and recovery of hydrocarbons thrown in the environment to mitigate the possible negative impacts.

In this work, a review of existing methods and sorbents for the elimination of hydrocarbons from the water surface is performed, the characteristics and operating

V. Karev et al. (Eds.): PMMEEP 2017, SPRINGERGEOL, pp. 17–27, 2018.
https://doi.org/10.1007/978-3-319-77788-7_3

Table 1. Oil spills in history [3].

Location of the spill	Year	Vessel/source of spill	Volume of spill, 10^3 tons
Gulf of Mexico	2010	Deepwater Horizon, platform	up to 800
Kerch Strait	2007	Volgoneft-139	2
Arabian Sea	2003	Tasman Spirit	60
Coast of the Philippines	2002	Solar 1	1,8
Bay of Biscay	2002	Prestige	64
Gulf of Aden	2002	Limburg	300
Guanabara Bay	2000	Petrobras pipeline	up to 1000
Atlantic coast of Angola	1991	ABT Summer	260
Ligurian Sea	1991	M/T Haven	0,2
Kuwait coast	1990	Oil deposit terminal	up to 1500
Alaska coast	1989	Exxon Valdez	40
New Scotland coast (Canada)	1988	Odyssey	0,2
Indian Ocean, RSA coast	1983	Castillo de Bellver	252
Persian Gulf	1983	Nowruz, platform	250
Gulf of Mexico	1980	Ixtoc I, platform	467
Ionian Sea	1980	Irenes Serenade	20
Caribbean Sea	1979	Atlantic Empress and Aegean Captain	290
Atlantic coast of France	1978	Amoco Cadiz	223
Gulf of Oman	1972	Sea Star and Horta Barbosa	115
Seven Stones Reef, Scilly Isles	1967	Torrey Canyon	119

principles of the existing devices are described. The authors carried out experimental study of the sorption capacity of different materials in relation to oil products and water, and applicability for liquidation of pollution of water surface with hydrocarbons. Sheep's wool is proposed for usage as a sorbent, as this material absorbs minimum 6 times their weight in oil and other hydrocarbons (diesel fuel, aviation and sunflower oil); a device to assemble liquid hydrocarbons (oil, diesel fuel, etc.) from the water surface is designed and its laboratory tests are undertaken.

2 Sorbents and Devices for Water Surface Purification

2.1 Sorbents for Water Surface Purification Review

There are several methods for liquidation of oil pollution of waters. Conventionally, the process can be divided into two stages: the localization of the spill, and then its amassment and disposal. The most widely used method is the mechanical collection of oil products after spill localization with the booms, which provide sufficient oil-layer thickness for the efficient collection. Spill spot, surrounded by oil booms, may be towed to a safer location which is also more comfortable to work [4]. Along with the mechanical method, thermal method of elimination (oil combustion) is widely used, but its scope is limited to a short period of time directly after the accident – when the oil layer has sufficient thickness and before the formation of water-oil emulsions (Fig. 2).

Physico-chemical method is based on the use of dispersants and sorbents and is implied when mechanical oil collection is impossible (at small layer thickness, in the presence of an immediate threat to ecologically vulnerable areas). With the use of dispersants natural dispersion of oil is activated to facilitate its removal from the water surface. Sorbents gather oil and form clumps that can be mechanically removed. The most delicate water purification is conducted by biological method: specific microorganisms process oil and oil products [5]. This method is often used after the applicability of physico-chemical, mechanical and other methods have already been exhausted. Nowadays nearly two hundreds of sorbents are produced and used for oil spill liquidation in the whole world. The sorbents can be attributed to four basic types: synthetic, inorganic, mineral-organic and organic. The efficiency and quality of the sorbents is determined by the basic characteristics: capacity to absorb oil, degree of hydrophobicity, buoyancy after the sorption, the ability for oil desorption, the availability of mechanisms for regeneration or disposal of the sorbent.

The experimental studies of sorption capacity and other properties of various materials in relation to oil products are performed, on prospect to use for the elimination of water surface pollution with hydrocarbons [6]. Table 2 shows that the following types of sorbents are presented on offer: sorbents of mineral origin (inorganic and mineral-organic),

Table 2. Basic characteristics of commercial sorbents.

Sorbent material	Capacity to absorb oil *, kg/kg	Capacity to absorb water, kg/kg	Oil desorption rate, %	Admission to absorb 1 ton of oil, kg/ton
Polypropylene (fiber)	13–25	3–6	70–80	40
Carbamide foam plastic	40–60	5–10	60–80	25–30
Urea-formaldehyde resin	30–50	4,6–10,0	70–80	33
Vermiculite	8–12	2–17	–	100–120
Turf	6–7	1,6	0	110
Moss	2–8	2	10–25	213
Graphite	40–60	0,2	–	25–30
Wool	6–9	1–4	65–80	150–270
Cellulose	7–9	5–7	–	100–200
Bacteria	5–10	–	–	200

* – According to the manufacturers

of animal or vegetable origin (organic and mineral-organic), synthetic and polymer sorbents (synthetic) [7].

In the conducted experiments it is revealed that 1 g of untreated sheep's wool provides to absorb 6 g of oil and other hydrocarbons (diesel fuel, sunflower oil). Consecutive images of the process of absorption of oil and the graph of sorption dynamics are presented in Fig. 1 (oil volume on the surface $V_o = 50$ ml).

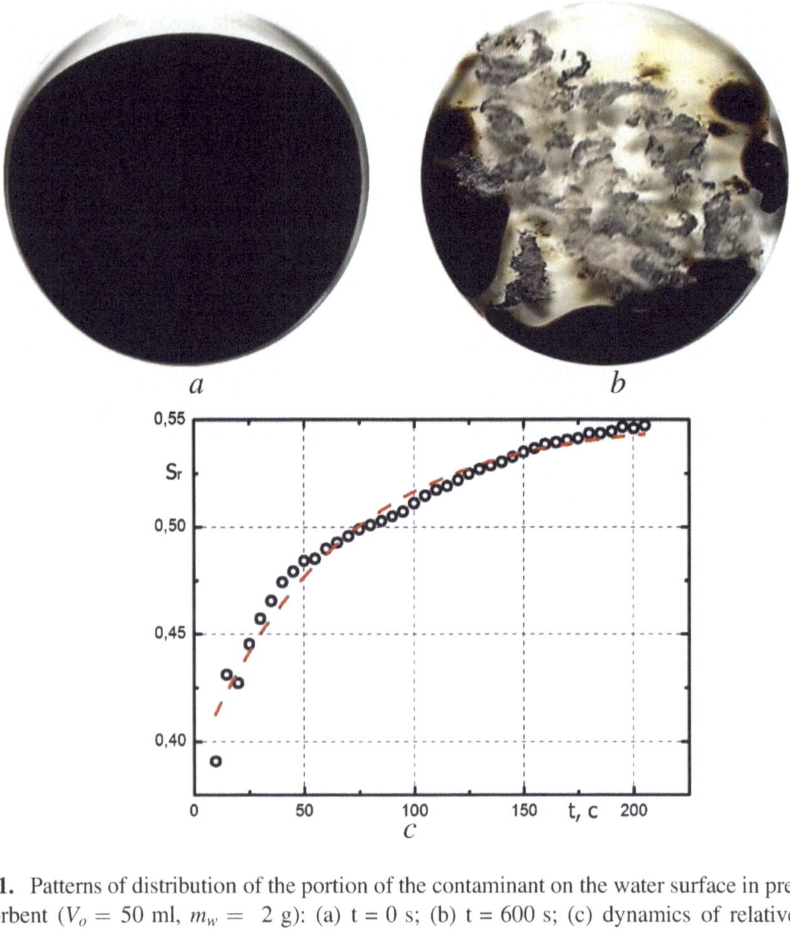

Fig. 1. Patterns of distribution of the portion of the contaminant on the water surface in presence of sorbent ($V_o = 50$ ml, $m_w = 2$ g): (a) t = 0 s; (b) t = 600 s; (c) dynamics of relative area occupied with oil patch.

After placing the fixed quantity of sorbent material on the surface of the water (in the experiment illustrated in Fig. 1a and b $m_w = 2$ g of natural sheep's wool), the sorption process disrupts the uniformity of thickness of the free surface oil patch.

The deposition of the sorbent on the surface of the two-layer oil-water system leads to the formation of flows and currents, under the action of which the distribution of the contaminating admixture changes, the shape of the oil spot on the surface changes. The evolution of the sorption process leads to a significant change in shape of the surface oil patch, it gets split into separate fragments. With time and exhaustion of the sorption capacity of the material (or rather its parts located in direct contact with the admixture/contaminant) a stationary pattern of distribution of the remaining portion of the contaminant on the surface is formed (Fig. 1b).

Oil droplets are detached from the initial patch. The observed velocity of droplets allows to evaluate the velocities of currents induced by sorption. In the sorption process the velocities of several detached drops of oil are estimated. The calculated values of the displacement velocities for the selected drops are shown in Fig. 2. It should be mentioned that individual droplets have different sizes (visible transverse size of the drops from 0.6 up to 3.3 mm).

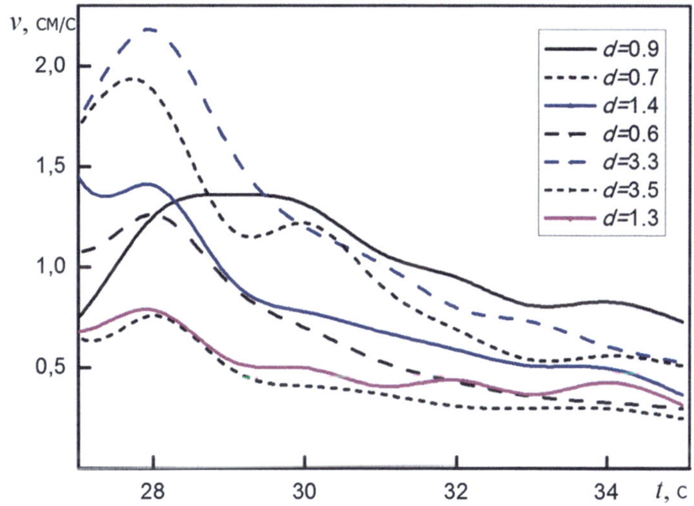

Fig. 2. Oil droplets measured displacement velocities in the sorption-induced currents.

Due to the hydrophobic properties and low density sheep's wool together with the absorbed substance is not immersed in the water depth, and therefore such sorbent can be easily removed from water surface mechanically. Thus, it is proposed to use relatively cheap raw material (sheep's wool which is partly destroyed by farmers along with other waste) as a sorbent. Wool is hydrophobic, resulting in little water sorption, and has a high sorption coefficient for pure hydrocarbons and their water emulsions. The captured oil is firmly associated with fibers of wool; the resulting compact volume can be easily mechanically extracted from the water surface.

2.2 Existing Devices for Water Surface Purification Analysis

The development and improvement of technical means for removal of water surface oil pollution have long engaged in the Russian and international institutions and commercial companies. For the elimination of oil spills various designs of collecting devices are used (most consist of a combination of different devices for hydrocarbons collecting).

Skimmers are designed for oil collection from water surface. Depending on the type and quantity of spilled oil and weather conditions different types of skimmers (design and operation principles) are effectuated. Oil-collecting devices are classified according to the method of movement or mounting principles: stationary mounted; self-propelled; portable or towed at various floating crafts. The principles of action are: threshold (the flow of the fluid surface layer through the barrier in a container with a lower liquid level), oleophilic (these skimmers is based on the ability of certain materials to expose the oil and oil products coating), hydrodynamical and vacuum.

Another device for removing oil from the water surface is the so-called rope-mop. It consists of an oil absorbent belt, which is made of polyurethane strands stretched as a circumferential pile in the radial direction through the suspension cable. The absorbent belt passes between two rotating rollers that squeeze the oil and it is drained into the tray from which it is pumped into the separate reservoir [8]. Other known device for removal of oil and oil products from the water or earth surface is a mat of oil adsorbent material [9]. Glass wool fibers coated with a mixture of silica and starch is used as an adsorbent material. Disadvantages of the device are the low adsorptive capacity and required high qualification to operate with the wool fibers system securely, which reduces performance.

The other known method of water and soil surface cleaning from petroleum products contamination assumes the implementation of devices made in the form of a cotton-comprising sorbent mat, the outer side of which is covered with a thin layer of transformer or machine oil [10]. The sorbent layer in the mat is fixed between the layers of synthetic or rare netting cotton fabric or cotton mesh. The distance between the threads of the fabric or the size of cells must be less than size of sorbent particles. The disadvantage of this device is the low adsorptive capacity. In addition, cotton is easily moisturized and mats are immersing down the water surface, before the oil sorption capacity is exhausted.

There also exists a device for removal of oil and oil products from water surface, comprising a rotating drum with a hydrophobic surface and the surrounding system for the removal and collection of the product [11]. The disadvantage of this device is a significant influence of ambient temperature on the performance of the device, – as when it is operating at night, in the cold autumn and spring periods and during winter, – due to the increase of the oil product viscosity the absorptive capacity of the hydrophobic shell is reduced, which reduces the device performance.

Also, for liquidation of oil spills, a device containing an outer shell of wool felt, which surrounds a core of finely divided dried pine bark, is used [12]. A disadvantage of the device is the low adsorptive capacity.

3 Preliminary Sample of the Device for Collecting Liquid Hydrocarbons from the Water Surface

As a result of the conducted experiments, the preliminary sample of the device for collecting liquid hydrocarbons from the water surface is developed. The device can be applied to eliminate various surface hydrocarbon spills on water objects. Features and advantages of the announced device are explained by the example of the implementation and illustration, which schematically shows a general view of the device (Fig. 3).

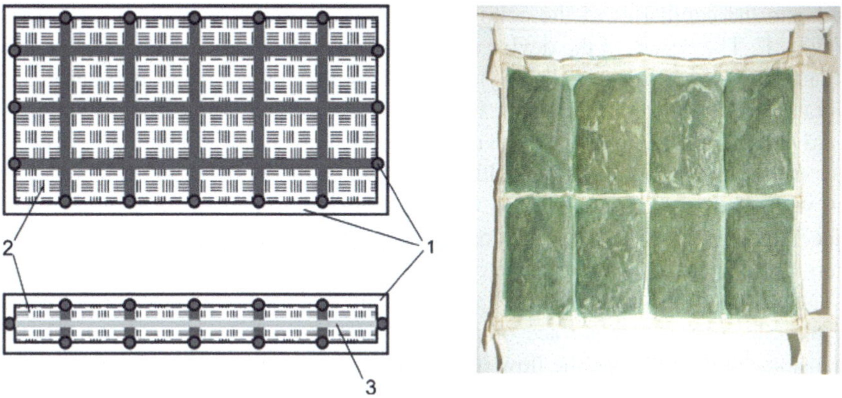

Fig. 3. Preliminary sample of the device for collecting liquid hydrocarbons from the water surface: (a) – scheme, (*b*) – photo.

A device for separating from the water and collecting liquid hydrocarbons includes a frame made of hollow elements or mesh 1, filled with natural sheep's wool 2 (washed or unwashed). The frame can be made of various geometric shapes in the form of a mat in the form of a rectangular parallelepiped (as shown in Fig. 3) or any other. The amount of wool 2 is separated by a waterproof barrier 3, located perpendicular to the direction of the smallest size of the frame [13].

Waterproof barrier can be made of polyethylene or fluoroplastic, the outer frame of the device is made removable. In one part of the device, separated by the barrier, cellulose fiber can be supplemented in an amount of from 10 to 70% of the wool volume. Properties of wool – its hydrophobicity and high sorption ability to hydrocarbons, including mixtures with water and even emulsions – serve as a basis for the use of wool as a sorbent for elimination of pollution by oil and oil products.

Table 3. Basic technical parameters of the device for collecting liquid hydrocarbons.

Overall dimensions, m	0,4 × 0,8
Mass of item, g	515 ± 36
Mass of basic sorbent, g	320 ± 21
Mass of supplementary sorbent, g	33 ± 5

The partition of the volume of wool in the device by the barrier perpendicular to the direction of the smallest size of the frame, aimed to the thorough use of the fibers volume in the process of sorption of hydrocarbons. The consistent use of the device from different sides eliminates the water sorption by the entire volume of sorbent, which increases the buoyancy of the device and its efficiency after flipping (Table 3).

In particular cases, to improve the efficiency of collecting liquid hydrocarbons, it is advisable to add cellulose fiber to the basic sorbent (sheep's wool). The cellulose fiber has a greater sorption capacity (cellulose fiber absorbs up to 9–10 times its weight in hydrocarbons). Along with the increase in the sorption capacity the supplementation of cellulose fibers extends the temperature range of the device application, as water infiltrated into the capillaries of the cellulose fibers, freezes at temperatures significantly below 0 °C. The design of the device including removable frame reduces consumption of materials, which are not directly involved in the process of water purification, facilitates and optimizes the processes of device storage and transportation.

Advantage of the developed device for collecting liquid hydrocarbons from the water surface is the possibility of multiple (up to three times) use: after filling the frame cells with hydrocarbons the device can be removed from the contaminated water surface, and then put back to operation after the removal of the collected contaminant material (oils, petroleum products, etc.), for example, by pressing.

The device can be used, depending on the conditions of the occurred pollution, as a mat placed on the water surface or partially submerged barrier or screen. The unit dimensions allow to reduce the flow of pollutants in outer area, when used as a barrier or screen to outline the tainted region.

Such devices could be used for cleaning and removing oil-containing liquids, when heavy equipment is not recommended to use, or in conjunction with it for increasing the purification rate. The device provides the possibility of operation in two modes: from a board of surface craft and from waterfront. Also, the developed device can be used as a preventive measure for potentially hazardous facilities: tank farms, transport companies, companies for air and sea transportation and any industry which is associated with the use of liquid hydrocarbons, accompanied by risks of potential leaks and spills. The cycle of the device usage allows removal of the collected contaminant material and repetitive use (at least three times).

The staple advantage of the developed device for collecting liquid hydrocarbons from the water surface is relatively low cost and environmental safety of its use, as well as the possibility of multiple (up to three times) use.

4 The Tests of the Developed Device for Collecting Liquid Hydrocarbons (Oil, Diesel Fuel, etc.) from the Water Surface in Laboratory Conditions

4.1 Measurements of the Hydrocarbon Sorption Capacity of the Device

The procedure of measurement of the sorption capacity of the device (mat) in relation to the hydrocarbons includes placement of mat on the surface of liquid hydrocarbons in a reservoir. The exposure time of the mat in the reservoir with hydrocarbons on each side was 10 min. Thereafter the mat was removed and weighed. The sorption capacity coefficient in relation to hydrocarbons have been determined by the formula:

$$K_{sorb} = m_{sorb}/m_{init} \tag{1}$$

where m_{sorb} is the mass of the absorbed substances (petroleum, oils, diesel fuel), m_{init} is the initial mass of the mat, which is calculated from a simple ratio $m_{sorb} = M - \mathrm{m}_{init}$ in which M is the total mass of the mat together with the collected substance. Table 4 presents the values of the sorption capacity coefficient in relation to hydrocarbons for the tested samples.

Table 4. Sorption capacity coefficient for the tested samples.

Sunflower oil	Aviation oil	Diesel fuel (summer)	Oil
7,4	6,9	4,9	7,7

4.2 Measurements of the Water Sorption Capacity of the Device

To determine the coefficient of water sorption samples the device of different sizes, each of the mats was weighed and placed in an inflatable reservoir half-filled with water (water volume 4800 L). The exposure time of each mat was 60 min. Then the samples are removed, weighed and the coefficient of water absorption is calculated based on the obtained data:

$$K_{water} = M_w/M_{init} \tag{2}$$

where M_{init} is the initial mass of the sample (mat), $M_w = M_{sorb} - M_{init}$ is the mass of absorbed water, M_{sorb} is the total mass of the sample with water.

4.3 Evaluation of the Results of Hydrocarbons Collecting by the Developed Device

For the analysis of water samples taken at various stages of the process of water purification from oil pollution the method developed in the laboratory of "Laser spectroscopy of water media and laser biophotonics" of the M.V. Lomonosov Moscow State University is used. The method is based on the procedure of normalization of the oil and oil products fluorescence intensity to the intensity of the spectral band of Raman (combination) scattering of light by molecules of the media – in our case, hexane or water [14, 15]. The results of the purification rate measurements are presented in Table 5.

Determination of the residual oil concentration is carried out to assess the efficiency of waters purification from oil pollution and also to study the sorbent properties [15, 16].

In most practically important situations, the required parameter to measure is the amount of oil and other hydrocarbons remaining in the water after cleaning by every particular sorbent. The sorption capacity measurements (see formula (1)) can be compromised by the additional substances absorbed along with the hydrocarbons in

Table 5. The results of the purification rate measurements by the extraction method.

Type of contaminant/admixture	Initial concentration, mg/l	Purification rate, %
Diesel fuel (summer)	25,4	35
	41,0	36
Diesel fuel (winter)	12,8	40
	20,6	37
Siberian Light oil	20,2	55
	32,3	74

uncontrolled proportions. This parameter (amount of remaining oil and other hydrocarbons) can be measured more accurately.

A laboratory experiment allows to use as a reference the same admixture which the water is contaminated with before the beginning of the purification process. The technique is called the method of internal mark, and can significantly improve the accuracy of fluorescence measurements and to represent the fluorescence intensity in unified quantities. The experiments have shown a high sensitivity of the method and the ability to provide acceptable measurement accuracy due to the application of the procedure of normalization.

In field conditions the exact type of hydrocarbon pollution has hit water masses and how the properties of an admixture have changed during the contact with water. The favorable situation for the concentration evaluation (to know the precise type and composition of admixture/contaminant) occurs very rarely, so in every particular case a specific choice of the calibration parameters is required.

5 Conclusions

The experimental studies of the sorption capacity of various sorbents and their applicability for the elimination of pollution of the water surface by hydrocarbons are made. The following materials are studied as sorbents: sheep's wool, cellulose fiber, peat, biological product "DOP-UNI", as hydrocarbons crude oil, diesel fuel, aviation and sunflower oil are used.

It is found experimentally that raw sheep's wool absorbs at least 6 times its weight in oil and other hydrocarbons (diesel fuel, aviation and sunflower oil). The preliminary sample of device for collecting liquid hydrocarbons from the water surface that can be applied to eliminate surface spills of various hydrocarbons on water objects is developed.

During the preliminary testing of the sample device for collecting liquid hydrocarbons (oil, diesel fuel, etc.) with surface water under laboratory conditions, it is shown that the sample effectively collects hydrocarbon contamination from the water surface due to the high sorption capacity of wool and its hydrophobicity. The device can be successfully used to eliminate surface spills of various hydrocarbons on water objects in two modes: from a board of surface craft and from waterfront [13].

The main advantages of the developed device for collecting liquid hydrocarbons from the water surface are relatively low prime cost and environmental safety of its usage.

Acknowledgement. The work was supported by the Foundation of assistance to development of small forms of enterprises in scientific-technical sphere (Foundation for the promotion of innovation, FASIE), Contract № 1463ГС1/22762.

References

1. Analysis of the situation of environmental pollution by oil products. Group of companies "Expert". http://www.expertyug.ru/analitics/33-analiz-situacii-zagrjaznenija/
2. Stepanova, E.V., Chaplina, T.O.: Modeling oil spreading in the vortex flow of water and oil spill response by using fleece. Theoret. Appl. Ecol. **2**, 108–115 (2015)
3. https://neftegaz.ru/analisis/view/7509-Krupneyshie-razlivy-nefti-v-istorii-chelovechestva
4. Prevention and liquidation of oil spills and oil products, Protection of the environment, nature management and environmental security. St. Petersburg (2003)
5. Belik, E.S., Rudakova, L.V., Kalashnikova, M.E.: Estimation of the effectiveness of the application of a biosorbent based on carbonizate in the process of degradation of petroleum hydrocarbons. Theoret. Appl. Ecol. **1**, 22–26 (2013)
6. Chaplina, T.O., Voloshina, O.V., Stepanova, E.V., Fadeev, V.V.: The fluorescent quality control of purification of water polluted by oil based on sheep wool sorbent. Process. GeoMedia **2**, 81–92 (2016)
7. Veprikova, E.V.: Features of water purification from petroleum products using oil sorbents, filtering materials and active coals. J. Sib. Fed. Univ. Chem. **3**, 285–304 (2010)
8. Cormac, D.: Fighting marine pollution with oil and chemicals. Transport Moscow (1989)
9. Bernhardt, B.: Agent for binding oil present in liquids. Patent DE4140247
10. Antropova, O.N., Leu, S.L.: Method of surface cleaning water and soil contamination by oil and petroleum products. Patent RU 2091539
11. Karavaev, I.I.: Device for removing oil from the water surface. Patent RU 1618836
12. Geles, I.S.: Method of cleaning the surface of water from oil and petroleum products. Patent RU 2115468
13. Chaplina, T.O., Stepanova, E.V.: Device for collecting liquid hydrocarbons. Patent RU 169140
14. Fadeev, V.V.: Possibility of standardisation of normalized fluorescent parameters as a measure of organic admixtures concentration in water and atmosphere. In: Proceedings of the International Symposium "Envirosense", Munich, Germany, Proc. SPIE, vol. 3821, pp. 458–466 (1999)
15. Rajakovic, V.: Efficiency of oil removal from real wastewater with different sorbent materials. J. Hazard. Mater. **143**(1), 494–499 (2007)
16. Suni, S.: Use of a by-product of peat excavation, cotton grass fibre, as a sorbent for oil-spills. Mar. Pollut. Bull. **49**(11), 916–921 (2004)

Investigations of Internal Waves in the Seas of Russia and in the Central Atlantic

K. S. Grigorenko[1(✉)] and S. M. Khartiev[2]

[1] Institute of Arid Zones Southern Scientific Center RAS,
41 Chekhova street, Rostov-on-Don 344006, Russia
klim_grig@mail.ru

[2] Southern Federal University, 105/42 Bolshaya Sadovaya Street,
Rostov-on-Don 344006, Russia

Abstract. The hydrodynamic regime of the seas and oceans plays an important role for mankind's economic activity and the formation of the Earth's climate. Internal waves are observed everywhere in the ocean and permeate all of its thickness, and therefore are an important component of hydrodynamics of water masses. Internal waves have a great importance for solving a number of applied problems, such as underwater navigation, transportation and extraction of hydrocarbons, hydraulic engineering, fishing, as well as for biology, sedimentation and optics of the ocean. The purpose of the investigation is to study the influence of various hydrological conditions of the northern and southern seas of Russia on the formation of modes of internal waves and their comparison with ocean conditions by the example of the hydrological situation of the central Atlantic.

Keywords: Internal waves · Kinematic characteristics · Dispersion curves
Underwater navigation

1 Introduction

Internal waves investigation interest arose during the heyday of oceanology in the 70's. XX century., At the same time, basic mathematical models, describing the dynamics of internal waves, were derived, from which the theoretical directions of investigations of internal waves originate. Recent years, due to the rapid development of information technology mathematical modeling and instruments for measuring the parameters of the marine environment, classical models receive a new tool base for development.

Waves have a wide frequency range of existence - from the inertial frequency to the Väisälä-Brent frequency. Hydrological conditions have an exceptional influence on the formation of various modes of existence of internal waves. Thus, the main issue of this work is how the hydrological conditions of different parts of the World Ocean affect the lengths and periods of internal waves.

The following tasks were accomplished:

1. Software development based on the equations of hydrodynamics, which allows to calculate the kinematic characteristics and to construct the modal composition of internal waves directly in the expeditionary conditions on board the vessel.

V. Karev et al. (Eds.): PMMEEP 2017, SPRINGERGEOL, pp. 28–34, 2018.
https://doi.org/10.1007/978-3-319-77788-7_4

2. Investigation of the average long-term and inter-seasonal variability of the hydrological conditions for the formation of internal waves in the Barents and the Black Seas, comparison of the conditions of the northern (Barents) and southern (Black) seas of Russia with the conditions of the central part of the Atlantic Ocean.
3. Approbation of the developed methods and reconstruction of the vertical density profile by surface manifestations of internal waves.

2 The Mathematical Statement of the Problem

Free internal waves are investigated within the framework of continuously stratified sea and in the presence of horizontal density diffusion. Equations and boundary conditions describing the dynamics of internal waves in a dimensionless form, with allowance for the "hard cover" approximation, in the case of an ideal fluid are described in [1–3].

$$\frac{\partial \mathbf{V}}{\partial t} = -\frac{1}{\rho_0}\nabla p + \mathbf{F}\, div\mathbf{V} = 0, \tag{1}$$

$$\frac{\partial p}{\partial t} + \frac{d\rho_0}{dz}v_z = 0 \tag{2}$$

$$v_z = 0 \text{ when } z = -1;\ v_z = 0 \text{ when } z = 0, \tag{3}$$

where the dimensionless values are related to the dimensional values as:

$$(\overline{x}, \overline{y}, \overline{z}, k) = (x, y, z, \overline{k})\, H, \quad (\overline{t}, \sigma, f) = \sqrt{H/g}(t, \overline{\sigma}, \overline{f}), \tag{4}$$

Here $\mathbf{V} = \{v_x, v_y, v_z\}$ - velocity vector of wave perturbations. $\mathbf{F} = \{fv_y, -fv_x, -\rho/\rho_0\}$ - body forces vector, p, ρ - wave perturbations of pressure and density, respectively; $\overline{f} = 2\Omega \sin\varphi$ - Coriolis parameter; Ω - angular velocity of rotation of the Earth; φ - the latitude; g - acceleration of gravity; $\overline{\sigma}$ - wave frequency; $\overline{k}$ - horizontal wave number.

The solution of the homogeneous boundary value problem (1)–(3) can be represented in the form:

$$[v_x, v_y, v_z, p, \rho] = [U(z), V(z), W(z), P(z), R(z)] \exp(i(kx - \sigma t)) \tag{5}$$

Substituting the expression (5) into Eqs. (1), (2) and the boundary conditions (3), we obtain for a given frequency σ the eigenvalue problem of the following form (the prime denotes the derivatives with respect to z).

$$W'' - N^2(z)W' - k^2\frac{\sigma^2 - N^2(z)}{\sigma^2 - f^2}W = 0 \tag{6}$$

$$W = 0 \text{ when } z = -1 \tag{7}$$

3 Expeditionary Investigations and Work with Electronic Databases

The initial hydrological data on the vertical density stratification of the Black Sea were obtained during the expeditions of the research vessel "Deneb" of the Southern Scientific Center of the Russian Academy of Sciences in 2011. The total number of processed stations was 189. The work was carried out in the Azov Sea, the Russian sector of the Black Sea and the territorial waters of the Abkhazia. The thermohaline characteristics of water masses were investigated by oceanographic CTD probes SBE-19 and SBE-19 + V2. The obtained data was processed by the software of the manufacturer of the probe Sea-Bird Electronics Inc. Cartographic data was carried out by ArcGIS geoinformation package software.

Calculations of the internal waves kinematic characteristics in the Barents Sea are based on the data of the Climatic Atlas of the Seas of the Arctic 2004 [4]. The initial hydrological data are divided into one-degree squares for which averaging was performed and one vertical density profile for the given month for the entire history of observations in this square. 199 vertical density distribution profiles were constructed for all seasons. Dispersion curves of internal waves were constructed for all seasons based on vertical density distributions, total - 22 one-degree squares, respectively 88 sets of dispersion curves characterizing the kinematic characteristics of internal waves. 32 sets of internal wave decay decrements for eight one-degree squares for all seasons were calculated.

Expedition in the Atlantic Ocean was carried out in cooperation between the Institute of Arid Zones of the SSC RAS, the Institute of Oceanology named after P.P. Shirshov RAS Institute of Hydrodynamics named after M.A. Lavrentyev the RAS Siberian branch and the Institute of Applied Physics RAS. The expedition in October-November 2012 was conducted on the vessel of the Russian Academy of Sciences "Academik Sergei Vavilov". The works were carried out with the help of the ship's standard equipment: SBE – 19 + probe and the LADCP RDI Workhorse 300 kHz current profiler. Three investigation areas were performed along the route of the vessel in the Kane Pass, in the Romash fault (1 °S) and in the Veema channel. In 2015, during the 39th and 40th cruises of the R/V "Academik Sergei Vavilov", measurements were made in the fracture zones of the Mid-Atlantic Ridge Veema and Dolrams [5].

Investigations of the internal waves kinematic characteristics in the Sea of Okhotsk were carried out on the basis of data on the thermohaline structure of water masses in the cross-section of the cape Aniva - cape Dokuchaev in 2003, obtained during SakhNIRO expeditions [6].

4 Seasonal Variability of Hydrological Data and Kinematic Characteristics of Internal Waves

The hydrophysical situation of the Viasal-Brent frequency and density fields and in the Black Sea during the expedition period of 2011 was slightly different from the long-time average annual, described in [7], it led to the formation of another mode of internal waves in the north-eastern part of the Black Sea water area. The absolute values of the Väisälä-Brent frequency in the density transition zone in June are higher by 10–25 cycles/h, in October a similar to a summery density stratification is observed, respectively, the frequency maxima are also higher than the long-time average annual of 10–15 cycles/h. In the main density transition zone, the observed values of the buoyancy frequency maxima are also higher: by 7–15 cycles/h in June and 2–7 cycles/h in October. A more stable stratification of water masses leads to the development of higher-frequency internal waves. Resonance zones are situated on internal waves dispersion curves, whereas on long-time average annual dispersion curves [8], such zones are absent (See Fig. 1.)

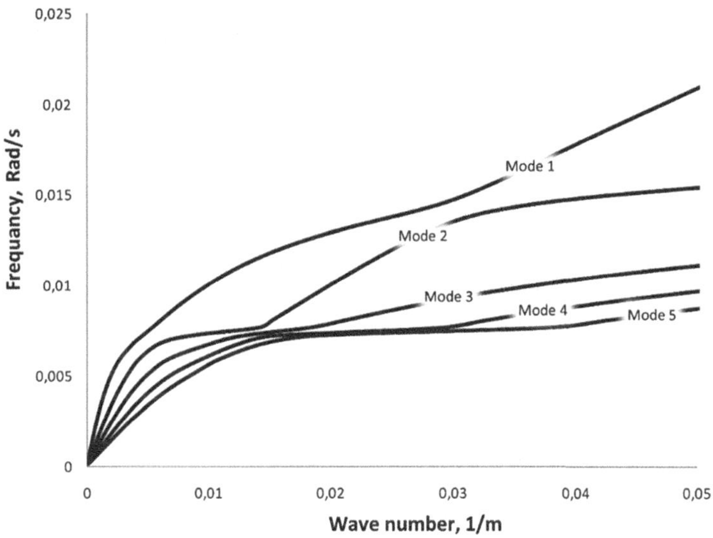

Fig. 1. Internal waves dispersion curves, based on density stratification data, obtained in June 2011, onboard research vessel "Deneb" in the Black sea.

The Barents Sea is extremely heterogeneous in the distribution of internal wave characteristics during the year and in space, however, some trends in the spatial distribution of the kinematic characteristics of internal waves can be identified. Obviously, the higher the density gradient and the maximum of the Väisälä-Brent frequency extends the frequency range of the existence of internal waves. Therefore, in the summer season, the internal wave oscillation frequency will be the largest in the square of 76 °N. and 30 °E, where warm Atlantic water masses spread maximally to the north. At summer seasonal density transition zone forms, density gradients and internal waves frequencies increase. In off-season time the highest Väisälä-Brent frequencies are obtained on the depths of border between Atlantic and Arctic water masses, and have smaller values than at summer in the seasonal density transition zone. At that time of year anomalies appear near the critical frequencies, which are called resonance zones [9]. Internal waves dispersion curves from the Barents sea, 71 N 51 E, depth 170 m at summer are shown on Fig. 2.

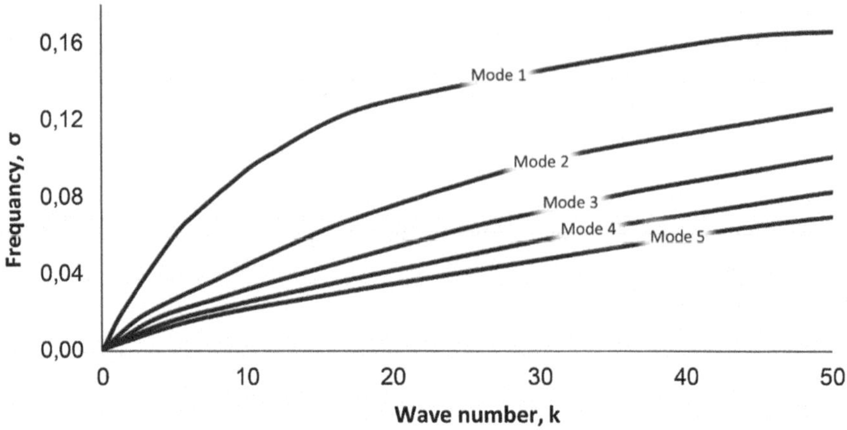

Fig. 2. Internal waves dispersion curves in dimensionless form, based on density stratification data in the square 71 N 51 E, at summer, calculated on data from the Climatic Atlas of the Seas of the Arctic 2004

Internal waves existing in regions with a stable vertical density structure containing a density transition zone with large gradients will decay more rapidly than in regions with a smooth density increasing with depth, however, for the same lengths, internal waves in areas with stable stratification will have more high frequencies.

At the 74th degree of the northern latitude the inertial period of the internal waves approaches the period of a semidiurnal tide, therefore the existence of characteristic internal waves with low frequencies in the Barents Sea is impossible. This fact directly affects the vertical modal structure of internal waves, even at the lowest frequencies close to inertial, the maxima of the horizontal and vertical velocity components are tied to horizons with maximum density gradients

Smooth increasing of dispersion curves without resonance zones is characteristically for dispersion curves, constructed on density stratification data in the Sea of Okhotsk. Despite the small seasonal density stratification changes we obtain principled differences in the wave motion modes, in November more internal wave modes have periodical decay mode than in June.

Surface area in the Cane Gap investigation area in 2012 was much sweetened because of the weather conditions, in the Veema Canal surface layer was much mixed, and density gradients were smaller than in Cane Gap and Romanche Fracture zone, and vertical density stratification profile was similar to an exponential function. Internal waves dispersion curves in Cane and Romanche investigation areas are similar, and internal waves in Veema canal have much smaller frequencies with equal lengths. Dispersion curve of the internal wave of the first mode in Veema canal has similar increasing character with the second mode of the internal wave in Cane and Romanche [10]. Internal waves dispersion curves, based on density stratification data from station 2477, obtained in 2012, onboard research vessel “Akademik Sergei Vavilov” in the Atlantic ocean are shown on Fig. 3.

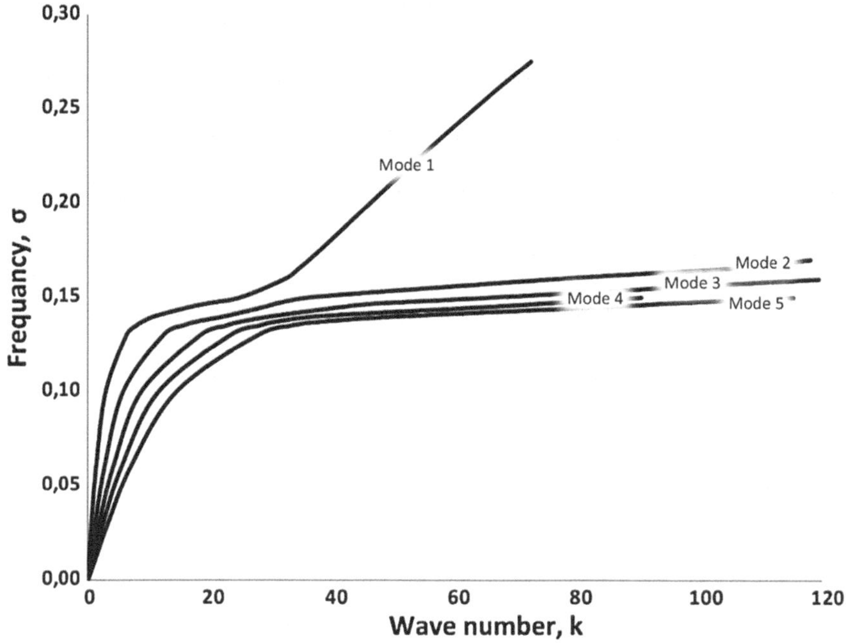

Fig. 3. Internal waves dispersion curves in dimensionless form, based on density stratification data from station 2477, obtained in 2012, onboard research vessel "Akademik Sergei Vavilov" in the Atlantic Ocean.

According to the results of expeditions in 2015 it was found out that between the stations located at a distance of 200–250 km along the meridian, there is a spatial variability of hydrological conditions and, respectively, modes of existence of internal waves. The southern stations are located closer to the axis of the zone of intertropical convergence (the equator), due to which in this region there is a higher amount of precipitation, desalination of the surface layer of the ocean. This leads to higher density gradients, the main pycnocline is located deeper and thicker, which in turn allows the development of high-frequency internal waves within it.

Acknowledgments. The project was carried out within the RFBR project № 17-08-00085 "Processes in the deep channels connecting the East and West Atlantic basins", Research project "Present-day situation and long-term variability of the coastal ecosystems of the southern seas of Russia" State registration CITIS № 01201363187 and SFedU Internal Project-07/2017-4 "Development of methodological bases and recommendations for integrated management of the coastal zone of the Azov Sea in conditions of growth of dangerous exogenous processes, recreational stress, climate variability".

References

1. Monin, A.S., Nauka, M. (eds.) Physics of the Ocean: Ocean Hydrodynamics, vol. 2, p. 456 (1978)
2. Cherkesov, L.V., Ivanov, V.A., Khartiev, S.M.: Introduction into Hydrodynamics and Wave Theory. 264 p. Gidrometeoizdat, St. Petersburg (1992)
3. Soloviev, A.N., Khartiev, S.M., Solovieva, A.A., Grigorenko, K.S., Matishov, D.G.: Internal gravity and acoustical waves parameters analysis based on "the climatic Atlas of the arctic regions seas 2004". SSC RAS Bull. **6**(1), 24–32 (2010)
4. Matishov, G.G., Zuev, A.N., Golubev, V.A., Adrov, N.M., Timofeev, S.F., Karamushko, O. V., Pavlova, L.V., Braunshtejn, A.A., Fadjakin, O.J., Buzan, A.P., Moiseev, D.V., Smoljar, I., Lokarnini, R., Tatushko, R., Bojer, T., Levitus, S.: Climatic Atlas of the Seas of the Arctic 2004. 148 p. Silver Spring, MD (2004)
5. Morozov, E.G., Demidova, T.A., Grigorenko, K.S., Gricenko, A.M., Zajachkovskij, A.O., Makarenko, N.I., Tarakanov, R.Y., Ul'janoval, M.O., Churin, D.A.: Measurements of bottom currents in Atlantic submarine channels in the 36 cruise of "Akademik Sergei Vavilov" research vessel. Oceanology **53**(6), 851–853 (2013)
6. Shevchenko, G.V., Chastikov, V.N.: Seasonal and interannual variations of oceanological conditions in the southwestern part of the Sea of Okhotsk from observations on a standard cross-section the Aniva cape – the Dokuchaev cape. Meteorol. Hydrol. **3**, 69–85 (2007)
7. Bukatov, A.E., Babij, M.V., Belokopytov, V.N., Pavlenko, E.A.: Buoyancy frequency, free internal wave vertical structure, and temperature of surface of the Black sea spatial-temporal variability. In: Eremeev, V.N., Konovalov, S.K. (eds.) Sustainability and Evolution of the Oceanological Characteristics of the Black Sea Ecosystem, pp. 126–142. NAS Ukraine, Marine Hydrophysical Institute, Sevastopol (2012)
8. Bukatov A.E., Cherkesov L.V.: Waves in an inhomogeneous sea, p. 224. Naukova Dumka, Kiev (1983)
9. Goncharov, V.V.: On some features of internal waves in the ocean. Tsunami and internal waves, pp. 87–96. MHI AS USSR, Sevastopol (1976)
10. Grigorenko, K.S., Khartiev, S.M., Solovieva, A.A., Ermoshkin, A.V.: Kinematic characteristics of internal waves in the central Atlantic investigation according to CTD-profiling data. Ecol. Bull. Res. Cent. Black Sea Econ. Cooperation **1**, 41–50 (2015)

Critically Stressed Fractures and Their Relation to Elastic Moduli

Nikita Vladislavovich Dubinya(✉) and Ilya Vladimirovich Fokin

The Schmidt Institute of Physics of the Earth of the Russian Academy of Sciences, Moscow, Russian Federation
Dubinya.NV@gmail.com

Abstract. The analysis of critically stressed fractures' relationship to elastic moduli of the rock is carried out. Series of triaxial loading tests were performed for rock samples. The behavior of differential stress vs. volumetric strain curve in the vicinity of elasticity limit of the rock is particularly analyzed. It is revealed that the slope of this curve changes considerably when the principal stresses exceed the specific values determined by the criterion of critically stressed fractures appearance. Thus the relationship between fracture initiation Griffith criteria and critical stress state of fractures in the medium is established. It is shown that it is possible to define a specific elastic modulus which is influenced by the fraction of critically stressed fractures in the rock. This modulus may be used to solve the inverse problem of stress state estimation from independent observations related to fractures. Moreover, a formalized approach for estimation of the parameters of rock failure criterion from a cyclic loading test is proposed.

Keywords: Rock mechanics · Stress state · Critically stressed fractures

1 Introduction

The main scope of the study is the mechanical behavior of rock while subjected to dynamically changing stress field. The point where the stress state of the rock corresponds to its elasticity limit is of particular interest: the aim of the study is finding the proper approach to determine the elasticity and plasticity limit of the rock. Naturally fractured rocks are of particular interest as the stresses acting on the planes of the fractures are analyzed considering the hypothesis of critically stressed fractures [1].

Critically stressed fractures existing in a naturally fractured rock play an important role in the mechanical and filtration properties of the considered medium. The hypothesis of critically stressed fractures having an impact on the permeability of the rock was introduced in [1]: the data from a borehole televiewer in crystalline rock from the Cajon Pass borehole on the orientations of the fractures existing in the vicinity of the well were analyzed using the Mohr diagram. Using the additional data on hydraulic permeability of the medium it was shown that the following rule takes place for the majority of fractures: in case the fracture is critically stressed it tends to be hydraulically conductive. The critical stress state of the fracture may be determined if normal and

V. Karev et al. (Eds.): PMMEEP 2017, SPRINGERGEOL, pp. 35–44, 2018.
https://doi.org/10.1007/978-3-319-77788-7_5

shear stress alongside with the internal friction angle of the medium are known. The fracture is considered as critically stressed if the following inequality takes place:

$$\tau_n \geq \sigma_n \cdot \mathrm{tg}\varphi, \tag{1}$$

where τ_n and σ_n are shear and normal stresses acting on the fracture's plane respectively; φ is the internal friction angle of the rock containing the fracture.

The hypothesis of high hydraulic permeability of the critically stressed fractures was developed in the further studies [2–4]. This hypothesis appeared to be useful for estimating the in-situ stress state of the medium surrounding the wellbore from the interpretation of various well logs [5, 6]. Nevertheless, there are some obstacles [7], preventing this hypothesis from being directly applied for hydrodynamic modeling as the fractures' permeability may originate not only from the critical stress mechanism but also from geological processes and history of the stress state.

In the current study the mechanical properties of the rock are analyzed instead of the filtration properties. The hypothesis introduced in [8] was formulated as following: if a naturally fractured rock is subjected to stresses providing that the most of fractures become critically stressed, the elastic moduli of this medium tend to decrease. This statement was used for stress profiles reconstruction along the trajectories of wells located in the vicinity of Nankai Trough, Japan. The single obtained in the study proof for the hypothesis was the match between the stress profiles reconstructed from the fractures' data and the stress state estimations given in independent studies [9].

One of the important aims of the current study is obtaining the solid substantiation for the phenomenological statement that medium with critically stressed fractures has lowered elastic moduli compared to the same rock in a stress state not leading to critically stressed fractures.

The description of the features of the rock behavior above the elasticity limit was introduced in [10]. In this study three different stress-strain curves describing the rock behavior during gradual loading process were analyzed: axial stress vs. axial strain; axial stress vs. lateral strain; axial stress vs. volumetric strain. On the first stage of loading the pre-existing fractures in the medium are closing and all the curves are non-linear. Than a linear stage takes place and the strain remains elastic. Further deformation leads to fracture initiation when the closed fractures start to propagate. The axial stress, at which this process starts, is associated with fracture initiation Griffith criteria. The analysis of the curves shows that the lateral and volumetric strains become non-linear while the axial strain curve remains linear. The process of fracture propagation continues with part of elastic energy is released to extend the crack surfaces. After the critical level of released energy is achieved the unstable fracture propagation takes place, which is revealed on the stress-strain curves: the curvature of volumetric strain-stress curve changes its sign and the axial strain-stress curve departs from linearity.

In the current study an attempt to establish the relationship between fracture initiation Griffith criteria and critically stressed fractures concept is made. If the inequality (1) is related to the strain curves' bending, one can assume whether the fractures in the rock are critically stressed or not just from the observations of its specific elastic moduli.

2 Experimental and Theoretical Study Basis

The series of cyclic triaxial loading tests were performed for different rock samples. The core samples of cylindrical shape were adjusted to gradually changing axial load providing axial stress σ_a and varying radial stresses σ_r (see Fig. 1). Differential stress $d\sigma = \sigma_a - \sigma_r$ ($d\sigma \geq 0$) was used as the governing parameter. For a given value of radial stress the differential stress had been gradually increasing with all strains being controlled. When curvature sign of the observed volumetric strain vs. differential stress curve was drawn upon zero, the test was moved to the unloading stage: differential stress had been decreasing to zero. Then several loading – unloading cycles were performed for different radial stresses. On the last cycle the unloading step was performed only after the stress state achieves the failure criterion. Mohr-Coulomb failure criterion was used to describe the rock behavior.

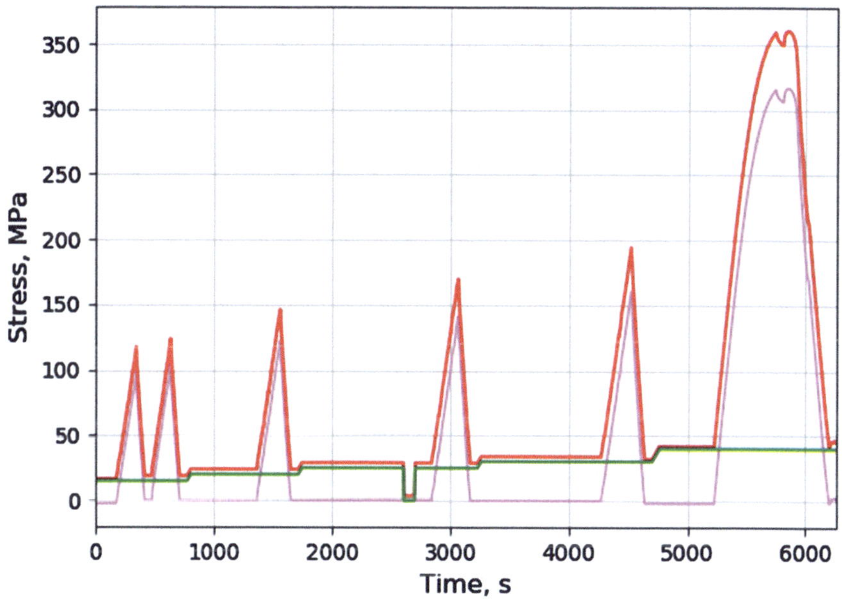

Fig. 1. Principal scheme of the loading test. Red curve represents axial stress; pink curve represents differential stress; green curve represents radial stress.

For the given conditions the axial and radial stresses are considered to be principal. Following the general rule used in geomechanics the compressive stresses are considered to be positive. For three principal stresses σ_1, σ_2, and σ_3 the following rule takes place by definition: $\sigma_1 \geq \sigma_2 \geq \sigma_3$. In the considered process maximum principal stress σ_1 is equal to axial stress σ_a, and two least principal stresses σ_2 and σ_3 are equal to σ_r.

When applying the critically stressed fractures concept one has to determine the normal and shear stresses acting on a plane of known spatial orientation. These stresses are related to principal stresses as following:

$$\sigma_n = \sigma_2 + l^2(\sigma_1 - \sigma_2) + n^2(\sigma_3 - \sigma_2), \tag{2}$$

$$\tau_n = \sqrt{\left(\frac{\sigma_2 - \sigma_3}{2}\right)^2 + l^2(\sigma_1 - \sigma_2)(\sigma_1 - \sigma_3) - \left(\sigma_n - \frac{\sigma_2 + \sigma_3}{2}\right)^2}. \tag{3}$$

Here l and m are cosines of the angles between the normal to the considered plane and the first and the third principal axes of the stress tensor respectively. Assuming $\sigma_2 = \sigma_3$ one can change Eqs. (2) and (3) to simpler form:

$$\sigma_n = \sigma_2 + l^2(\sigma_1 - \sigma_2), \tag{4}$$

$$\tau_n = \sqrt{l^2(\sigma_1 - \sigma_2)^2 - (\sigma_n - \sigma_2)^2}. \tag{5}$$

It can be shown that the orientation of the plane at which the fracturing initiation begins at the lowest values of external stresses is governed by the friction angle φ of the rock. In particular, the angle between normal to this plane and first principal axis is equal to $\pi/4 + \varphi/2$. Then Eqs. (4) and (5) may be substituted into inequality (1) assuming the last one is equation. Then a simple relationship between principal stresses σ_{1cr} and σ_{1cr} providing at least one critically stressed fracture in the rock may be obtained:

$$\frac{\sigma_{1cr}}{\sigma_{3cr}} = \frac{1 + \sin\varphi}{1 - \sin\varphi}. \tag{6}$$

In case when radial stress does not change during the test, one can easily obtain the following equation for differential stress $d\sigma_{cr}$ corresponding to the state when the possible fracture oriented along the weakest plane reaches the critical stress state:

$$d\sigma_{cr} = \sigma_r \frac{2\sin\varphi}{1 - \sin\varphi}. \tag{7}$$

The differential stress vs. volumetric strain curves were analyzed for different rocks and different radial stresses. The behavior of this curve at the point corresponding to critical stress state was particularly analyzed. It is expected that the critical value $d\sigma_{cr}$ is related to the point of fracture initiation or, following [10], Griffith criteria.

The slope of this curve in the elastic zone (before differential stress exceeds the critical value determined using Eq. (7)) is governed by a special elastic modulus. The slope of the curve itself serves as an indicator of critically stressed fracture appearance, although it is worth considering, how it is related to elastic moduli widely applied in practice. In order to determine this relation an isotropic homogeneous medium is considered. The differential forme of the Hooke's law is cylindrical coordinate system provides the following equations:

$$\begin{cases} \delta\sigma_r = (\lambda + 2\mu)\delta\varepsilon_r + \lambda(\delta\varepsilon_\varphi + \delta\varepsilon_z), \\ \delta\sigma_\varphi = (\lambda + 2\mu)\delta\varepsilon_\varphi + \lambda(\delta\varepsilon_r + \delta\varepsilon_z), \\ \delta\sigma_z = (\lambda + 2\mu)\delta\varepsilon_z + \lambda(\delta\varepsilon_r + \delta\varepsilon_\varphi). \end{cases} \quad (8)$$

Here $\delta\sigma_r$, $\delta\sigma_\varphi$, and $\delta\sigma_z$ are the changes in radial, hoop and axial stresses respectively, and $\delta\varepsilon_r$, $\delta\varepsilon_\varphi$, and $\delta\varepsilon_z$ are changes in corresponding strains. λ and μ are Lamé parameters of the medium.

According to the principal scheme of the experiment (Fig. 1), the changes in hoop and radial stress are equal to zero during one loading cycle. The substitution of this condition $\delta\sigma_r = \delta\sigma_\varphi = 0$ into the first two equations in (8) provides a simple relationship between change in axial stress (change in differential stress $\delta d\sigma$ is the same) and change in axial strain:

$$\delta\sigma_z = \frac{\mu(3\lambda + 2\mu)}{\lambda + \mu}\delta\varepsilon_z. \quad (9)$$

The similar relation may be established between $\delta d\sigma$ and change in volumetric strain $\delta\varepsilon_V$ using the Lamé parameters or the pair of Young's modulus E and Poisson's ratio ν; or bulk modulus K:

$$\delta d\sigma = (3\lambda + 2\mu)\delta\varepsilon_V = \frac{E}{1 - 2\nu}\delta\varepsilon_V = \frac{K}{3}\delta\varepsilon_V. \quad (10)$$

It is clear that the coefficient between differential stress and volumetric strain is related to the bulk modulus of the rock which is evident from the test conditions.

Thus the following tendency is expected: whenever at least one fracture becomes critically stressed, the slope of differential stress vs. volumetric strain starts changing. Moreover this bending point is expected to be related to critical value of differential stress $d\sigma_{cr}$ determined from Eq. (7).

3 Experimental Results

Cyclic loading experiments were performed for a number of different rock samples. A typical diagram of the loading process can be found at Fig. 1: a preliminary loading cycle was performed at the beginning of each test; than several cycles were performed without exceeding the plasticity limit of the rock. The behavior of differential stress vs. volumetric strain curve served as an indicator of the loading stage finish: the unloading stage started after the volumetric strain curve drew upon vertical. The loading stages were performed for a set of radial stresses remaining constant during the cycle. The radial stresses were chosen according to preliminary estimations of in-situ stress state of the rock. The last loading cycle was performed up to the plasticity limit – the classical differential stress vs. axial strain curve behavior provided necessary data: maximum value of differential stress was associated with failure criterion.

The typical behavior of differential stress vs. volumetric strain curve is shown at Fig. 2. Here the pink line the pink curve represents the measured stress and volumetric strain; dotted line is the polynomial approximation of the measured curve; black line represents the linear approximation of the volumetric strain-stress curve in the linear zone; and finally, the horizontal line represents the condition $d\sigma = d\sigma_{cr}$. The value of $d\sigma_{cr}$ was calculated from Eq. (7). In order to determine $d\sigma_{cr}$ it was necessary to estimate the internal friction angle of the medium, which was performed after combining the results of several loading cycles: it appeared that the differential stresses at which the curve became vertical provide a linear relationship between the radial stresses σ_r and axial stresses σ_a at different loading cycles. Following [11] the linear coefficient between this stresses was considered to be related to the internal friction angle. In particular, this coefficient was chosen to be equal to the tangent of internal friction angle.

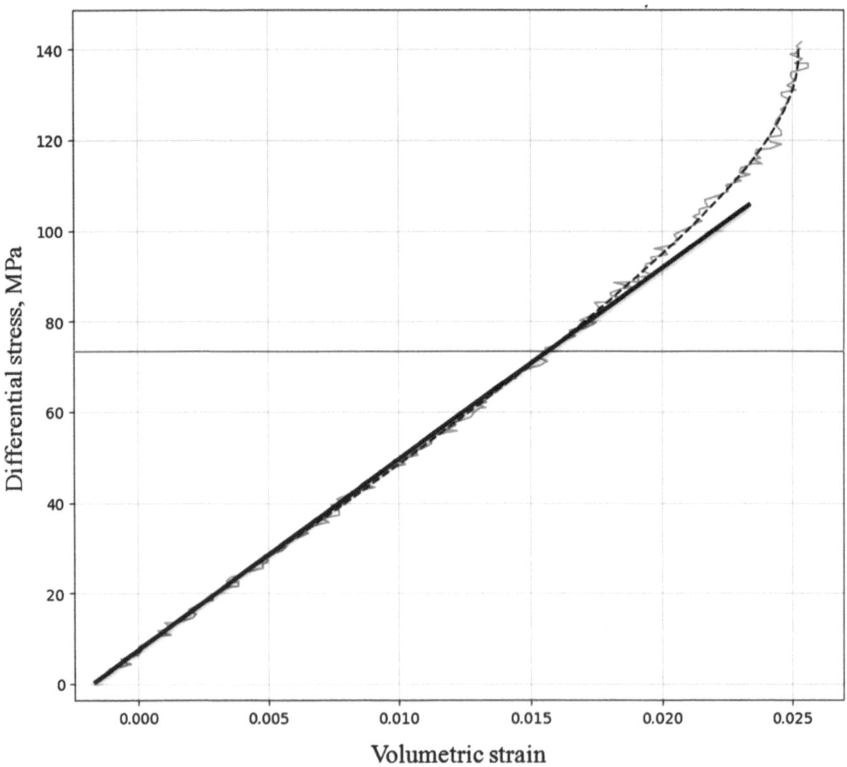

Fig. 2. Differential stress vs. volumetric strain curve for one cycle of loading the granite rock sample

The linear trend of the curve was constructed using the data on stress and strain in the zone of elastic deformation. The coefficient for this slope was related to the bulk modulus of the rock K.

It can be clearly seen from Fig. 2 that the analyzed curve may be considered as linear while $d\sigma < d\sigma_{cr}$ and nonlinear when the inequality $d\sigma > d\sigma_{cr}$ becomes true. The simultaneously analyzed lateral and axial strain curves proved that these curves remain linear with gradual differential stress increase during the stage of stable fracture propagation following [10].

The same results were obtained for other samples: the differential stress vs. volumetric strain curves were losing their linearity after achieving the critical differential stress value calculated from Eq. (7). While the results shown at Fig. 2 were obtained for a granite sample, the same studies were carried out for samples of other nature. For example the differential stress vs. volumetric strain curve for all the cycles are shown at Fig. 3 for carbonate rock.

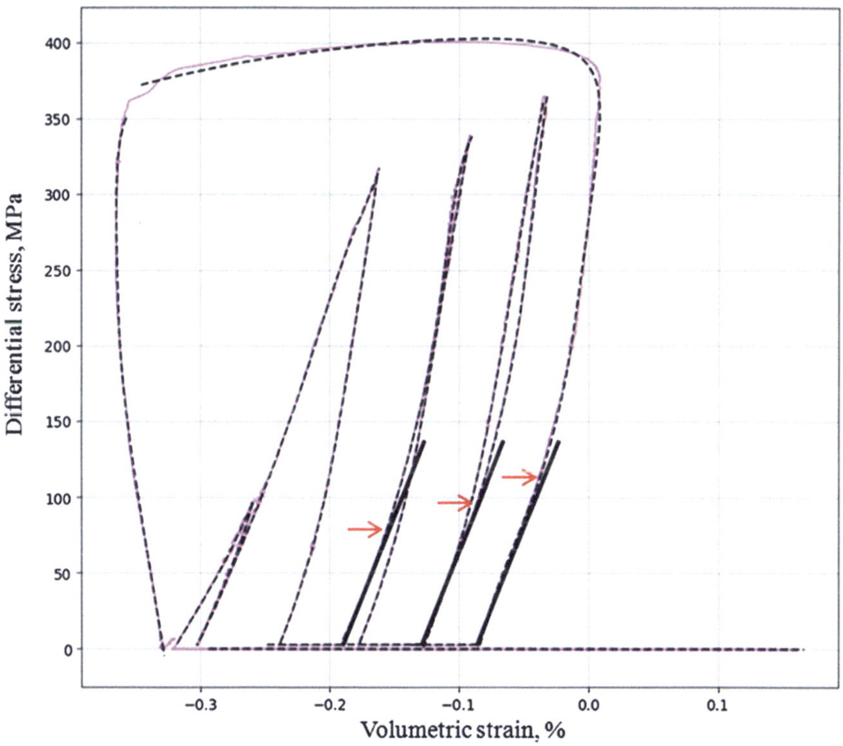

Fig. 3. Differential stress vs. volumetric strain curve for all cycles of carbonate rock loading

Red arrows point at the stress-strain test at which differential stress is determined from Eq. (7). The corresponding strain is obtained from the loading curve. Black lines still represent the linear approximations at the elastic zone. It is clear from the figure that the start of stress-strain curve bending is in good match with the estimation of $d\sigma_{cr}$.

4 Discussion

The obtained results provide an experimental basis for the statement that Griffith fracture initiation criterion is related to the external loading providing the fractures existing in the rock to become critically stressed.

The same behavior was observed for rocks of different mineral composition, inner structure and physical-mechanical properties. For some samples, nevertheless, the later cycles were characterized with differential stress vs. volumetric strain curve bending earlier than it was expected. There are two possible explanation of this effect. First, the usage of linear Mohr-Coulomb criterion may lead to a worse description of the rock's failure for all possible values of principal stresses. Then the internal friction angle observed for lower radial stresses may not be the same as this angle measured at higher radial stresses. In other words the critically stressed fractures appear at lower relation σ_1/σ_3 for higher values of σ_3 which is equal to σ_r for the considered process.

On the other hand the loading process may only be considered as elastic while the differential stress remains lower than critical value determined from Eq. (7). Despite that it was mentioned above that the unloading step starts not after the critical differential stress is achieved but when the differential stress vs. volumetric strain curve becomes vertical. That means that some of the fracture propagation processes may be characterized by a portion of irreversible deformation thus total unloading does not provide zero change in volumetric deformation. This may result in change of the rock's microstructure and thus in the change of critical stress value corresponding to Griffith criteria.

There are two important applications for the experimental results obtained in the present study. First of all, if one carries out a series of triaxial tests with differing radial stresses and obtains the array of $d\sigma_{cr}$ values corresponding to volumetric strain curves' bendings in the assumption that linear Mohr-Coulomb failure criterion describes the rock behavior decently, than this criterion may be obtained after achieving one plasticity limit. In this case the internal friction angle may be obtained from the array of $d\sigma_{cr}$ values and cohesion may be estimated from the plasticity limit.

Another important result is connected to stress analysis: it provides that if rock has increased elastic modulus characterizing the slope of differential stress vs. volumetric strain curve (but not Young's modulus or Poisson's ratio considering the isotropic medium), than the majority of existing fractures are expected to be critically stressed. This result might be further used to estimate stress state following the approach [5, 6, 8].

The last statement may have a significant influence on practically important problems, especially in the field of hydrocarbon reservoir engineering. Geophysical logging carried out in production wells provides decent amount of data regarding the natural fractures distribution existing in the rocks surrounding the wellbore. Moreover, information about elastic properties of the rock can also be obtained through corresponding studies, although only dynamic moduli can be estimated without laboratory experiments. Nevertheless these two sources provide some help in estimating the reservoirs: if some special behavior of the elastic moduli is observed near the fractures, these fractures may be considered from the positions of critically stressed fractures hypothesis. In particular, if the local maximum of bulk modulus is observed on the corresponding log and a fracture appears in the rock surrounding the wellbore at the same depth, this fracture is expected to be critically stressed according to results of the present study. Assuming the critically stressed fractures hypothesis is true one should expect the improved filtration properties of the rock at the corresponding depth.

5 Conclusion

The conducted study was aimed at the investigation of the rock behavior while subjected to gradually changing loading. It was shown that the fractures existing in the rock may have influence on the elastic moduli of the medium. This effect leads to the bendings of the stress-strain curves characterizing the rock.

Although the axial stress vs. axial strain curve may remain linear, the volumetric and radial strain curves may be bended. The moment when these two curves' curvatures lose linearity is associated with the critically stressed fractures existing in the rock: it appears that when the fractures oriented along the weakest plane in the medium become critically stressed, the gradual loading leads to nonlinear behavior of volumetric strain curve. The elastic modulus describing this behavior is closely related to the bulk modulus: this process is characterized by significant increase in bulk modulus.

In this study the experimental results of loading tests alongside with the theoretical basis were provided. Further investigation may be carried out from the position of mathematical modeling of the elastic properties using the approaches developed in effective medium theory.

The experimental complex used in the study was created with the financial support of Russian Science Foundation, grant #14-17-00658.

References

1. Barton, C.A., Zoback, M.D., Moos, D.: Fluid flow along potentially active faults in crystalline rocks. Geology **23**, 683–686 (1995). https://doi.org/10.1130/0091-7613(1995)023<0683:FFAPAF>2.3.CO;2
2. Hickman, S.H., Barton, C.A., Zoback, M.D., Morin, R., Sass, J., Benoit, R.: In situ stress and fracture permeability along the stillwater fault zone, dixie valley, nevada. Int. J. Rock Mech. Mining Sci. Abs. **34**, 3–4 (1997). https://doi.org/10.1016/S0148-9062(97)00169-1
3. Rogers, S.: Critical stress-related permeability in fractured rocks. In: Ameen, M. (ed.) Fracture and In Situ Stress Characterization of Hydrocarbon Reservoirs, vol. 209, pp. 7–16. Geological Society, London (2002). https://doi.org/10.1144/GSL.SP.2003.209.01.02
4. Townend, J., Zoback, M.D.: How faulting keeps the crust strong. Geology **28**, 399–402 (2000). https://doi.org/10.1130/0091-7613(2000)28%3c399:HFKTCS%3e2.0.CO;2
5. Ito, T., Zoback, M.D.: Fracture permeability and in situ stress to 7 km depth in the KTB scientific drillhole. Geophys. Res. Lett. **27**, 1045–1048 (2000). https://doi.org/10.1029/1999GL011068
6. Dubinya, N.V., Ezhov, K.A.: In-situ horizontal stress estimation based on the geometrical properties of fractures in well vicinity. Geophys. Res. **18**(2), 5–26 (2017). https://doi.org/10.21455/gr2017.2-1

7. Sathar, S., Reeves, H.J., Cuss, R.J., Harrington, J.F.: The role of stress history on the flow of fluids through fractures. Mineral. Mag. **76**, 3165–3177 (2012). https://doi.org/10.1180/minmag.2012.076.8.30
8. Dubinya, N.: Horizontal stress profiles reconstruction based on elastic properties and natural fractures' characteristics. NanTroSEIZE case study. In: Japan Geosciences Union Meeting Proceedings, SIT24-P05 (2017)
9. Wu, H.-Y., Chan, C.-Y., Kinoshita, M., Saito, S.: Stress field observation and modeling from the NanTroSEIZE scientific drillings in the nankai trough system, SW Japan. Tectonophysics **600**, 99–107 (2013). https://doi.org/10.1016/j.tecto.2013.04.009
10. Bieniawsky, Z.T.: Propagation of Brittle Fracture in Rock. Republic of South Africa Council of Science and Industrial Research: Report MEG 664 (1968)
11. Kim, M.M., Ko, H.-Y.: Multistage triaxial testing of rocks. Geotech. Test. J. **2**(2), 98–105 (1979)

Mechanical Properties of Thin Films of Coals by Nanoindentation

Elena Kossovich[1](✉), Svetlana Epshtein[1], Nadezhda Dobryakova[1], Maxim Minin[2], and Darya Gavrilova[1]

[1] National University of Science and Technology "MISiS", 4, Leninsky Prospekt, Moscow 119049, Russian Federation
e.kossovich@misis.ru

[2] Ural Federal University named after the first President of Russia B.N. Yeltsin, 19, Mira Street, Ekaterinburg 620002, Russian Federation

Abstract. In the current work, depth-sensing indentation tests were applied at different coals microcomponents in order to characterize their mechanical properties. To this end, three coals were chosen differing by their rank and properties. Thin transparent films of such coals were prepared as samples in order to use transmitted light microscopy to achieve exact positioning of indent probe in the zone occupied by specific coal microcomponent. Experimental procedures allowed obtaining load-displacement curves at different microcomponents. These curves revealed differences between microcomponent vitrinite behavior among coals of different types. Measured values of elastic moduli and hardness also proved that they variate with coal rank. It was concluded that matrix of coals (vitrinite microcomponent) transforms its mechanical properties with rank, whereas the inclusions properties do not change as significantly.

Keywords: Coal · Mechanical properties · Microcomponent · Matrix Inclusion · Depth-sensing indentation · Elastic modulus Hardness

1 Introduction

Coals propensity to destruction at mining, storage and utilization could be one of the possible reasons for loss of products quality and hazards occurrence such as dust formation and its further explosion, as well as methane outbursts. Coals propensity to destruction and fine dust formation is hard to be characterized by traditional means of mechanical testing [1]. Among the possible reasons for such phenomena one may mention the fact that coals could be considered as natural organic composites with a complex structure not only at macro-, but also at nano- and microscales [2]. And this complex structure (consisting of various lithotypes including vitrinite as matrix and different organic and inorganic

V. Karev et al. (Eds.): PMMEEP 2017, SPRINGERGEOL, pp. 45–50, 2018.
https://doi.org/10.1007/978-3-319-77788-7_6

inclusions) may reason to formation of defects and fractures at low-scales, i.e. the scales where the initial stages of destruction are found.

In order to reach for more information on coals mechanical properties, it is necessary therefore to use approaches that are traditional to composites. Such approaches include, among others, those of depth-sensing indentation (DSI) at different scales (nano- and micro) [3]. At the moment, DSI methods are widely used for characterization of various mechanical properties and defects formation of thin tribological coatings [4], various "matrix-inclusions" systems of composite materials [5], etc.

One of the most common difficulties that arise at attempts of application of DSI techniques at coals samples is identification of the location, sizes and shapes of inclusions as well as matrix. It was shown previously that utilization of reflected light microscopy for coals microcomponents identification is complicated due to relatively low reflectance of their surfaces, even polished ones [6]. A novel approach was proposed recently for characterization of coals microcomponents mechanical properties by depth-sensing nanoindentation with microscopy using transmitted light. To this end, specially prepared samples of coals were considered shaped as transparent thin films glued to glass substrate (see Fig. 1) [6]. Such samples allow exact identification of separate coals microcomponents. This, in turn, allows more strict positioning of the zone where indentation is to be performed.

Fig. 1. Transparent thin film of coal sample

2 Materials and Methods

In the current work, indentation experiments were performed at thin films of coals prepared in accordance with the previously developed by our co-authors technique [6,7]. Samples included three coals of different type, rank and metamorphism degree, namely, lignite, low-rank and medium-rank bituminous coal. Coals films thickness was 14–15 µm.

During indentation tests, the following microcomponents were considered for each of the coal sample: vitrinite (matrix) and inertinite (inclusion).

Main experiments were performed at Hysitron TI750 UBI precision installation for DSI. Load-control regime was used for indentation of samples. Localization of objects was provided by transmitted light microscopy. Each object was subject to not less than 20 indentation measurements in order to achieve statistically significant results.

3 Results

After experimental procedures, a set of load (P) - displacement (h) curves was obtained as well as a number of calculated parameters such as values of elastic moduli and hardness evaluated according to Oliver and Pharr method [8] and Bulychev-Alekhin-Shorshorov dependence [9]. Typical load-displacement curves for coals microcomponents are shown in Fig. 2.

It could be seen from Fig. 2 that lignite's matrix behaves rather plastic in comparison with matrix of low-rank bituminous coal. On the other hand, behavior of inclusions is rather similar for both coals. Moreover, it should be mentioned that load-displacement curves obtained for medium-rank bituminous coal were qualitatively similar to those of low-rank one, therefore they were not included in Fig. 2.

Table 1 shows measured elastic moduli (E) and hardness values (H) for each of the coal sample, along with data on standard deviation (StDev).

Table 1. Measured values of elastic constants and hardness of coals samples microcomponents

Coal sample	Object (matrix or inclusion)	E, GPa	StDev E, GPa	H, MPa	StDev H, MPa
Lignite	Matrix	5.55	0.74	303.15	43.33
	Inclusion	6.31	0.32	518.47	44.75
Low-rank bituminous coal	Matrix	5.58	0.09	608.23	21.26
	Inclusion	5.21	0.26	481.26	43.40
Medium-rank bituminous coal	Matrix	5.05	0.38	504.71	63.34
	Inclusion	5.18	0.91	334.10	57.45

It could be observed from Table 1 that the elastic moduli measured at coals matrix were of similar values slowly decreasing from lignite to medium-rank bituminous coal. At the same time, elastic moduli of inclusions descended significantly from lignite to bituminous coals. As of the hardness values measured for the matrix, one may observe an opposite situation, namely, leap from lignite to bituminous coals. On the other hand, hardness of inclusions also decreased from lignite to both bituminous coals.

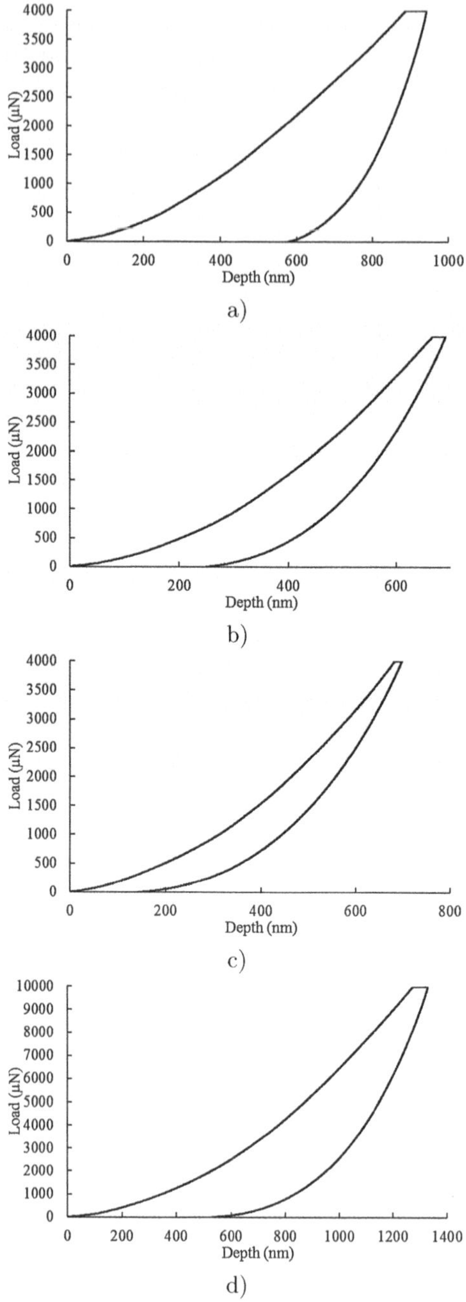

Fig. 2. Typical load-displacement curves after DSI of coals microcomponents: (a) matrix of lignite; (b) inclusion of lignite; (c) matrix of low-rank bituminous coal; (d) inclusion of low-rank bituminous coal

It is interesting to note that the highest differences between the measured elastic moduli and hardness values for matrix and inclusion are observed for lignite. Also, it was found that such differences become less pronounced with rank growth.

It could be concluded that according to our observations, matrix of coals (vitrinite microcomponent) transforms its mechanical properties with rank, whereas the inclusions properties do not change significantly.

It should also be mentioned that the measured values of elastic moduli of coal films are actually the ones for the system "sample-substrate", as it was shown in [10], and additional investigation is required for evaluation of their actual values according to the proposed scheme [10].

4 Conclusions

In the current work, depth-sensing indentation was used for characterization of coals microcomponents mechanical properties.

Thin transparent films of three coals of different types and rank were prepared as samples in order to use transmitted light microscopy to achieve exact positioning of indent probe in the zone occupied by specific coal microcomponent.

As a result, load-displacement curves were constructed characterizing mechanical behavior of different microcomponents of coals. These curves revealed changes of microcomponent vitrinite behavior with rank. Measured values of elastic moduli and hardness also proved that they variate with coal rank. It was concluded that matrix of coals (vitrinite microcomponent) transforms its mechanical properties with rank, whereas the inclusions properties do not change as significantly.

Acknowledgements. The work was supported by the Russian Science Foundation (grant No. 16-17-10217).

References

1. Jiang, H., Du, C., Dong, J.: Investigation of rock cutting dust formation and suppression using water jets during mining. Powder Technol. **307**, 99–108 (2017). https://doi.org/10.1016/j.powtec.2016.11.029
2. Balokhonov, R.B., Romanova, V.A., Makarov, R.V., Voroshilov, S.P.: Effect of complex interface geometry on deformation of a coal composite. Numerical simulation. Fiz. mezomechanica **10**, 75–79 (2007)
3. Palacio, M.L.B., Bhushan, B.: Depth-sensing indentation of nanomaterials and nanostructures. Mater. Charact. **78**, 1–20 (2013). https://doi.org/10.1016/j.matchar.2013.01.009
4. Bull, S.J.: Nanoindentation of coatings. J. Phys. D. Appl. Phys. **38**, R393–R413 (2005). https://doi.org/10.1088/0022-3727/38/24/R01
5. Konopka, K.: Crack propagation in composites with ceramic matrix. In: IUTAM Symposium on Multiscale Modelling of Damage and Fracture Processes in Composite Materials, vol. 135, pp. 255–262 (2006). https://doi.org/10.1007/1-4020-4566-2_30

6. Epshtein, S.A., Borodich, F.M., Bull, S.J.: Evaluation of elastic modulus and hardness of highly inhomogeneous materials by nanoindentation. Appl. Phys. A Mater. Sci. Process. **119**, 325–335 (2015). https://doi.org/10.1007/s00339-014-8971-5
7. Borodich, F.M., Bull, S.J., Epshtein, S.A.: Nanoindentation in studying mechanical properties of heterogeneous materials. J. Min. Sci. **51**, 1062–7391 (2015). https://doi.org/10.1134/S1062739115030072
8. Oliver, W.C., Pharr, G.M.: An improved technique for determining hardness and elastic modulus using load and displacement sensing indentation experiments. J. Mater. Res. **7**, 1564–1583 (1992). https://doi.org/10.1557/JMR.1992.1564
9. Bulychev, S.I., Alekhin, V.P., Shorshorov, M.K., Ternovskij, A.P., Shnyrev, G.D.: Determination of Young modulus by the hardness indentation diagram. Zavod. Lab. **41**, 1137–1140 (1975). (in Russian)
10. Kossovich, E.L., Borodich, F.M., Bull, S.J., Epshtein, S.A.: Substrate effects and evaluation of elastic moduli of components of inhomogeneous films by nanoindentation. Thin Solid Films **619**, 11–119 (2016). https://doi.org/10.1016/j.tsf.2016.11.018

Using the Variational Approach and Adjoint Equations Method Under the Identification of the Input Parameter of the Passive Admixture Transport Model

Sergey Germanovich Demyshev, Vladimir Sergeevich Kochergin, and Sergey Vladimirovich Kochergin(✉)

Marine Hydrophysical Institute RAS, Sevastopol 299029, Russian Federation
vskocher@gmail.com

Abstract. Method of the adjoint equations, estimation method and variational algorithm of the input parameters identification of the passive admixture transport model is considered in that work. Modification of the variational algorithm of the measurements data assimilation, allows saving the calculation resources under the algorithms numerical realization considerably. Problem of the contamination point source power estimation on the base of the evaluation method and working out the obtained estimations is solved. The numerical experiments for estimation the concentration fields according to the initial data and solving the corresponding adjoint problem are conducted. Variation algorithm of the initial pollution spot identification is realized. Possibility to use assimilation variational methods and measurements data filtration for the wide range of problems connected with investigated basin ecological state was shown in the present work. Modified scheme for realization of the variational algorithm identification of the model input parameters is suggested. Conditions, under which the suggested algorithm has advantages over the standard approach, were obtained. The conducted numerical experiments have shown the proven work of the identification algorithm of the contamination source parameters applying the passive admixture transport model in the Azov Sea. Good converging of the iteration process and accuracy of the concentration field initial distribution determination was obtained for the variational algorithm of the passive contaminant concentration initial field identification.

Keywords: Transport model · Passive admixture · Variational algorithm Adjoint problem · Identification · Azov sea

1 Introduction

Analysis of the potential impact of the different origin admixture is necessary under solving the problems, connected with water basins ecological state estimation. Such analysis of the concentration fields can be realized on the base of mathematical modeling. Two inter-related problems are solved at that. The first one is solved on the

V. Karev et al. (Eds.): PMMEEP 2017, SPRINGERGEOL, pp. 51–61, 2018.
https://doi.org/10.1007/978-3-319-77788-7_7

base of optimal transport model construction and its numerical realization, taking into consideration its behavior in the water environment.

The second problem is connected with identification of the model input parameters and measurements data adoption. Coefficients, velocity fields and initial data are the model input parameters. Calculations for the Azov Sea hydrodynamic fields modeling under the different wind impact were set for the realization of the variation algorithm of the measurements data adoption. The received current fields and coefficients of the turbulent diffusion were used as the input parameters for the main and adjoint problem.

Using the adjoint equations and variation principals for solving the problems connected with ocean dynamics modeling became widespread [1, 2]. It happened so, that variation approach was used for solving problems of the of the atmosphere dynamics [3, 4] and for solving the oceanological problems hereafter [5–8]. The adjoint problem solution and quadratic functional of the prediction quality are in the basis of such algorithms. The model serves as a spatial-temporal interpolant and that is the important peculiarity of such algorithms [4]. The obtained solution is coordinated with the model itself as well as with the measurements data.

Identification of the input parameters is realized at the expense of the prediction quality quadratic functional minimization. That functional is convex and it have only one minimum. Transport model of the passive admixture is linear, that is why it's using as a connections (limitations) under chosen functional minimization don't change its convexity. That allows fulfilling search of the model input parameters under which its solution corresponds the obtained measurements data in the best way.

Solution of the adjoint problem is the "impact function" [9], which gives the opportunity to estimate the model integration different regions impact on the characteristics which are of our interest. It is necessary to follow some functional values in the regions of our interest, but not the contamination concentration, when we solve some ecological problems. For instance, that can be average or total concentration of some admixture in the researched area. That functional values can be measured directly according to the concentration values in the knots of the calculation grid. At that, values for the integration final moment can be found by solving the transport model with the given initial data and functions of the contamination sources.

Thus, giving different initial data and sources functions we get series of the concentration fields, for which the necessary functional in the pointed area is calculated. New spatial-temporal calculation should be done for different initial data and sources power. The most optimal calculation, which does not exceed some its allowed values by the functional, can be chosen among these calculations. On the other hand, such problems can be solved easily, if they do that on the base of adjoint equations. At that, corresponding to [9], the adjoint problem can be solved once and the investigated functional is solved without spatial-temporal calculation under different initial fields and sources functions.

2 Method of the Adjoint Equations, Estimation Formula

Let's consider model of the passive admixture transport [10, 11] in the σ - coordinates

$$\frac{\partial DC}{\partial t} + LC = 0, \tag{1}$$

with conditions on the profile boundaries

$$\Gamma : \frac{\partial C}{\partial n} = 0, \tag{2}$$

with conditions on the surface and on the bottom

$$\begin{cases} \sigma = 0 : \frac{\partial C}{\partial \sigma} = Q \\ \sigma = -1 : \frac{\partial C}{\partial \sigma} = 0 \end{cases}, \tag{3}$$

where Q is $Q = Q_S \cdot \delta(t-0) \cdot \delta(x - x_0) \cdot \delta(y - y_0)$ and initial data

$$C(x, y, \sigma, 0) = C_0(x, y, \sigma). \tag{4}$$

In (1)–(4) $L = \frac{\partial DU}{\partial x} + \frac{\partial DV}{\partial y} + \frac{\partial W}{\partial \sigma} - \frac{\partial}{\partial x} A_H \frac{\partial D}{\partial x} - \frac{\partial}{\partial y} A_H \frac{\partial D}{\partial y} - \frac{\partial}{\partial \sigma} \frac{K}{D} \frac{\partial}{\partial \sigma}$; $t \in [0, T]$; is time; x_0, y_0 are coordinates of the point source; D is dynamic depth; Q is capacity of the instant point sources; C is admixture concentration; U, V, W are velocity fields components; A_H and K are coefficients of the horizontal and vertical turbulent diffusion correspondingly; n is normal to the side boundary, Γ is boundary of the domain M; $M_t \in M \times [0, T]$.

Let's set the adjoint problem:

$$-\frac{\partial DC^*}{\partial t} + L^* C^* = 0, \tag{5}$$

$$\Gamma : \frac{\partial C^*}{\partial n} = 0, \ \sigma = 0 : \frac{\partial C^*}{\partial \sigma} = 0, \ \sigma = -1 : \frac{\partial C^*}{\partial \sigma} = 0, \tag{6}$$

$$t = T : C^* = h, \tag{7}$$

where L^* is formally adjoint operator to the operator L.

Multiplying (1)–(4) to C^* and integrating partially considering analogue of the equation of continuity and (5)–(7) we got under $Q = 0$

$$\int_M hCdM = \int_M C_0 C^* dM. \tag{8}$$

Let's choose h as

$$h = \begin{cases} \frac{1}{m(G)}, (x, y, \sigma) \in G \\ 0, (x, y, \sigma) \notin G \end{cases}, \tag{9}$$

where m is a measure of some areas $G \in M$. At that, we get average concentration $\overline{C}_T$ in the area G for the final time moment in the left part of the Eq. (8).

Choosing G as a cell of the calculation grid, we get

$$\overline{C}_T = \int_M C_0 C^* dM. \tag{10}$$

Thus, using solution of the adjoint problem (5)–(7) according to the formula (10) we can estimate concentration of the admixture $\overline{C}$ in the given cell. Numerical experiments, conducted in the work [12], have shown high accuracy of the concentration field simulation according to the initial data and solution of the adjoint problems series without basic transport model.

3 Estimation of the Point Source Capacity

Under identification of the contamination point source parameters on the sea surface, we consider that $C = 0$, choosing C^* as solution of the next adjoint problem

$$-\frac{\partial DC^*}{\partial t} - \frac{\partial DUC^*}{\partial x} - \frac{\partial DVC^*}{\partial y} - \frac{\partial WC^*}{\partial \sigma} - D\frac{\partial}{\partial x}A_H\frac{\partial C^*}{\partial x} - D\frac{\partial}{\partial y}A_H\frac{\partial C^*}{\partial y} - \frac{\partial}{\partial \sigma}\frac{K}{D}\frac{\partial C^*}{\partial \sigma} = 0 \tag{11}$$

$$\Gamma : \frac{\partial C^*}{\partial n} = 0,\ \sigma = 0 : \frac{\partial C^*}{\partial \sigma} = g,\ \sigma = -1 : \frac{\partial C^*}{\partial \sigma} = 0 \tag{12}$$

$$t = T : C^* = 0. \tag{13}$$

We get:

$$\int_S Q \cdot C^* \cdot dS = \int_S C \cdot g \cdot dS, \tag{14}$$

where S is the sea surface, $g = \delta(t - T) \cdot \delta(x - x_n) \cdot \delta(y - y_n)$, $n = 1, \ldots, N$, N is the total number of the measurement points for the final time moment. Considering Q and g from the formula (14) we get:

$$Q_S C_n^*(0, x_0, y_0, 0) = C(T, x_n, y_n, 0),\ n = 1, \ldots, N. \tag{15}$$

Formula (14) is similar to the formula, obtained in the method of estimating [12] the concentration field according to the different initial data and is the double presentation [9] of the concentration through the capacity and solution of the corresponding adjoint problem. Mind, that N of the adjoint problem for N points are independent and realize in parallel mode by different processors.

For determining the emission coordinate point (x_0, y_0) we solve N of the adjoint problem (11)–(13) and find area $G = \cap L_n$, $n = 1, \ldots, N$, where L_n is areas of the C_n^* considerable meanings. The conducted preliminary numerical experiments have shown that priory information concerning the pollution spot location and its boundaries is important under area G determination. Location of (x_i, y_i) points on the pollution spot boundary allows to narrow down area G, considerably. And using measurements data in the points of the maximal concentration leads to the better stipulation of Q_S finding problem. Choosing $(x_i, y_i) \in G$, $i = 1, \ldots, M$ one can make M systems for the different coordinates of the pollution source.

The best estimation Q_S will correspond the real location of the source. For $(x_i, y_i) = (x_0, y_0)$ values $Q_S^n \approx Q_S$, $n = 1, \ldots, N$. Values of Q_S can be found from the system (15) using filtration method [9], which proved to be very helpful for the oceanological problems solution [10]. Filtration of the measurements data considering the whole information focused in the N equations, in the case of newly defined system (15) and mistakes under C_n measuring. As a result, equations of the new system are reordered and not-informative equations are deleted. And equation with the best determinacy, from which value Q_S can be found, appears in the first line of the system.

4 Variational Method of Identification

Task of assimilating measurements data can be conducted at the expense of different input parameters identifications on the base of minimization the quadratic functional of the prediction quality.

$$I_0 = \frac{1}{2}(P(C - C_m), (C - C_m))_M \tag{16}$$

where C_m is the data measurements, P is the operator of zeros filling the prediction disparity fields under absence of these measurements data. Following [4] for the concentration initial field identification, we have:

$$\begin{aligned} I = I_0 &+ \left(\frac{\partial C}{\partial t} + LC, \lambda^*\right)_{M_t} + \left(\frac{\partial C}{\partial n}, \lambda^*\right)_{\Gamma_t} + (C - C_0, \lambda^*) \\ &+ \left(\frac{\partial C}{\partial \sigma}, \lambda^*\right)_{\sigma_t^0} + \left(\frac{\partial C}{\partial \sigma}, \lambda^*\right)_{\sigma_t^{-1}} \end{aligned} \tag{17}$$

where scalar product is determined in a standard way. Writing down functional variation (17) and integrating partially, taking into consideration boundary conditions and analogue of the continuity equation, we will choose solution of the following problem as Lagrange multipliers

$$-\frac{\partial D\lambda^*}{\partial t} + L^*\lambda^* = 0 \tag{18}$$

$$\Gamma : \frac{\partial \lambda^*}{\partial n} = 0,\ \sigma = 0 : \frac{\partial \lambda^*}{\partial \sigma} = 0,\ \sigma = -1 : \frac{\partial \lambda^*}{\partial \sigma} = 0 \tag{19}$$

$$t = T : \lambda^* = P(C_m - C). \tag{20}$$

Then, under the initial field boundary identification, we've got:

$$\nabla_{C_0} I = \lambda^*|_{t=0}. \tag{21}$$

Iteration emptying is carried out according to the formula:

$$C_0^{n+1} = C_0^n + \tau \nabla_{C_0} I, \tag{22}$$

where τ is the iteration parameter, which is being found taking into account solution of the task in variations. Algorithm for identifying other input parameter (contamination capacity [13], for instance) can be made analogically.

5 Modification of the Variation Method of the Measurements Data Assimilating

The measurements data are not observed in all knots of the integrating area very often. That is why number of the demanded adjoint problems for the assessment formula (10) is decreased considerably, because assessment of the concentration field values can be carried out in the points of measurements only. One must solve k adjoint problem, where k is the measurements points number, for to asses $\overline{C}$ concentration. Analogically, we can estimate solution of the problem in variations according to these solutions of the adjoint problem:

$$\delta \overline{C} = \int_M \lambda_0 C_0^* dM, \tag{23}$$

where C_0^* is the solution for the problem (5)–(7), λ_0 is the decision for the problem (18)–(20). Thus, under the main task solving, we carry out assessment according to the formula (10), variation is assessed according to the correlation (23) and adjoint problem (18)–(20) is integrated for the constructing the prediction quality functional gradients in the parameter space.

Integrating the main, adjoint problems and problem in variations for each iteration takes all the time under realization of the identification variation algorithm. Mind that the more is the model integrating time interval, the more iterations we need for achievement the functional minimum under other equal conditions. Let t_p be the processor time for the model integrating for $[0, T]$, then the total time of standard identification algorithm realization is $t_s = 3Jt_p$, where J is the total iterations number, needed for the functional minimum achievement. Total time of the counting is $t_m = t_p(J + N/R)$ under realization of the modified algorithm. Argument N is the number of measurements data and R is the number of the used processors. Let $t_m < t_s$, so we have:

$$R = \frac{N}{2J}. \tag{24}$$

Correlation (24) should be carried out for the number of the used processors R if we want modified procedure of the variation identification to have advantage comparing with the standard one. Such approach allows realizing search of the optimal modeling input parameter using multiprocessor computers under the assimilation variation algorithm realization. That improves efficiency of the admixture fields prediction values obtaining considerably.

6 Numerical Experiments Results

Numerical experiments were carried out with the model [10, 11] for the Azov Sea aquatorium. The model currents field under the east wind 10 m\s was calculated for testing the concentration field estimation algorithm. Spatial distribution of the used coefficients A_H and K was obtained besides the velocity fields, as the result of the calculations on the model. The obtained velocity and turbulent diffusion fields were used as input parameters under integrating the model of passive admixture transport for 5 days. At that, the time step is $\Delta t = 240$ s, and spatial step is $\Delta x = 0.78$ km, $\Delta y = 1.125$ km. The calculation grid in σ – coordinates with 15 horizons is used on vertical in the model. The following norms are given for the assessment of the reproducing accuracy

$$NC = \frac{\max_{x \in D} \left| \overline{C} - C \right| \times 100}{\max_{x \in D} |C|} \tag{25}$$

$$NL = \frac{\sum_{x \in D} \left| \overline{C} - C \right| \times 100}{\sum_{x \in D} |C|} \tag{26}$$

Initial concentration field was given as equal to unit in the area marked by the square at the Fig. 1. Result of the concentration calculation on the sea surface according to the model (1)–(4) is represented at the same figure. Maximal concentration value is $C = 0.4192$ The reconstructed concentration field for this horizon according to the formula (10) is visually the same as the represented at the Fig. 1 maximal value $\overline{C} = 0.4098$, norm values (26), (25) are the following: $NC = 2.2438$ and $NL = 1.7562$. Thus, formula (10) gives good accuracy of the concentration field reconstruction according to the calculated beforehand solutions of the adjoint problem and can be used under field estimation according to the different initial data.

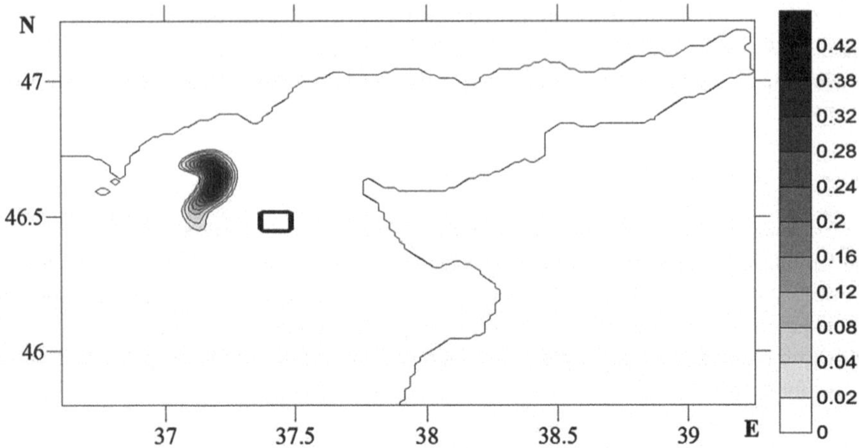

Fig. 1. Location of the initial contamination spot and modeling result.

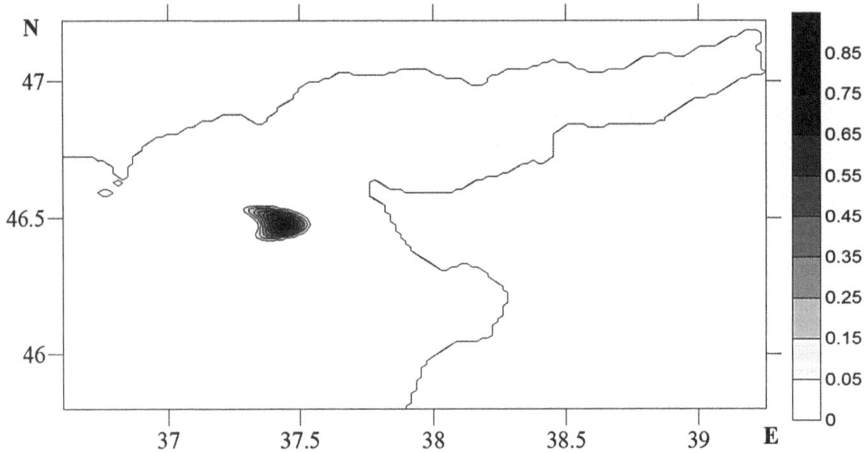

Fig. 2. Identification result (20-th iteration).

The concentration field at the Fig. 1 calculated using model (1)–(4) was used as measurements data for identification of the initial distribution on the base of the basic assimilation method. Figure 2 characterizes spatial distribution of the initial concentration field for the 20-th iteration. Location of the found initial contamination spot corresponds with the coordinates of the initial given (Fig. 1). Total concentration in the contamination spot corresponds to the given values for the initial time moment.

Calculation of the model currents field under the constant north eastern wind with the velocity of 10 m/s was carried out for testing the algorithm of the source capacity identification. Spatial distribution of fields, velocities and coefficients A_H and K which were used as an input information under integrating the model of passive admixture transport for 5 days was obtained as a modeling result.

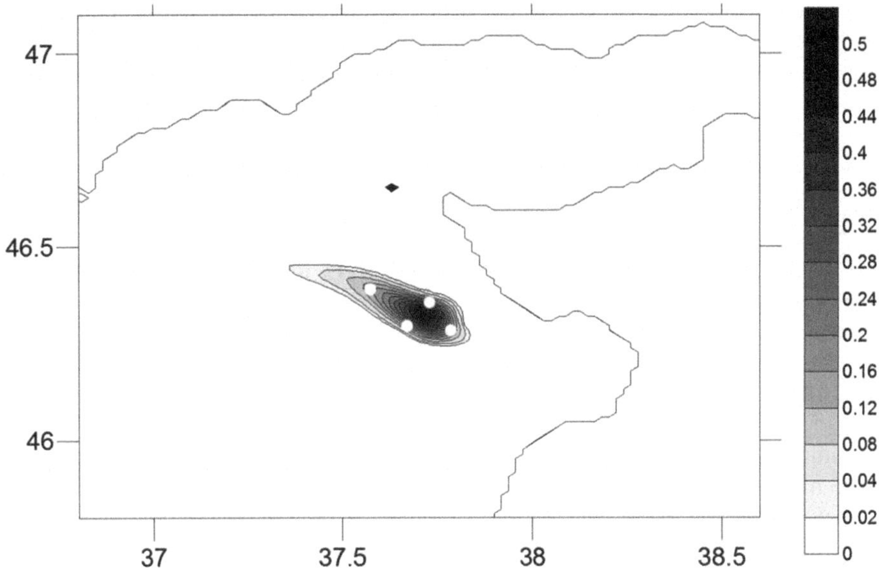

Fig. 3. Location of the instant point source, normalized concentration field and measurements points.

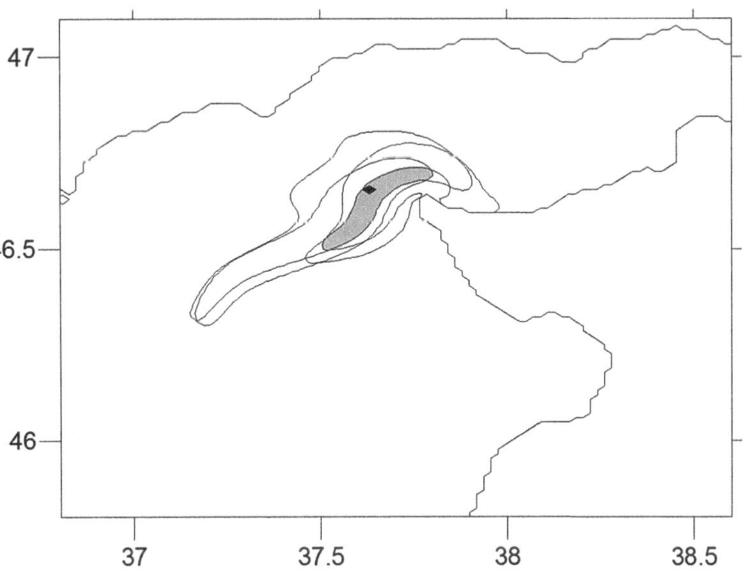

Fig. 4. Solution of the adjoint problem, area G and location of the contamination sources.

Concentration model field was calculated under $Q_S = 1$ and calculation results were standardized according to the maximal value. Real location of the instant point source, which is marked by the black point is represented at the Fig. 3. Four measurements points ($N = 4$) are represented at this figure besides the model concentration field. Result of four adjoint problem integrating is given at the Fig. 4. Area G, which corresponds to these adjoint problem is marked at this figure by the grey color. It is obvious that real location of the source is in the G area. For this point solution of four equations of the system (10) has the following values: $Q_S^1 = 1.032$, $Q_S^2 = 0.9977$, $Q_S^1 = 0.9531$, $Q_S^1 = 1.1446$. We can find $Q_S = 1.0115$ for the point 1 on the base of filtration method [14] from the redefined system (15).

7 Conclusions

- Numerical experiments revealed that identification result depends considerably on the measurements points location. The most accurate reproducing of the real value of the pollution source capacity can be obtained in case when measurements are carried out in the area of concentration field maximal values which would lead to the better conditioning of the solving problem.
- Possibility of using variation assimilation methods and measurements data filtration for solving the wide range of problems connected with estimating the research basin ecological state was represented in this work. The modified schemes for realizing the variation algorithm of the model input parameters identification was suggested. Conditions under which the suggested algorithm has some advantages to the standard approach were obtained.
- The conducted numerical experiments have shown the reliable work of the algorithm of pollution source parameters identification, concerning the passive admixture transport model in the Azov Sea. Good convergence of iteration process and accuracy of determining the concentration field initial distribution for the variation algorithm of identifying the passive admixture concentration initial field was obtained. The results can be used for solving the different problems of ecological origin under studying the anthropogenic contamination sources impact in the Black and Azov seas.

The given work was conducted in frames of the state task on the topic № 0827-2014-0010 “Complex interdisciplinary investigations of the oceanological processes, determining functioning and evolution of the Black and Azov Seas ecosystems on the base of the modern methods of controlling marine environment and grid technologies”.

References

1. Malanotte-Rizzoli, P., Holland, W.R.: Data constraints applied to models of the ocean general circulation. Part II: the transient, eddy-resolving case. J. Phys. Oceanogr. **18**(8), 1093–1107 (1988)
2. Yu, L., O'Brien, J.J.: Variational estimation of the wind stress drag coefficient and the oceanic eddy viscosity profile. J. Phys. Oceanogr. **21**, 709–719 (1991)
3. Marchuck, G.I.: Basic and adjoint equations of the atmosphere and ocean dynamics. Meteorol. Hydrol. (2), 17–34 (1974)
4. Penenko, V.V.: Methods of the Numerical Modeling the Atmosphere Processes. Gidrometeoizdat, Leningrad, 350 p. (1981)
5. Demyshev, S.G., Kochergin, S.V., Kochergin, V.S.: Construction of the functions of impact in the passive admixture transport model. In: Ecological Safety of the Coastal and Shelf Zones and Complex Using the Shelf Recourses. MHI of NAS of Ukraine, Sevastopol, no. 19, pp. 228–233 (2009)
6. Demyshev, S.G., Kochergin, S.V., Kochergin, V.S.: Initialization of the initial data in the three-dimension model of passive admixture transport in the Black Sea. In: Ecological Safety of the Coastal and Shelf Zones and Complex Using the Shelf Recourses. MHI of NAS of Ukraine, Sevastopol, no. 15, pp. 236–241 (2007)
7. Eremeev, V.N., Demyshev, S.G., Kochergin, S.V., Kochergin, V.S.: Identification of the initial data in the three-dimension model of passive admixture transport in the Black Sea. In: Marine Ecological Magazine, Севастополь, no. 3, pp. 36–46 (2007)
8. Demyshev, S.G., Eremeev, V.N., Kochergin, S.V., Kochergin, V.S.: Using the variation approach and solutions of the adjoint problem under identification of the input parameters of the passive admixture transport model in the Black Sea. In: Monography, Stability and Evolution of the Oceanologic Characteristics of the Black Sea Ecosystem. ECOSI-GIDROPHISICA, Sevastopol, 357 p. (2012)
9. Marchuk, G.I.: Mathematical Modeling in the Problem of the Environment. Nauka, Moscow, 320 p. (1982)
10. Ivanov, V.A., Fomin, V.V.: Mathematical Modeling the Dynamic Processes in the Sea – Land Zone. ECOSI-Gidrophisica, Sevastopol, 363 p. (2008)
11. Fomin, V.V.: Numerical model of the Azov Sea waters circulation. In: Scientific Works of UkrNIGMI, no. 249, pp. 246–255 (2002)
12. Kochergin, V.S.: Determination of the passive admixture concentration field according to the initial data on the base of adjoint problem solving. In: Ecological Safety of the Coastal and Shelf Zones and Complex Using the Shelf Recourses, no. 25, vol. 2, pp. 270–376. ECOSI-GIDROPHISICA, Sevastopol (2011)
13. Kochergin, V.S., Kochergin, S.V.: Identification of the pollution source capacity in the Kazantip gulf on the base of variation algorithm. In: Physical Oceanography, no. 2, pp. 69–76 (2015)
14. Strakhov, V.N.: Method of the linear algebraic equations filtration is the base for solution of the linear problems of gravimetry and magnetometry. Report, Academy of Science of USSR, vol. 320, no. 3, pp. 595–599 (1991)

Mechanisms Accounting for Interannual Variability of Advective Heat Transport in the North Atlantic Upper Layer

A. B. Polonsky and P. A. Sukhonos(✉)

Institute of Natural and Technical Systems,
Lenina Street, 28, Sevastopol 299011, Russia
pasukhonis@mail.ru

Abstract. The article describes physical mechanisms accounting for interannual variability of horizontal heat advection in the upper mixed layer (UML) in the North Atlantic in January and July. The data from ocean reanalyses ORA-S3, GFDL and GODAS over 1980–2011 are used for this analysis. The relative contribution of currents' intensity, horizontal temperature gradients and their mutual influence into interannual variations of the advective heat transport in the UML is examined. In the most part of the North Atlantic basin, the variations of currents' intensity are crucial factor accounting for the UML anomalies. The interannual heat advection anomalies in the Guiana current and the Gulf Stream vicinity (before veering off the continental slope) in January and July are caused by temperature gradients variations. In general, the influence of horizontal temperature gradient anomalies transported by abnormal currents in the North Atlantic is small.

Keywords: Horizontal heat advection · Temperature gradient
Current velocity · Upper mixed layer · North Atlantic

1 Introduction

Interannual-to-interdecadal temperature oscillations in the upper ocean layer are determined by natural variability of the ocean-atmosphere system. These temperature oscillations reliably stand out against significant warming of the North Atlantic. It is well known, that this quasi-periodic variability is a joint result of the local atmospheric forcing, the ocean-atmosphere interaction and the internal dynamics of the ocean [1, 2].

The spatio-temporal distribution of interannual-to-interdecadal temperature anomalies in the upper ocean layer reveals an alternation of positive and negative anomalies within all main currents of the North Atlantic. There are various points of view on the prevailing mechanisms of sea surface temperature anomalies (SSTA) generation on interannual-to-interdecadal scale. However, many authors emphasize the importance of heat advection in the formation of observed variability of upper ocean layer characteristics.

A model study [3] has revealed an anticyclonic character of circulation of large-scale temperature anomalies of the upper ocean layer around the North Atlantic

V. Karev et al. (Eds.): PMMEEP 2017, SPRINGERGEOL, pp. 62–70, 2018.
https://doi.org/10.1007/978-3-319-77788-7_8

Subtropical Gyre (STG). These authors showed that the advection of mean temperature by anomalous currents and the advection of temperature anomalies by mean currents are responsible for this movement. The advective character of the SSTA propagation on the decadal scale from the southern part of the Gulf Stream to the North Atlantic Subpolar Gyre (SPG) is recorded from both observations [4–6] and numerical modeling of the ocean dynamics [7, 8]. At the same time, there is another point of view, according to which only a small part of the surface waters of the Gulf Stream reaches the SPG, and the waters that manage to reach the SPG propagate under the mixed layer [9]. Above-mentioned authors found that interannual SSTA do not propagate from the southern part of the Gulf Stream to the SPG and heat transport between the STG and SPG, which is manifested as a part of the Atlantic meridional overturning circulation, is realized in the subsurface layer. Therefore, they noted that the forecasting the SSTA within the SPG by tracking the SSTA along the Gulf Stream is impossible. Therefore, the study of causes of emergence of SSTA in the upper ocean layer of the North Atlantic and their propagation is an actual task.

The presence of high-amplitude decadal oscillations of horizontal heat advection in the upper ocean layer confirms the crucial role of advective heat transport in the North Atlantic for the generation of decadal variability [10]. The advection time of thermohaline anomalies from the Tropical Atlantic to subarctic latitudes determines the phase shift between long-period changes of the heat fluxes at the ocean-atmosphere boundary in the high latitudes of the North Atlantic and the meridional heat transport in the Subtropical Atlantic. This phase shift is one of the possible mechanisms for maintaining the interdecadal variability [11]. A model study of an interaction of the long-term SSTA and the currents' velocity field in the North Atlantic at the scales from months to decades has revealed the negative spatio-temporal feedback of the main modes of temperature and velocity oscillations in the midlatitudes of the North Atlantic [12]. These authors pointed out that this connection might be an important part of the interdecadal variability. At the present, the question of importance of effect of temperature and velocity anomalies on the intra- and inter-annual scale for the longer-term changes of advective heat transport in the upper layer of the North Atlantic Ocean remains open. Therefore, the evaluation of interannual variability of the upper ocean layer temperature due to the anomalies of advective heat transport is important for understanding the mechanisms of climatic variability not only in the North Atlantic, but in the Arctic also [13]. At the same time, the quantitative evaluation of contribution of various processes to the overall variability of horizontal heat advection requires further clarification.

The goal of this paper is to estimate the role of physical mechanisms forming the interannual anomalies of horizontal heat advection within the upper mixed layer (UML) of the North Atlantic in January and July. The estimation is based on the homogeneous and rather long-term data of ocean reanalyses. The use of several independent data sets will provide more reliable conclusions about the features of interaction between the currents and the temperature gradients in the upper layer of the North Atlantic Ocean.

2 Data and Methods

The monthly data of ocean temperature (°C), zonal and meridional components of the ocean current velocity vector (m/s) and mixed layer depth (m) from ocean reanalyses ORA-S3 for 1959–2011 [14], GFDL for 1961–2015 [15] and GODAS for 1980–2016 [16] were used. The area of the North Atlantic is restricted by the following coordinates: 0–60° N, 10–80° W. In order to calculate the mixed layer depth in the ORA-S3 reanalysis, a scheme based on the semi-empirical theory of turbulence is applied. The main idea of the scheme is to calculate the Richardson number in the form by [17]. The mixed layer depth is assumed to be equal to the depth at which the Richardson number reaches a critical value. The mixed layer depth in the GFDL reanalysis is defined by fixed density criterion; namely as a depth where density increases compared to the density at 10 m depth by 0.03 kg/m^3. The mixed layer depth in the GODAS reanalysis is defined as a depth where the temperature deviation from the surface temperature is less than 0.8°C.

According to the above-mentioned data, the zonal (U) and meridional (V) components of the current velocity vector, zonal (T_X) and meridional (T_Y) temperature gradients, zonal (UT_X), meridional (VT_Y) and total horizontal ($UT_X + VT_Y$) heat advection were calculated within variable UML for each month of the entire period.

The present study covers the period from 1980 to 2011, which is common for all the ocean reanalyses. Data for January and July were for the entire analyzed period. Further, the separation of advective heat transports in the regular spatial grid onto average values and anomalies is performed using the following procedure.

The horizontal temperature gradients and the components of current velocity vector are the sum of mean climatic value in January and July ($\overline{\ }$) and the interannual deviations (′):

$$U = \overline{U} + U', V = \overline{V} + V', T_X = \overline{T}_X + T'_X, T_Y = \overline{T}_Y + T'_Y \quad (1)$$

The zonal and meridional heat advection in the UML are as follows:

$$UT_X = \left(\overline{U} + U'\right) \cdot \left(\overline{T}_X + T'_X\right) = \overline{U}\,\overline{T}_X + \overline{U}T'_X + U'\overline{T}_X + U'T'_X, \quad (2)$$

$$VT_Y = \left(\overline{V} + V'\right) \cdot \left(\overline{T}_Y + T'_Y\right) = \overline{V}\,\overline{T}_Y + \overline{V}T'_Y + V'\overline{T}_Y + V'T'_Y. \quad (3)$$

It follows from (2) and (3) that the anomalies of zonal and meridional heat advection are determined by the superposition of following mechanisms responsible for the UML temperature anomalies: the transport of anomalous temperature gradients by mean currents; the transport of mean temperature gradients by anomalous currents; the transport of anomalous temperature gradient by anomalous currents.

After that, the variance for each grid point is calculated using the time series of components of interannual anomalies of zonal and meridional heat advection. Then the ratio of variance of individual component to the total variance of anomalies of zonal and meridional heat advection in the UML was determined. The magnitude of this ratio for each term allows to determine the factors responsible for the generation of interannual anomalies of advective heat transports in the upper layer of the North Atlantic Ocean in winter and summer.

3 Analysis of Results

The interannual variations of transport of mean temperature gradients by anomalous zonal currents $\left(U'\overline{T}_X\right)$ dominate in the UML in January in the eastern part of the Equatorial Atlantic and the STG, the Gulf Stream recirculation zone and the inner part of the SPG. Moreover, the ratio of variance of these anomalies to the total variance of advective heat transport in zonal direction in the region of the West African upwelling and the inner part of the SPG exceeds 80% for all the data sets. The interannual fluctuations of transport of anomalous zonal temperature gradients by mean zonal currents in the UML $\left(\overline{U}T'_X\right)$ are responsible for the development of anomalies of zonal heat advection in January in most of the North Atlantic basin. They are also characterized by the largest variance in the vicinity of intensive currents: the Guiana current, the North Atlantic current, the East Greenland current, the Azores current and Gulf Stream. The ratio of variance of interannual anomalies of zonal advective heat transport in the UML in January formed by temperature gradient anomalies carried by anomalous currents ($U'T'_X$) to the total dispersion of anomalies of zonal heat advection is small in the most part of the North Atlantic. The Sargasso Sea and the central part of the Equatorial Atlantic are exceptions. The relative contribution of this mechanism reaches 80% in these regions (Fig. 1a, b, c).

The contribution of variance which characterizes the interannual fluctuations of transport of mean temperature gradients by anomalies of velocity of meridional currents $\left(V'\overline{T}_Y\right)$ to the total variance of meridional heat advection anomalies makes is at a maximum (>80%) in January in the West African upwelling vicinity, the West and East Greenland currents and the latitudinal band 15–50° N. The vicinity of the Gulf Stream and the Canary Current are exceptions. The variance of interannual variations in the transport of anomalous meridional temperature gradients by mean meridional currents $\left(\overline{V}T'_Y\right)$ in January is high in the central part of the North Atlantic Tropical Gyre and the vicinity of the Labrador Current. Its ratio to the total variance of anomalies of advective heat transport in the meridional direction is more than 50%. The contribution of this mechanism to the total dispersion of meridional heat advection anomalies in the Guiana current and the Gulf Stream area before its veering from the continental slope exceeds 80% for all January data sets. The ratio of variance of interannual anomalies of meridional temperature gradients transported by anomalous currents ($V'T'_Y$) to the total variance of interannual anomalies of meridional heat advection in January is characterized by high values (more than 50%) in the North Equatorial Countercurrent, the Gulf Stream recirculation zone and the inner part of the SPG. In the rest of the North Atlantic basin, the ratio of variance of this component to the total variance of interannual anomalies of meridional heat advection in the UML in January is small (Fig. 1d, e, f).

The interannual variations of transport of mean temperature gradients by anomalous zonal currents $\left(U'\overline{T}_X\right)$ in the UML in July prevail in the inner part of the STG and SPG. The ratio of the variance of these anomalies to the total variance of zonal heat advection anomalies in the inner part of the SPG exceeds 80% for all the data sets.

The spatial distribution of regions with a high variance of transport of anomalous zonal temperature gradients by mean zonal currents $\left(\overline{U}T'_X\right)$ in the UML in July is the

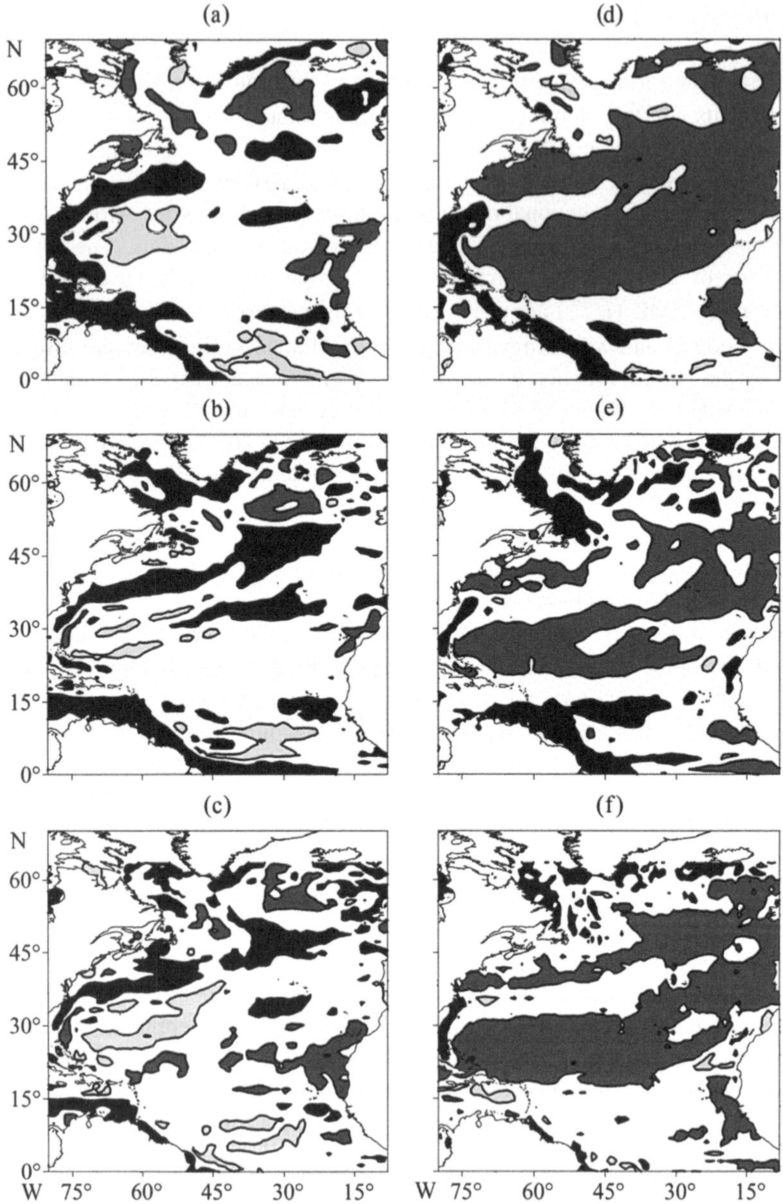

Fig. 1. The ratio (in %) of dispersion of components of heat advection anomalies to the total variance of zonal (a, b, c) and meridional (d, e, f) heat advection anomalies in January in the upper mixed layer of the North Atlantic for the period 1980–2011 according to ORA-S3 (a, d), GFDL (b, e), and GODAS (c, f) data. The type of color filling characterizes the main mechanism: black – the transport of anomalous temperature gradients by mean currents, dark gray – the transport of mean temperature gradients by anomalous currents and light gray – the transport of anomalous temperature gradient by anomalous currents. The isoline characterizes the value of 75%.

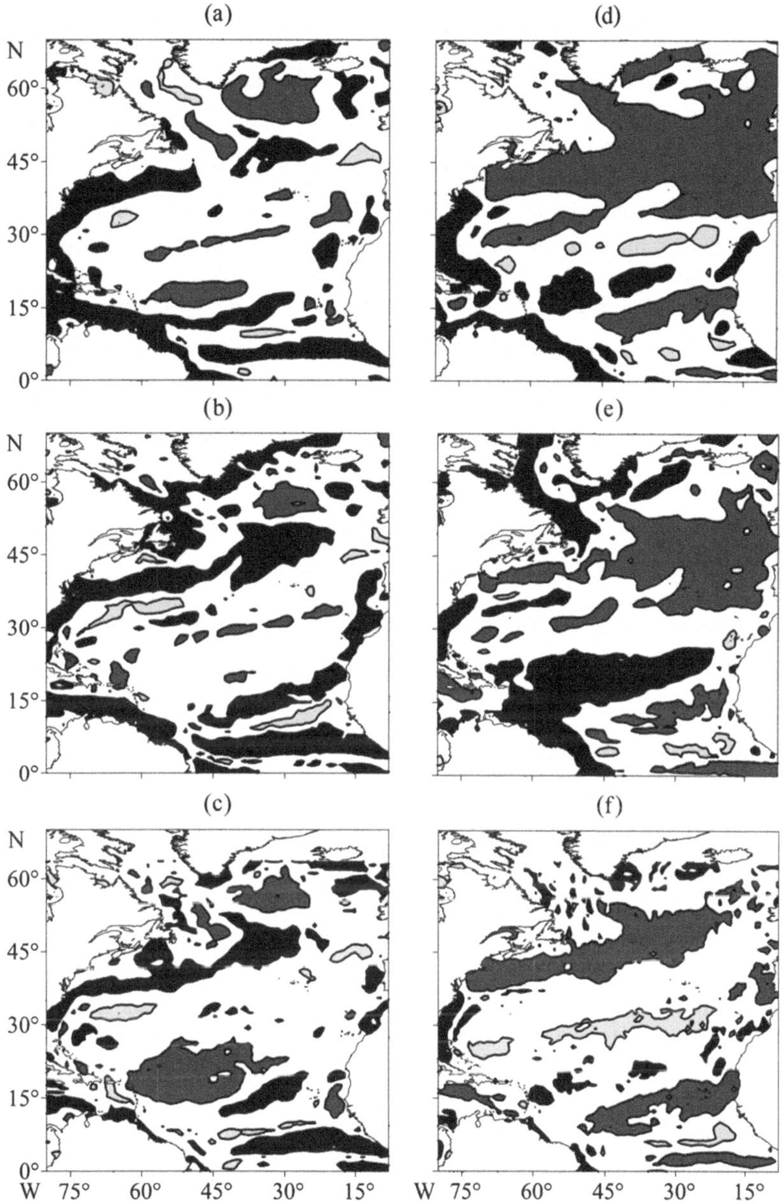

Fig. 2. The ratio (in %) of dispersion of components of heat advection anomalies to the total variance of zonal (a, b, c) and meridional (d, e, f) heat advection anomalies in July in the upper mixed layer of the North Atlantic for the period 1980–2011 according to ORA-S3 (a, d), GFDL (b, e), and GODAS (c, f) data. The type of color filling characterizes the main mechanism: black – the transport of anomalous temperature gradients by mean currents, dark gray – the transport of mean temperature gradients by anomalous currents and light gray – the transport of anomalous temperature gradients by anomalous currents. The isoline characterizes the value of 75%.

same as for January. The ratio of variance of interannual temperature gradient anomalies transported by anomalous zonal currents ($U'T'_X$) to the total dispersion of anomalous zonal heat advection in July is small in the most part of the North Atlantic basin. The small areas in the Sargasso Sea and along 10° N are exceptions. The relative contribution of this mechanism reaches 80% in these regions (Fig. 2a, b, c).

The interannual fluctuations of transport of mean temperature gradients by anomalous meridional currents $\left(V'\overline{T}_Y\right)$ in the UML in July dominate in the eastern part of equatorial zone, the inner part of the North Atlantic tropical gyre, the East Greenland Current and the latitudinal band 30–50° N. The vicinity of the Gulf Stream and the Labrador Current are exceptions. The variance of interannual variations of transport of anomalous meridional temperature gradients by mean meridional currents $\left(\overline{V}T'_Y\right)$ in July is high in the latitudinal band 15–25° N. In the Guiana and Labrador currents as well as in the Gulf Stream region, before its veering from the continental slope, the contribution of this mechanism to the total variance of meridional heat advection anomalies in July is more than 80% for all the data sets. The ratio of variance of interannual anomalies of meridional temperature gradients transported by anomalous meridional currents ($V'T'_Y$) to the total variance of meridional heat advection anomalies in July is large (more than 50%) in the area of the North Equatorial Countercurrent, the Gulf Stream recirculation zone and the inner part of the STG. The ratio of the variance of this component to the total variance of meridional heat advection anomalies in the UML in July is small in the rest of the North Atlantic area (Fig. 2d, e, f).

4 Discussion and Conclusion

It is well known, the heat advection caused by ocean currents is one of the main factors determining the thermal state of the UML. Moreover, in the tropical Atlantic the heat transport completely occurs within the UML [18]. The advection of mean temperature gradients by anomalous currents represents the most important contribution to the recent decadal trend reversal in the SPG heat content [19]. This conclusion is completely confirmed by our results. Variations in the Atlantic water inflow define the main part of interannual variability of climatic parameters in the Barents Sea in the cold season [20]. However, in winter and summer the character of interannual changes of advective heat transports in the UML is different. The greatest contribution of variations of horizontal heat advection to the UML temperature change in the North Atlantic at the interannual scale is observed in winter, and the smallest – in summer [21].

Thus, the quantitative evaluations of contribution of various mechanisms, accounting for the interannual anomalies of horizontal heat advection in the UML in the North Atlantic in winter and summer, are analyzed from the ORA-S3, GFDL and GODAS oceanic reanalyses data for 1980–2011.

The following results are obtained. The interannual anomalies of horizontal heat advection in January and July in the regions of the Guiana Current and the Gulf Stream until its veering from the continental slope are largely formed by variations in the temperature gradients. In winter (summer) the interannual anomalies of zonal heat advection in the inner part of SPG (in the inner parts of STG and SPG) and the anomalies of meridional heat advection in the inner part of STG (in the vicinity of Inter-Tropical Convergence Zone and the North Atlantic Current) are due to changes in the currents intensity. The magnitude of interannual anomalies of horizontal temperature gradient transported by anomalous currents in the UML is generally small in the North Atlantic. The interannual anomalies of zonal heat advection in the Sargasso Sea and the central part of the Equatorial Atlantic are exception in January. The interannual anomalies of zonal heat advection in the Gulf Stream recirculation zone and the interannual anomalies of meridional heat advection in the North Equatorial Countercurrent and the inner part of the STG are exceptions in July.

Acknowledgements. This work was partially supported by research project No. 15-05-02019 of the Russian Foundation for Basic Research (RFBR).

References

1. Levitus, S., Antonov, J.I., Boyer, T.P., Baranova, O.K., Garcia, H.E., Locarnini, R.A., Mishonov, A.V., Reagan, J.R., Seidov, D., Yarosh, E.S., Zweng, M.M.: World ocean heat content and thermosteric sea level change (0–2000 m), 1955–2010. Geophys. Res. Lett. **39**(10), L10603 (2012). https://doi.org/10.1029/2012GL051106
2. Polonsky, A.B.: Oceans, Global Warming Hiatus and Regional Climate Variability. Lambert Academic Publishing, Saarbrucken, 192 p. (2015)
3. Dong, B., Sutton, R.T.: Variability in North Atlantic heat content and heat transport in a coupled ocean-atmosphere GCM. Clim. Dyn. **19**(5), 485–497 (2002)
4. Hansen, D.V., Bezdek, H.F.: On the nature of decadal anomalies in North Atlantic sea surface temperature. J. Geophys. Res. **101**(C4), 8749–8758 (1996)
5. Sutton, R.T., Allen, M.R.: Decadal predictability of North Atlantic sea surface temperature and climate. Nature **388**(6642), 563–567 (1997)
6. Chepurin, G.A., Carton, J.A.: Subarctic and Arctic sea surface temperature and its relation to ocean heat content 1982–2010. J. Geophys. Res. **117**(C6), C06019 (2012). https://doi.org/10.1029/2011JC007770
7. Hakkinen, S.: Decadal air–sea interaction in the North Atlantic based on observations and modeling results. J. Clim. **13**(6), 1195–1219 (2000)
8. Dong, S., Kelly, K.A.: Heat budget in the Gulf Stream region: the importance of heat storage and advection. J. Phys. Oceanogr. **34**(5), 1214–1231 (2004)
9. Foukal, N.P., Lozier, M.S.: No inter-gyre pathway for sea-surface temperature anomalies in the North Atlantic. Nat. Commun. **7**, 11333 (2016)
10. Polonsky, A.B., Kuzmin, A.S.: Decadal variability of hydrometeorological elements in the North Atlantic. Russ. Meteorol. Hydrol. **9**, 51–63 (2000)
11. Polonsky, A.B.: Interdecadal variability in the ocean-atmosphere system. Russ. Meteorol. Hydrol. **5**, 37–44 (1998)
12. Bagno, A.V., Dianskii, N.A., Moshonkin, S.N.: Interaction of the ocean surface temperature anomalies with the circulation in the North Atlantic. Oceanology **36**(5), 652–661 (1996)
13. Panin, G.N., Diansky, N.A.: Climatic variations in the Arctic, North Atlantic, and the Northern Sea Route. Dokl. Earth Sci. **462**(1), 505–509 (2015)
14. Balmaseda, M.A., Vidard, A., Anderson, D.L.T.: The ECMWF ocean analysis system: ORA-S3. Mon. Wea. Rev. **136**(8), 3018–3034 (2008)

15. Chang, Y.-S., Zhang, S., Rosati, A., Delworth, T.L., Stern, W.F.: An assessment of oceanic variability for 1960–2010 from the GFDL ensemble coupled data assimilation. Clim. Dyn. **40**(3–4), 775–803 (2013)
16. Behringer, D.W., Xue, Y.: Evaluation of the global ocean data assimilation system at NCEP: the Pacific Ocean. In: Proceedings of the Eighth Symposium on Integrated Observing and Assimilation Systems for Atmosphere, Oceans, and Land Surface, Seattle, WA, pp. 11–15. American Meteorological Society (2004)
17. Pacanowski, R.C., Philander, S.G.H.: Parameterization of vertical mixing in numerical models of tropical oceans. J. Phys. Oceanogr. **11**(11), 1443–1451 (1981)
18. Sukhovey, V.F., Camara, T.: Thermal advection in the tropical Atlantic upper layer. Phys. Oceanogr. **6**(6), 399–410 (1995)
19. Piecuch, C.G., Ponte, R.M., Little, C.M., Buckley, M.W., Fukumori, I.: Mechanisms underlying recent decadal changes in subpolar North Atlantic Ocean heat content. J. Geophys. Res. Oceans **122**, 7181–7197 (2017). https://doi.org/10.1002/2017jc012845
20. Alekseev, G.V., Glok, N.I., Smirnov, A.V., Vyazilova, A.E.: The influence of the North Atlantic on climate variations in the Barents Sea and their predictability. Russ. Meteorol. Hydrol. **41**(8), 544–558 (2016)
21. Polonsky, A.B., Sukhonos, P.A.: Interannual variations in the components of heat budget in the upper layer of the North Atlantic in different seasons. Izv. Atmos. Oceanic Phys. **53**(4), 459–466 (2017)

Convective Jets: Volcanic Activity and Turbulent Mixing in the Boundary Layers of the Atmosphere and Ocean

Alexander Vulfson[1,2(✉)], Oleg Borodin[1], and Petr Nikolaev[3]

[1] Institute of Oil and Gas Problems, Russian Academy of Sciences, Moscow 117701, Russia
vulfson@ipng.ru

[2] National Research University Higher School of Economics, Moscow 101000, Russia

[3] National University of Science and Technology MISIS, Moscow 119049, Russia

Abstract. In the article, it is considered a modification of an integral model of unsteady turbulent jet with a presence of pressure force. Stationary solutions of the presented model is compared with well-known analytical results of classical models. It is shown that the inclusion of pressure forces changes dynamic parameters of a jet by about 15%. An analytical solution of a steady forced buoyant jet that corresponds to a volcanic outburst is deduced. An analytical solution for the spontaneous jet of convective surface layer is presented. The simplest model of an ensemble of the buoyant jets of convective surface layer is built. A hydrodynamic formation mechanism of vertical profiles of the turbulent diffusivity and the turbulent statistical moments of the atmospheric surface layer related to the ascent of the jets' system, is formulated.

Keywords: Convective thermal · Convective jet · Forced convective jet Ensemble of convective jets · Eddy diffusivity · Turbulent moments

1 Introduction

Integral models of convective jets (plumes) are widely used for a description of volcanic activity and turbulent mixing within the boundary layers of the atmosphere and ocean.

Modern integral models of unsteady jets are presented in [1–7]. Most of the well-known integral models of buoyant jets and forced plumes use the approximation of the vertical boundary layer. This approximation enables the pressure force in the equation of motion to be neglected.

The present paper discusses modifications of the integral models of a convective jet by including the pressure force. The modifications have exact solutions that can be compared against the well-known analytical solutions of classical models. This comparison allows to estimate the effect of a pressure forced, that changes the vertical velocity and the buoyance of a jet by about 15%.

V. Karev et al. (Eds.): PMMEEP 2017, SPRINGERGEOL, pp. 71–83, 2018.
https://doi.org/10.1007/978-3-319-77788-7_9

In geophysical environments, we can observe both isolated convective plumes [8] and ensemble of buoyant jets [9]. Isolated convective jets form during volcanic outburst into atmosphere. A steady forced convective jet and its model may be considered as a simplest model of this outburst.

An ensemble of convective thermals (bubbles) forms in the atmosphere over a heated surface of the land or ocean. A steady spontaneous jet and its analytical solution may be considered as a simplest model of an isolated element of the thermals' ensemble.

It is shown that an ensemble of the dynamically identical buoyant jets forms the vertical profile of turbulent eddy diffusivity and the profiles of the turbulent moments of velocity and buoyance up to fourth order.

2 Integral Model of a Convective Jet

Consider an unsteady jet (plume) rising in a neutrally stratified atmosphere. Traditional models use the approximation of the vertical boundary layer and some other hypotheses.

Suppose that a convective jet forms over a point source of heat and buoyancy. Experimental measurements in a neutrally stratified environment show that the plume has a near-conical shape with the radios R_w. The shape of the cone is defined by $\partial R_w/\partial z = \alpha_R$, where $\alpha_R \approx 0.1$ is a constant coefficient. The hypothesis about a conical shape of a jet is a part of the model [4].

Consider that the vertical velocity and buoyancy in each horizontal cross section of the jet are assumed to have stepwise Π-shaped profiles with amplitudes $\hat{w}$ and $g\hat{\theta}$. The amplitude equations of the integral model [4] in the boundary-layer approximation with the hypothesis about a conical shape of a jet are

$$\begin{cases} \frac{\partial}{\partial t}\hat{w}R_w^2 + \frac{\partial}{\partial z}\hat{w}^2R_w^2 = g\widehat{\theta}R_w^2, \\ \frac{\partial}{\partial t}\hat{\theta}R_w^2 + \frac{\partial}{\partial z}\hat{\theta}\hat{w}R_w^2 = 0, \quad \frac{\partial}{\partial z}R_w = \alpha_R \end{cases} \tag{1}$$

The equations of motion and heat influx in (1) suggest that the jet does not exchange by vertical momentum and heat with its surroundings.

System (1) must therefore be supplemented with boundary conditions. The boundary conditions for the point sources of heat and momentum are

$$\lim_{z\to 0}\hat{w}^2R_w^2 = P_*(t), \quad \lim_{z\to 0}\hat{\theta}\hat{w}R_w^2 = S_*(t) \tag{2}$$

Here, $P_* = P_*(t) > 0$ and $S_* = S_*(t) > 0$ are the time dependent strengths of momentum and heat fluxes, $[P_*] = m^4/s^2$, $[S_*] = m^3/s$.

The comparison between the numerical calculation by the model [4] and other known integral models [2, 3, 5], that was made in [5], is shown in Fig. 1. For notational convenience, in these graphs we use normalized vertical velocity $\tilde{w}$, buoyance $g\tilde{\theta}$, and altitude $\tilde{z}$, which description provides in models [2–5].

Let us consider an alternative way of deducing the amplitude equations of unsteady convective jet that does not use the approximation of the vertical boundary layer.

According to [10], in the structure of a turbulent jet we can distinguish a coherent system consisting of a sequence of moving eddied that can be stylized as a chain of spherical thermals (see Fig. 2).

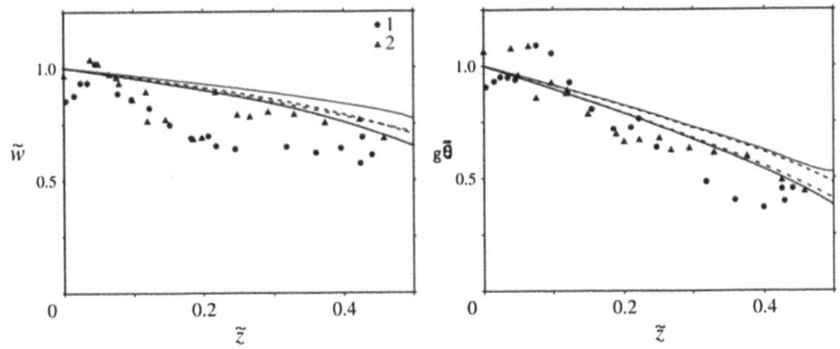

Fig. 1. A comparison between integral models [2–5] and experimental data [2]. (a) The normalized velocity $\tilde{w}$ field; (b) the normalized buoyancy $g\tilde{\theta}$ field. Circles and triangular dots are experimental data. dash-and-dot line – model [1]; solid line – model [2]; thin solid line – model [3, 4]; dotted line – model [5].

Let us deduce an equation of the isolated thermal in the chain. $V_w = V_w(t)$ is the volume of the spherical thermal with the radios $R_w = R_w(t)$. Let us introduce the V_w-averaged mean velocity $\widehat{w}$ and dimensionless potential-temperature fluctuation $\widehat{\theta}$ and the height of the centre of mass of a thermal $\widehat{z} = \widehat{z}(t)$.

In [11], the equation of the motion of a chain spherical thermal was suggested

$$\frac{d}{dt}\widehat{w} = \frac{2}{3}g\widehat{\theta} - \frac{\alpha_R}{R_w}\widehat{w}^2, \quad \frac{d}{dt}\widehat{w} = \frac{2}{3}g\widehat{\theta} - \frac{1}{\widehat{z}}\widehat{w}^2, \tag{3}$$

The factor 2/3 in (3) is related to introduction of a pressure force, that is parametrized by the added-mass, (see [12]).

Based on the concept of a jet as a chain of thermals, we construct an alternative model of a convective jet. Instead of the classical Eq. (1), we consider a modify system, that include a pressure force, in a parametric form

$$\begin{cases} \frac{\partial}{\partial t}\hat{w}R_w^2 + \frac{1}{2}\frac{\partial}{\partial z}\hat{w}^2R_w^2 = \frac{2}{3}g\hat{\theta}R_w^2, \\ \frac{\partial}{\partial t}\hat{\theta}R_w^2 + \frac{\partial}{\partial z}\hat{\theta}\hat{w}R_w^2 = 0, \quad \frac{\partial}{\partial z}R_w = \alpha_R \end{cases} \tag{4}$$

with the boundary conditions (2).

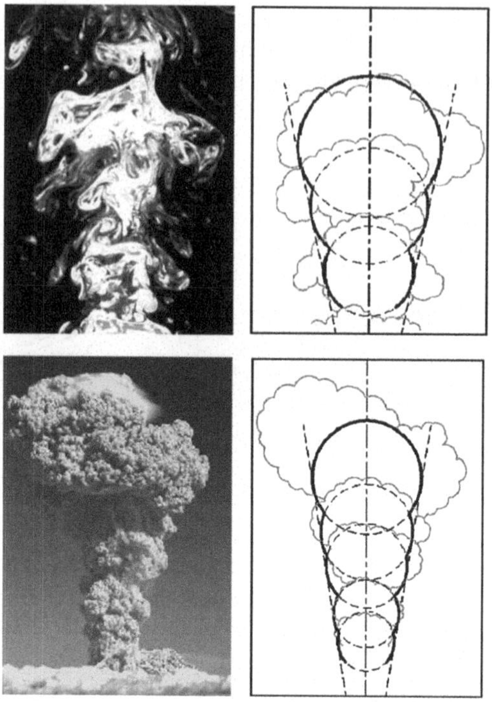

Fig. 2. Internal coherent structure of a turbulent jet and its geometric representation. (a) Submerged turbulent water jet effluent from a round orifice at the Reynolds number 4000 (reproduced from [10]); (b) geometric representation of a turbulent jet in the shape of a chain of spherical bubbles; (c) a convective turbulent ash plume from the eruption of Shiveluch volcano (Kamchatka), September 24, 2014; (d) geometric representation of a convective turbulent plume in the shape of a chain of spherical bubbles.

Let us clarify the physical meaning of the equation of motion in modified system (4). Suppose that $d/dt = \partial/\partial t + \hat{w}\,\partial/\partial z$ is an individual time derivative. Obviously, when $\widehat{w} = \hat{w}$ and $g\widehat{\theta} = g\hat{\theta}$, Eq. (4) turns into the form (3).

It is significant that the Eq. (4) includes an approximation of pressure force and cannot therefore be derived in the vertical boundary-layer approximation.

3 Steady Integral Model of a Convective Jet

The Euler form of the modified stationary Eq. (4) with the boundary conditions (2) is

$$\begin{cases} \frac{1}{2}\frac{d}{dz}\hat{w}^2 R_w^2 = \frac{2}{3} g\hat{\theta} R_w^2, \quad \frac{d}{dz} g\hat{\theta}\hat{w} R_w^2 = 0, \quad R_w = \alpha_R z \\ \lim\limits_{z\to 0} \hat{w}^2 R_w^2 = P_*, \quad \lim\limits_{z\to 0} g\hat{\theta}\hat{w} R_w^2 = gS_* \end{cases} \tag{5}$$

The Lagrange form of the modify steady Eq. (5) with the boundary conditions (2) is

$$\begin{cases} \frac{d}{dz}\hat{w}^2 = \frac{4}{3}g\hat{\theta} - \frac{2}{z}\hat{w}^2, \quad \frac{d}{dz}\hat{\theta}\hat{w} = -\frac{2}{z}\hat{\theta}\hat{w}, \quad R_w = \alpha_R z, \\ \lim_{z\to 0}\hat{w}^2 R_w^2 = P_*, \quad \lim_{z\to 0} g\hat{\theta}\hat{w}R_w^2 = gS_* \end{cases} \tag{6}$$

The steady equations system with the boundary conditions (6) can be transformed to the equivalent form

$$\begin{cases} \frac{1}{3}\frac{d}{dz}\hat{w}^3 = \frac{2}{3}g\hat{\theta}\hat{w} - \frac{1}{z}\hat{w}^3, \quad \frac{d}{dz}g\hat{\theta}\hat{w} = -\frac{2}{z}g\hat{\theta}\hat{w}, \\ \lim_{z\to 0}\hat{w}^2 R_w^2 = P_*, \quad \lim_{z\to 0} g\hat{\theta}\hat{w}R_w^2 = gS_* \end{cases} \tag{7}$$

The systems (5)–(7) can be considered on the unbounded domain $0<z<\infty$.

4 Steady Convective Jet with a Pressure Force

Consider a steady convective jet from a point source of buoyancy with $P_* = 0$ and $S_* > 0$.

The analytical solution to the classical model of a steady jet (1) is

$$R_w(z) = \alpha_R z, \;\; \hat{w}(z) = \left(\frac{3}{4}\right)^{\frac{1}{3}} \alpha_R^{-\frac{2}{3}} (gS_*)^{\frac{1}{3}} z^{-\frac{1}{3}}, \;\; g\hat{\theta}(z) = \left(\frac{3}{4}\right)^{-\frac{1}{3}} \alpha_R^{-\frac{4}{3}} (gS_*)^{\frac{2}{3}} z^{-\frac{5}{3}} \tag{8}$$

The analytical solution to the modify model of a steady jet (4) is

$$R_w(z) = \alpha_R z, \quad \hat{w}(z) = \alpha_R^{-\frac{2}{3}} (gS_*)^{\frac{1}{3}} z^{-\frac{1}{3}}, \quad g\hat{\theta}(z) = \alpha_R^{-\frac{4}{3}} (gS_*)^{\frac{2}{3}} z^{-\frac{5}{3}} \tag{9}$$

We now compare analytical solutions (8) and (9). Given that $(3/4)^{1/3} \approx 0.908$ and $(3/4)^{-1/3} \approx 1.100$, we may conclude that the model of a steady convective jet with the pressure force (5) and models of convective jet (1) with no pressure force give qualitatively close results. However, the inclusion of the pressure forces in the model decreases the amplitude of vertical velocity by about 10% and increases the amplitude of buoyancy.

The power-law dependences of the vertical velocity and buoyancy on the altitude are identical in the both models and correspond to the similarity law for the convective jet, presented in [13].

5 Steady Submerged Jet

Consider an isolated submerged jet from a point source of momentum and heat with $P_* > 0$ and $S_* > 0$. The symmetry axis of the submerged jet is horizontal, so the gravitational acceleration for such a flow is unimportant.

To describe the submerged jet, we use model (5) with $g = 0$. An analytical solution to (5) is

$$R_w(z) = \alpha_R z, \quad \hat{w}^2(z) = \frac{P_*}{\alpha_R^2 z^2}, \quad \hat{\theta}(z) = \frac{P_*^{-1/2}}{\alpha_R z} S_* \tag{10}$$

Relations (10) are consistent with the similarity law for a submerged turbulent jet [14].

6 Steady Buoyant Jet with a Pressure Force

Buoyant jets are of much interest in problems of volcanic outbursts. An extensive review of hydrodynamic models of volcanic outbursts was presented in [8].

An analytical solution to the volcanic outburst problem (5) is found base on an equivalent system (7). Then,

$$\begin{cases} R_w(z) = \alpha_R z, \quad \hat{w}^2(z) = \frac{1}{\alpha_R^2 z^2}\left[P_*^{3/2} + \alpha_R g S_* z^2\right]^{2/3} \\ g\hat{\theta}(z) = \frac{g S_*}{\alpha_R z}\left[P_*^{3/2} + \alpha_R g S_* z^2\right]^{-1/3} \end{cases} \tag{11}$$

We assume that $h_* = P_*^{3/4}/(\alpha_R g S_*)^{1/2}$ is a scaling parameter of height in a buoyant jet. Obviously, equalities (11) turn into (9) at $z >> h_*$; that is, the buoyant jet at high levels turns into an convective jet. When $z \ll h_*$, equalities (11) turn into (10); that is, the buoyant jet at low levels turns into a submerged jet.

A solution similar to (11) also exists in the model described in [15]. The jet radius is calculated from the equation of the kinetic energy of vertical motion

7 Spontaneous Jet with a Pressure Force

The free-convection layer in developed turbulence is the fluid or gas layer adjacent to a heated surface and having a thickness $0 < z/h < 0.1$, where h is the height of the convective layer. The free-convection layer is defined as the layer of a constant buoyancy flux per unit area $gS_\theta = const > 0$, where $[gS_\theta] = m^2/s^3$.

In the atmosphere, the layer of free convection is an unstable surface layer about 100-m thickness in the immediate vicinity of a heated underlying surface.

Let $\Gamma(z) = d\bar{\theta}/dz \le 0$ be a stratification parameter. Then,

$$g\Gamma(z) = g\frac{d\bar{\theta}}{dz} = -\lambda_\theta (gS_\theta)^{2/3} z^{-4/3} \tag{12}$$

where $0.9 \le \lambda_\theta \le 1.1$, according to the atmospheric field observations (e.g., [16, 17]).

Equality (12) follows from the classical Prandtl turbulence theory [18], or from the Obukhov similarity theory [19].

According to [20], spontaneous jets arise in unstable atmospheric layers. It is significant that spontaneous jets gain energy from the unstable layer and so do not require any additional heat and momentum sources at the underlying surface.

Extending the modified integral model (5) to the case of unstable stratification, we obtain equations for a stationary isolated spontaneous jet that include pressure. Assuming $\lambda_\theta = 1$ in the form (12), we have

$$\begin{cases} \frac{1}{2}\frac{d}{dz}\hat{w}^2R_w^2 = \frac{2}{3}g\hat{\theta}R_w^2, \quad \frac{d}{dz}g\hat{\theta}\hat{w}R_w^2 = (gS_\theta)^{2/3}z^{-4/3}\hat{w}R_w^2, \quad R_w = \alpha_R z, \\ \lim_{z\to 0}\hat{w}^2R_w^2 = 0, \quad \lim_{z\to 0}g\hat{\theta}\hat{w}R_w^2 = 0 \end{cases} \tag{13}$$

The analytical solution to (13) is

$$R_w(z) = \alpha_R z, \quad \hat{w}(z) = \frac{1}{2}(gS_\theta)^{1/3}z^{1/3}, \quad g\hat{\theta}(z) = \frac{1}{2}(gS_\theta)^{2/3}z^{-1/3} \tag{14}$$

Relations (14) are consistent with the similarity law for a spontaneous turbulent jet obtained in [20].

It is useful to compare exact solution (14) to that of the classical integral model of a spontaneous jet [20] in the boundary-layer approximation

$$\begin{cases} \frac{d}{dz}\hat{w}^2R_w^2 = g\hat{\theta}R_w^2, \quad \frac{d}{dz}g\hat{\theta}\hat{w}R_w^2 = (gS_2)^{2/3}z^{-4/3}\hat{w}R_w^2, \quad R_w = \alpha_R z, \\ \lim_{z\to 0}\hat{w}^2R_w^2 = 0, \quad \lim_{z\to 0}g\hat{\theta}\hat{w}R_w^2 = 0 \end{cases} \tag{15}$$

The solution to (15) is

$$R_w(z) = \alpha_R z, \quad \hat{w}(z) = \frac{\sqrt{3}}{4}(gS_2)^{1/3}z^{1/3}, \quad g\hat{\theta}(z) = \frac{1}{2}(gS_2)^{2/3}z^{-1/3} \tag{16}$$

Compare analytical solutions (14) and (16). The power-law dependences of the vertical velocity and buoyancy on the height are identical in both models. Given that $(\sqrt{3}/4) \approx 0.43$, we may conclude that the inclusion of the pressure forces in the model conserves the buoyancy and increases the amplitude of vertical velocity by about 15%.

The obtained results shows that introduction of the modify model (13) into the mathematical description of the ensemble of thermals may influence onto the turbulent moments of the surface layer in the framework of the mass-flux parametrization.

8 Ensemble of the Dynamically Identical Jets and Turbulent Diffusivity Within the Convective Boundary Layer

It is known that in the turbulent convective layer an ensemble of convective elements forms over a horizontally homogenous heated surface (see [9]).

An idea of what an ensemble of thermals looks like can be gained from laboratory modeling at large Rayleigh numbers. The results of laboratory experiments [21], presented in Fig. 3, clearly exhibit the fine structure of the surface layer.

Let h be a height of a convective layer. It should be noted that the ensemble of convective jets existed mainly within the lower part of a convective layer $0<z/h<0.5$. Within the upper part of a layer $0.5<z/h<1$, the amount of jets is low.

We construct a "generalized" convective jet based on the hypothesis that within the surface layer $0<z/h<0.1$ parameters of the "generalized" jet are defined as parameters of the spontaneous jet (14); and within the mixing layer $0.1<z/h<0.5$ they defined as parameters of the convective jet (9).

Using the results of laboratory experiments [21] and the ensemble model [22], we assume that the convective surface layer is packed with the system of identical "generalized" convective jets.

Let us consider the convective surface layer. Suppose that gS_θ is a buoyance flux at the underlying surface. According to [23], we introduce Deardorff's parameter of velocity and buoyance

$$w_D = h^{1/3}(gS_\theta)^{1/3}, \quad g\theta_D = h^{-1/3}(gS_\theta)^{2/3} \tag{17}$$

Based on the concept of statistical ensemble, we deduce a relationship for the heat eddy diffusivity. In the framework of the semi-empirical theory of turbulence by Prandtl [16], it is assumed that

$$<K_h> \;=\; <w> l_P, \quad \frac{l_P}{h} = k_v \frac{z}{h}\left(1 - \frac{z}{h}\right)^2 \tag{18}$$

Where $<w>$ is the average velocity of updraft flows; l_P is Prandtl's mixing length; $k_v = 0.4$ is von Karman constant.

Substituting the vertical velocity of the "generalized" convective jet into (18) and rewriting deduced relationships with regard to Deardorff's parameters (17) yield

$$\frac{<K_h>}{w_D\, h} \sim \begin{cases} (z/h)^{2/3}[1-(z/h)]^2 & when \quad 0.1<z/h<0.5. \\ (z/h)^{4/3} & when \quad 0<z/h<0.1 \end{cases} \tag{19}$$

Well-known approximations of the real heat eddy diffusivity K_h are

$$\frac{K_h}{w_D\, h} \sim \begin{cases} (z/h)[1-(z/h)]^2 & when \quad 0.1<z/h<0.5. \\ (z/h)^{4/3} & when \quad 0<z/h<0.1 \end{cases} \tag{20}$$

An approximation for K_h in the surface layer $0<z/h<0.1$ was deduced in [19] within the framework of similarity theory. An approximation for K_h in the mixing layer $0.1<z/h<0.5$ was deduced by numerical modeling in [24].

Fig. 3. Ensemble of thermals in the form of dense salt fingers descending in a water layer, according to [21]. The descending motion of thermals is visualized by adding fluorescein to the salt and illuminating fingers through a silt.

Similarity between approximations (19) and (20) means that the ensemble of the "generalized" convective jets is a main enabler of the turbulent diffusivity within the boundary layer.

9 Ensemble of the Dynamically Identical Spontaneous Jets and the Turbulent Moments of the Convective Surface Layer

Let us consider an ensemble of the identical spontaneous jets in the surface layer. Based on an integral model similar to (14), we deduce the second statistical turbulent moments of vertical velocity and buoyance. Then

$$\begin{cases} \dfrac{<\widehat{w}^2>}{w_D^2} = \dfrac{\sigma}{4}\dfrac{(gS_\theta)^{2/3}z^{2/3}}{w_D^2} = \dfrac{\sigma}{4}\dfrac{(gS_\theta)^{2/3}h^{2/3}}{w_D^2}\left(\dfrac{z}{h}\right)^{2/3} - \dfrac{\sigma}{4}\left(\dfrac{z}{h}\right)^{2/3} \\ \\ \dfrac{<(g\widehat{\theta})^2>}{(g\theta_D)^2} = \dfrac{\sigma}{4}\dfrac{(gS_\theta)^{4/3}z^{-2/3}}{(g\theta_D)^2} = \dfrac{\sigma}{4}\dfrac{(gS_\theta)^{4/3}h^{2/3}}{(g\theta_D)^2}\left(\dfrac{z}{h}\right)^{-2/3} = \dfrac{\sigma}{4}\left(\dfrac{z}{h}\right)^{-2/3} \end{cases} \tag{21}$$

Where $<\widehat{w}^2>$ and $<(g\widehat{\theta})^2>$ are statistical turbulent moments, related to a system of spontaneous jets and downdrafts; $\sigma_{ww}/4 = \sigma_{\theta\theta}/4 \approx 1.1$ is a dimensionless constant, which depends on the relative area of the updraft flows, the density of jets per area unit, and a description method of interference between jets.

The statistical moments (21) are defined by common coefficients $\sigma_{ww} = \sigma_{\theta\theta}$. If we use another integral model, the universality of this coefficient may not exist.

We assume that $\overline{w^2}$ and $\overline{\theta^2}$ are the real second order moments of vertical velocity and variance of potential temperature of the atmospheric surface layer. According to results of dimension theory [16], transformed by Deardorff's parameters (17), we deduce

$$\frac{\overline{w^2}}{w_D^2} = \lambda_{ww}\left(\frac{z}{h}\right)^{2/3}, \quad \frac{\overline{(g\theta)^2}}{(g\theta_D)^2} = \lambda_{\theta\theta}\left(\frac{z}{h}\right)^{-2/3} \tag{22}$$

According to [25], the empirical constant values are equal, $\lambda_{ww} = 1.8$ and $\lambda_{\theta\theta} = 1.8$.

Field data of the second moments of vertical velocity and buoyance [25] and their approximations (22) are presented in Fig. 4.

Similar results can be deduced for the third and fourth order moments of vertical velocity and buoyance. Relationships of the higher moments of vertical velocity, due to (14), are

$$\begin{cases} \frac{<\widehat{w}^3>}{w_D^3} = \frac{\sigma_{www}}{8}\frac{(gS_\theta)z}{w_D^3} = \frac{\sigma_{www}}{8}\frac{(gS_\theta)h}{w_D^3}\left(\frac{z}{h}\right) = \frac{\sigma_{www}}{8}\left(\frac{z}{h}\right) \\ \frac{<\widehat{w}^4>}{w_D^4} = \frac{\sigma_{wwww}}{16}\frac{(gS_\theta)^{4/3}z^{4/3}}{w_D^4} = \frac{\sigma_{wwww}}{16}\frac{(gS_\theta)^{4/3}h^{4/3}}{w_D^4}\left(\frac{z}{h}\right)^{4/3} = \frac{\sigma_{wwww}}{16}\left(\frac{z}{h}\right)^{4/3} \end{cases} \tag{23}$$

Where $<\widehat{w}^3>$ and $<\widehat{w}^4>$ are statistical convective turbulent moments, related to a system of spontaneous jets and downdraft flows; $\sigma_{www}/8 \approx 1.1$ and $\sigma_{wwww}/16 \approx 0.9$ are dimensionless constants, which depends on the relative area of the updraft flows, the density of jets per area unit, and the description method of interference between jets.

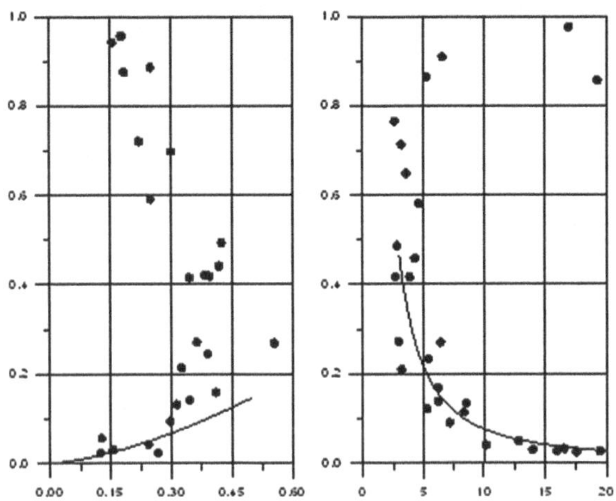

Fig. 4. Field data of the second moments of vertical velocity and buoyance and their approximations. (a) The dots are normalized field data [25]; the line is the approximation of the second moment $\overline{w^2}/w_D^2$ (22). (b) The dots are normalized field data [25]; the line is the approximation of the second moment $\overline{\theta^2}/\theta_D^2$ (22).

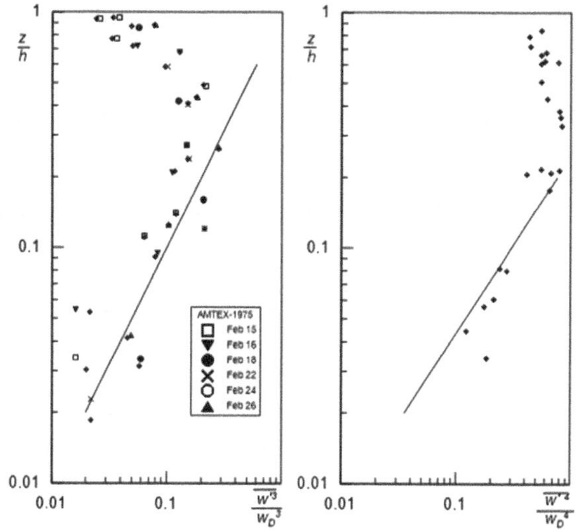

Fig. 5. Field data of the third and fourth moments of vertical velocity and their approximations. (a) The dots are normalized field data [25]; the line is the approximation of the second moment $\overline{w^3}/w_D^3$ (24). (b) The dots are normalized field data [27]; the line is the approximation of the fourth moment $\overline{w^3}/w_D^3$ (24).

Let us assume that $\overline{w^3}$ and $\overline{w^4}$ are the real third and fourth turbulent moments of vertical velocity of the atmospheric surface layer. Approximations of them, that deduced in the framework of dimension theory in [16, 26], are

$$\frac{\overline{w^3}}{w_D^3} = \lambda_{www}\left(\frac{z}{h}\right), \quad \frac{\overline{w^4}}{w_D^4} = \lambda_{wwww}\left(\frac{z}{h}\right)^{4/3} \tag{24}$$

where $\lambda_{www} \approx 1$ and $\lambda_{wwww} \approx 6.5$, according to [25, 27].

Field observation data of third and fourth moments of vertical velocity [25, 27] and their approximations by power-law profiles (24), are shown in Fig. 5.

According to both sets of relationships (21), (22) and (23), (24), the power-law dependences of statistical and real moments of vertical velocity and buoyance variations on the dimensionless altitude z/h are identical. Hence, an ensemble of spontaneous jets forms a dependence of momentum and heat diffusivity on the altitude within the convective surface layer.

When a model of the ensemble of thermals is well-balanced, a "principle of proportionality" is fulfilled. It means that a set of statistical turbulent moments $<\widehat{w}^{\,2}>$ and $<\widehat{(g\theta)}^{\,2}>$ must be proportional to a set of real turbulent moments $\overline{w^2}$ and $\overline{\theta^2}$. Naturally, the "principle of proportionality" must be true to the higher order moments. A wrong choice of an integral model of spontaneous jets leads to a failure of a proportionality between the turbulent moments.

10 Conclusion

Presented results show that the classical model of convective thermals and jets can be modified by including a pressure force.

It is demonstrated that the modified integral model of convective jets retain a power-law dependence of the vertical velocity and buoyance on the altitude, that deduced in the classical models. For convective jets, that rising within a neutrally stratified atmosphere, an introduction of a pressure force increases the amplitude of buoyance and decreases the amplitude of vertical velocity. The total varying of the amplitudes is about 10%. For spontaneous jets, that rising within an unstable stratified atmosphere, an introduction of a pressure force retain the amplitude of buoyance and increases the amplitude of vertical velocity by about 15%.

The relatively weak influence of the pressure force on the parameters of a jet enables the vertical boundary-layer approximation in the theory of buoyant jets to be regarded as a first approximation.

An isolated forced convective jets are effective for describing industrial emissions and volcanic outbursts into the atmosphere.

Ensembles of buoyant jets form vertical profiles of turbulent moments and eddy diffusivity within convective surface layers.

This study was supported by the Russian Foundation for Basic Research, grant no. 15-05-068491-a.

References

1. Delichatsios, M.A.: Time similarity analysis of unsteady buoyant plumes. J. Fluids Mech. **93**(2), 241–250 (1979)
2. Yu, H.-Z.: Transient plume influence in measurement of convective heat release rates of fast-growing fires using a large-scale fire products collector. Trans. ASME **112**, 186–191 (1990)
3. Vul'fson, A.N.: Convective-region top front propagation in a uniform medium under the action of point, linear, and plane heat and momentum sources. Fluid Dyn. **36**(3), 418–428 (2001)
4. Vul'fson, A.N., Borodin, O.O.: Self-similar propagation regimes of a nonstationary high-temperature convective jet in the adiabatic atmosphere. J. Appl. Mech. Tech. Phys. **42**(2), 255–261 (2001)
5. Scase, M.M., Hewitt, R.E.: Unsteady turbulent plume models. J. Fluid Mech. **697**, 455–480 (2012)
6. Craske, J., van Reeuwijk, M.: Generalised unsteady plume theory. J. Fluid Mech. **792**, 1013–1052 (2016)
7. Woodhouse, M.J., Phillips, J.C., Hogg, A.J.: Unsteady turbulent buoyant plumes. J. Fluid Mech. **794**, 595–638 (2016)
8. Costa, A., Suzuki, Y.J., Cerminara, M., Devenish, B.J., Ongaro, T.E., Herzog, M., Van Eaton, A.R., Denby, L.C., Bursik, M., Vitturi, M.D.M., Engwell, S.: Results of the eruptive column model inter-comparison study. J. Volcanol. Geoth. Res. **326**, 2–25 (2016)
9. Vulfson, A.N., Borodin, O.O.: System of convective thermals as a generalized ensemble of Brownian particles. Phys. Usp. **59**(2), 109 (2016)

10. Shraiman, B.I., Siggia, E.D.: Scalar turbulence. Nature **405**(8), 639–646 (2000)
11. Vulfson, A.N., Nikolaev, P.V.: Integral bubble and jet models with pressure forces. Izv. Atmos. Oceanic Phys. **53**(4), 419–427 (2017)
12. Sedov, L.I.: Mechanics of Continuous Media, 6th edn. LAN, St. Petersburg (2004)
13. Zel'dovich, B.Y.: Limiting laws of free-rising convective currents. Zh. Eksp. Teor. Fiz. **12**, 1463–1465 (1937)
14. Abramovich, G.N.: Theory of Turbulent Jets. Nauka, Moscow (1984)
15. Priestly, C.H.B., Ball, F.K.: Continuous convection from an isolated source of heat. Quart. J. Roy. Meteor. Soc. **81**(348), 144–157 (1955)
16. Priestly, C.H.B.: Turbulent Transfer in the Lover Atmosphere. University Chicago Press, Chicago (1959)
17. Kader, B.A., Yaglom, A.M.: Mean and fields and fluctuation moments in unstably stratified turbulent boundary layers. J. Fluid Mech. **212**, 637–662 (1990)
18. Prandtl, L.: Meteorogische anwendung der stromungslehre. Beitr. Phys. fr. Atmos. **19**(3), 188–202 (1932)
19. Obukhov, A.M.: Turbulence in thermally inhomogeneous atmosphere. Trudy Inst. Teor. Geofiz. Akad. Nauk SSSR **1**, 95–115 (1946)
20. Batchelor, G.K.: Heat convection and buoyancy effects in fluids. Quart. J. Roy. Met Soc **80**(345), 339–358 (1954)
21. Huppert, H.E., Turner, J.S.: Double-diffusive convection. J. Fluid Mech. **106**, 299–329 (1981)
22. Vulfson, A.N., Borodin, O.O.: An ensemble of dynamically identical thermals and vertical profiles of turbulent moments in the convective surface layer of atmosphere. Russ. Meteorol. Hydrol. **34**(8), 491–500 (2009)
23. Deardorff, J.W.: Convective velocity and temperature scales for the unstable planetary boundary layer and for Raylegh convection. J. Atmos. Sci. **27**(8), 1211–1212 (1970)
24. Troen, I., Mahrt, L.: A simple model of the atmospheric boundary layer: sensitivity to surface evaporation. Bound.-Layer Meteorol. **37**, 129–148 (1986)
25. Lenschow, D.H., Wyngaard, J.C., Pennel, W.T.: Mean field and second-moment budgets in a baroclinic, convective boundary layer. J. Atmos. Sci. **37**(6), 1313–1326 (1980)
26. Vulfson, A.N., Volodin, I.A., Borodin, O.O.: Local similarity theory and universal profiles of turbulent characteristics in the convective boundary layer. Russ. Meteorol. Hydrol. **10**, 1–10 (2004)
27. Gryanik, V.M., Hartman, J.: A turbulence closure for the convective boundary layer based on a two-scale mass-flux approach. J. Atmos. Sci. **59**(18), 2729–2744 (2002)

Theoretical and Experimental Evaluation of Formation Fluid Composition Influence on Filtration and Elastic Properties of Porous Media

Daniil Karmanskiy[1(✉)] and Andrey Maltsev[1,2]

[1] Department of Oil and Gas Fields Development and Operation, Saint-Petersburg Mining University, Saint-Petersburg, Russia
karmanskiy.da@yandex.ru, maltsev.postbox@gmail.com
[2] Gazpromneft NTC LLC, Saint-Petersburg, Russia

Abstract. Over the last few years interest in the multidisciplinary approach to modeling and evaluating processes associated with the development of oil and gas fields has incredibly increased. The authors of the article studied the effects of rock on the filtration-capacitive, physical and mechanical properties of the rock. The coefficients of porosity, permeability and cementation factor change were estimated taking into account water saturation of rock samples.

Keywords: Geomechanics · Poroelasticity · Oil and gas reservoir development
Oil and gas well production
Correlations of soil and rock properties in geotechnical engineering

1 Introduction

The existing integrated approaches for accounting of poroelastic effects of porous media in hydrodynamic modeling are based on the Bio equation [1]. But the mathematical model of rock compressibility is mainly used to assess the change in porosity. The company developer of one of the most common hydrodynamic simulator indicates an increase in the accuracy of adaptation and forecast when using a combined model of hydrodynamics and geomechanics [2]. In the simulator, this possibility is realized by introducing a permeability multiplier as a tabular function of the maximum and minimum main stress [3]. But this function can only be obtained after special core tests.

Authors of this article uses alternative mathematical model of Nikolaevsky [4], which describes the dependence of permeability on reservoir pressure. In equations [4] there is a parameter of rock cementation. Its change has been experimentally proved in a series of experiments described in this article.

Authors assume that in terrigenous reservoirs, the rock cementation parameter strongly depends on the fluid content. Clay is the most unstable component in the reservoir, but at the same time it makes up the cement of the rock, that creates bonds between the grains. Change in the properties of clay in the rock affects the entire solid mass as a whole. In the articles about modeling of low salinity water saturation the

V. Karev et al. (Eds.): PMMEEP 2017, SPRINGERGEOL, pp. 84–89, 2018.
https://doi.org/10.1007/978-3-319-77788-7_10

physical model is based on the swelling of clays and fines migration. The current understanding of dynamical processes of reservoir does not take into account changes in the physical and mechanical properties of the rock. A prerequisite for such an opinion is a series of experiments [5–7], where the dependence of destruction properties of oil rock on its water saturation was revealed.

Earlier the Berea sandstone samples were used in experiments, where researchers obtained the dependence of the breaking load, Young modulus and friction angle on the water saturation of the samples [8].

In this work experiments were carried out on samples of the BS-11 formation located in Western Siberia. Characteristics of the samples: sandstones with a predominant grain size of 0.06–0.12 mm; the terrigenous material is sorted into medium degree; quartz - 32–44%, feldspars - 41–45%, fragments of rocks - 10−20%, mica - 1−8%; cement consists of clay, distributed unevenly, contained in an amount of 5−10%; the clay composition is predominantly kaolinite, chlorite; according to the analogues, the presence of mixed clay formations and a small fraction of montmorillonite in the clay composition is observed; that clay formation can multiply swell when the mineralization of water changes [9].

2 Experimental Research

For the experiments, a special test procedure was developed:

- preliminary preparation of samples with saturation of their model oil and water in specified proportions;
- tests for determine the rock strength, elastic and deformation characteristics of saturated samples.

Prepared samples were tested using unconfined compressive strength (UCS) method for determine deformation characteristics, Young's modulus and Poisson's ratio [10, 11]. In total, 7 samples were tested with a ratio of diameter to height of 1:2, the diameter of the samples was 30 mm. The samples were saturated with mixtures of model oil (kerosene) and slightly mineralized water in proportions corresponding to a number of water saturation: 0%, 25%, 50%, 75%, 100%. Initially, the samples were saturated with kerosene, and then placed in the evacuated water. Water under the influence of capillary forces gradually penetrated into the sample. The degree of water saturation was determined by determining the total density and subsequent recalculation, corrected for a decrease in the pore volume. On the destroyed samples, additional tests were carried out using spherical indentors, which made it possible to determine the basic mechanical properties of the rock [12]. The test results are shown in the form of graphs in Figs. 1, 2, 3 and 4, Table 1.

In addition to modifying the elasticity modulus and the strength of the samples, depending on the water saturation, a gradual formation and an increase in the zone of plastic deformation is noted: Fig. 2.

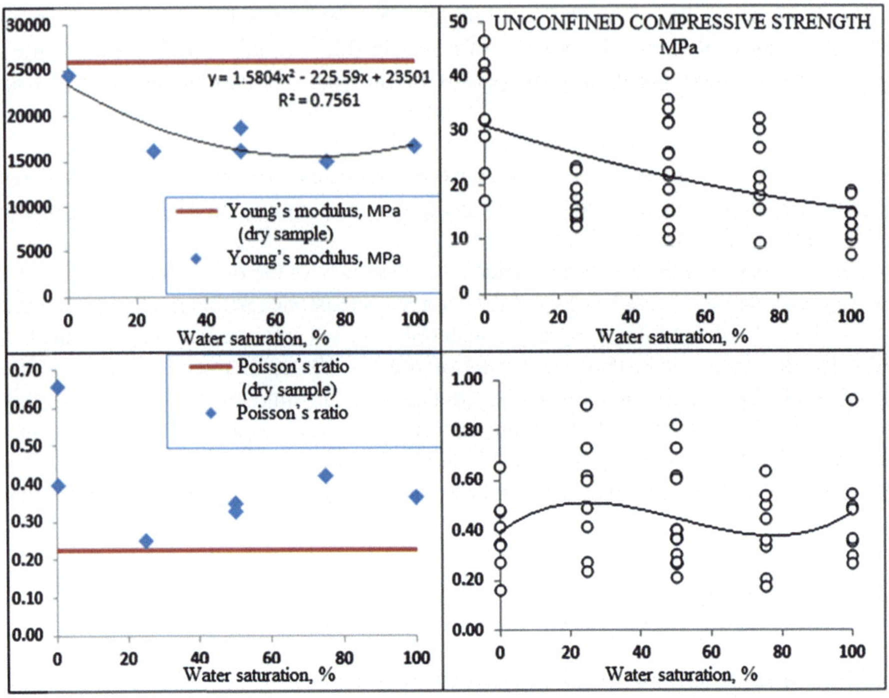

Fig. 1. The results of tests to determine the physical and mechanical properties of the rock.

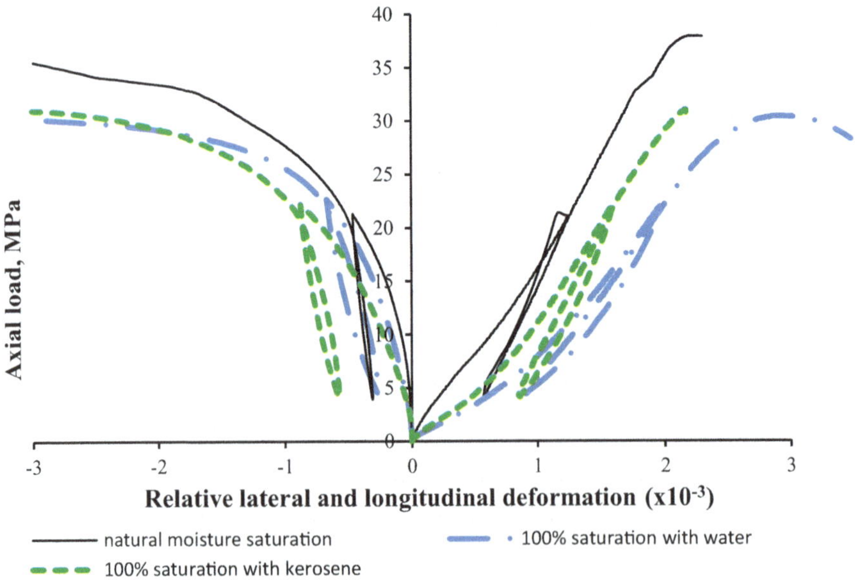

Fig. 2. The stress-strain graphs from a different kind of saturation.

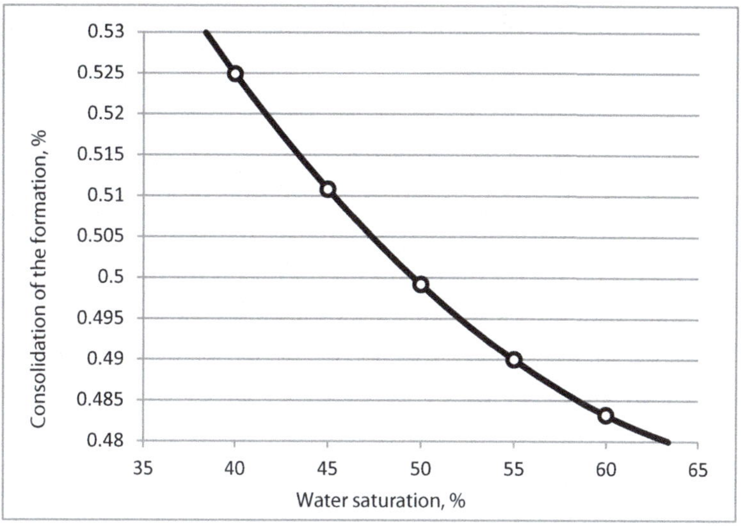

Fig. 3. Dependence of rock consolidation from water saturation.

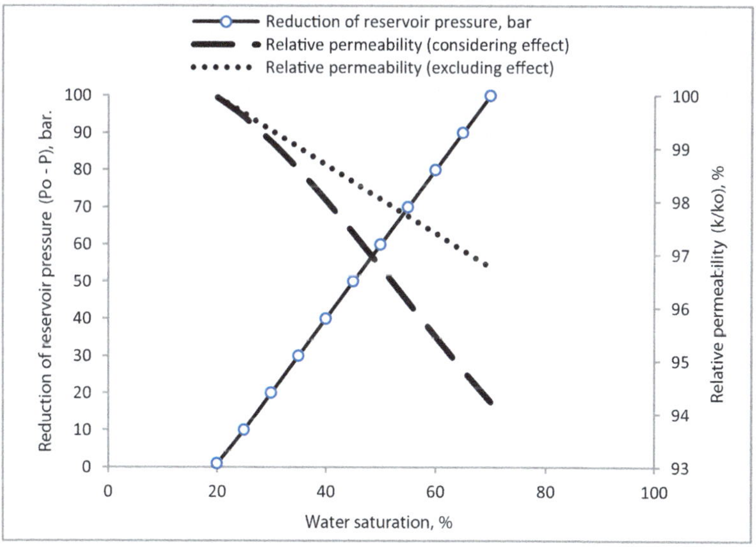

Fig. 4. Dependence of relative permeability and reservoir pressure from water saturation.

Table 1. Results of the UCS studies.

Sample №	Fracture load, N	Tensile strength, MPa	Modulus of deformation, MPa	Young's modulus, MPa	Poisson's ratio
2–11	26807	37.94	16459.60	25977.26	0.23
2–17	21950	31.07	16040.06	24459.38	0.40
2–23	18944	26.94	11000.34	16124.33	0.25
2–18	17027	24.10	11241.48	16134.32	0.35
2–14	18288	25.89	11717.07	14969.21	0.42
2–13	21462	30.40	13238.93	16592.70	0.37

3 Results and Discussion

The authors of the article assume that the prevailing effect of influence of water saturation on the changes in strength characteristics of the rock is a change of strength characteristics of the rock skeleton composed mainly of clays. The level of influence of the described effect is particularly high when clays includes formations with inclination to swelling. From the point of view of mathematical analysis, this effect is described by introducing a dependence of the cementation coefficient ($\beta_1 K$) on water saturation. The recalculation of the change in consolidation index from water saturation according to the test results is shown in Figs. 3 and 4.

To determine the change in effective permeability, relative phase permeability was taken using Corey model based on real data from filtration tests. To calculate the change in permeability, it was taken an equation describes the dependence of the change in porosity on permeability under mechanical influence [4].

$$k = k_0 \left(\frac{m}{m_0} \right)^{a_k / a_m} \tag{1}$$

where k_0, m_0 – permeability and porosity at initial effective pressure; a_k, a_m, – coefficients of change in permeability and compressibility of pores, 1/Pa.

Then the associated model (permeability/water saturation/porous pressure) was calculated in 2 versions: taking and not taking into account the change in consolidation of the rock: Figs. 3 and 4

$$p_{ef} = -\sigma_{11}^{f} + (1 - m_0)\beta_0 K p_0 = q - (1 - m_0)\beta_1 K p_0 \tag{2}$$

where p_{ef} – effective pressure, Pa; q – full applied load, N; m_0 – porosity, %; $\beta_0 K$, $\beta_1 K$ – cementation of the porous medium, Darsi/Pa; p_0 – initial formation pressure, Pa.

In conclusion, it should be mentioned that the change in depression to the formation may have both a positive and negative impact on permeability. The direction of change in permeability under conditions of extreme loading depends on the type of rock. Therefore it is extremely necessary to conduct mechanical core research at the stage of assessment and design [13].

4 Conclusions

- Using tests methods similar to the conducted studies, as well as using the poroelasticity equations, it is possible to increase the adaptation and prediction of reservoir characteristics when selecting well operation regimes and selecting an agent for waterflooding.
- Carrying out a series of tests taking into account the mineral composition of the rock will make it possible to single out the model dependences of cementation coefficient changing under influence of mineralogical composition and mineralization of the produced water. Analysis of the correlations will make it possible to evaluate the strength characteristics of rock fields with a limited number of core studies.
- The physical, mechanical and filtration properties of oil and gas reservoirs are affected by the ratio of oil and water in the rock, the mineralization of reservoir and injected water, the mineralogical composition of the reservoir.

References

1. Biot, M.A.: General theory of three-dimensional consolidation. J. Appl. Phys. **2**(12), 155–164 (1941)
2. Koutsabeloulis, N., Xing, Z.: 3D reservoir geomechanical modeling in oil/gas field production. In: SPE Saudi Arabia Section Technical Symposium, Al-Khobar, pp. 1–14 (2009)
3. Schlumberger ECLIPSE: Technical description. Schlumberger (2007)
4. Nikolaevsky, V.N., Basniev, K.S., Gorbunov, A.T.: Mechanics of Saturated Porous Media. Nedra, Moscow (1970)
5. Kadet, V.V., Chagirov, P.S.: Effects of ion-exchange processes in the joint flow of oil and electrolyte solutions in clay-containing reservoirs. Oilfield Rev. **1**, 68–72 (2016)
6. Erke, S.I., Volokitin, Y.E., Edelman, I.Y., et al.: Low salinity flooding trial at west Salym field. In: SPE Improved Oil Recovery Conference 2016, Tulsa, pp. 1–11 (2016)
7. Korrani, A.K.N., Jerauld, G.R., Sepehrnoori, K., et al.: Coupled geochemical-based modeling of low salinity waterflooding. In: 19th SPE Improved Oil Recovery Symposium, Tulsa, pp. 1–23 (2014)
8. Najmud, D., Hayatdavoudi, A., Ghalambor, A.: Laboratory investigation of saturation effect on mechanical properties of rocks. In: SPWLA 31st Annual Logging Symposium, Lafayette, pp. 1–23 (1990)
9. Vikulova, M.F., Burkov, U.K., Makedonov, A.V.: Facies types of clay rocks and their primary lithological features. Nedra, Leningrad (1973)
10. Maximova, M.I., Grishanova, N.S.: Methods for determination of axial compression strength, State Standard 21153.2-1984. Standartinform, Moscow (2000)
11. Koretnikova, L.V., Prusakova, V.N.: Methods for determination of deformation characteristics under uniaxial compression, State Standard 28985-1991. Standartinform, Moscow (2004)
12. Korshunov, V.A., Kartashov, U.M., Kozlov, V.A.: Determination of certificate of rock strength by the method of specimen destruction by spherical indentors. J. Mining Inst. **185**, 41–45 (2010)
13. Karev, V.I.: Influence of stress-strain state of rocks on the filtration process and well production rate. IPME RAS (2010)

Check for updates

Synchronous Changes of Geophysical Fields in the Earth's Near-Surface Zone

Svetlana Riabova(✉) and Alexander Spivak

Institute of Geospheres Dynamics of Russian Academy of Science, Leninsky Prospekt, 38, Building 1, Moscow, Russia
ryabovasa@mail.ru

Abstract. The research of Earth's physical fields and their variations is particularly important to establish causes and mechanisms of changes in the environment and in the climate. The results of the analysis of instrumental observation concerning the variations in electric and magnetic fields, as well as acoustic vibrations in the near-surface Earth zone are considered in relation to the variations of meteorological parameters. We used the results of synchronous observations concerning physical fields carried out in the period of 2011–2014 in conductions of Geophysical observatory "Mikhnevo" of Institute of Geosphere Dynamics of Russian Academy of Science (IDG RAS) (Russia, Moscow region, settlement Mikhnevo; 54.959° N, 37.766° E). Synchronism of the variations of mentioned geophysical fields and change in atmospheric parameters has been determined, and for the first time not only synchronous, but advanced manifestations of geomagnetic field perturbations are shown, we introduce a new parameter - "reverse" magnetic tipper, variations of which due to atmospheric disturbances are more strongly marked in comparison with the variations of magnetic tipper.

Keywords: Atmospheric parameters · Magnetic tipper · Thunderstorm activity Barometric micropulsations · Electric field variations

1 Introduction

The research of Earth's physical fields and their variations is particularly important to establish causes and mechanisms of changes in the environment and in the climate. Another equally important problem is to monitor processes associated with natural and man-made influences on the Earth's biosphere, to establish response of the environment to these influences, as well as to define boundaries of maximum permissible anthropogenic loads on both the environment and the wildlife, including a human body.

The Earth's physical fields play an important role in interactions between geospheres. Mutual ties between processes occurring in the Earth's hard shells and its upper geospheres are carried out due to transformation and interaction of physical fields [1]. The physical fields join together all geospheres by forming self-consistent geophysical system. Functioning of such a system, like any other complex multi-component system, is described by a quasi-stationary component, formed as a result of long-time interactions of individual components of this system with each other, and perturbations of

V. Karev et al. (Eds.): PMMEEP 2017, SPRINGERGEOL, pp. 90–98, 2018.
https://doi.org/10.1007/978-3-319-77788-7_11

different scales, which are caused by external and internal sources and which under certain conditions can disturb the dynamic equilibrium of the system.

Of greatest interest is the research of a perturbed component, the amplitude of which can exceed by several orders of background values. In doing so it provides an opportunity to research the nature and mechanisms of mass and energy exchange processes related, in particular, to energy transformation between physical fields of different nature. Here it is also important to note a special role of variations of physical fields in the formation of conditions for existence of living organisms (the Earth's biosphere) and, in particular, the human environment.

This paper presents the results of data analysis of synchronous observations concerning the variations in electric and magnetic fields and atmospheric pressure micropulsations in the Earth's near-surface zone in relation to the variations of meteorological parameters of the near-surface atmosphere. A synchronism of variations of mentioned fields and atmospheric parameters is shown. The quantitative relationships between amplitudes of electric field variations, acoustic vibrations, atmospheric parameters and also transfer function of geomagnetic variations as magnetic tipper are given.

2 Initial Data

In this paper we used the results of synchronous observations concerning physical fields carried out in the period of 2011–2014 in conductions of Geophysical observatory "Mikhnevo" of Institute of Geosphere Dynamics of Russian Academy of Science (IDG RAS) (Russia, Moscow region, settlement Mikhnevo; 54.959° N, 37.766° E), which are characterized by absence of industrial pollutions and by low electromagnetic interference level [2]. The measurements of the vertical component of the electric field and components of the magnetic field were carried out by the INEP fluxmeter (the frequency range 0–20 Hz) and the LEMI-018 magnetometer, respectively. The atmospheric pressure micropulsations are recorded by the MB-03 microbarometer, equipped with a wind reduction filter, in the frequency range of 0.001–10 Hz. The meteorological parameters of atmosphere were determined using the Davis Vantage Pro 2 Weather Station. All measurements were carried out in the near-surface layer of the atmosphere at a height of 3–5 m (the description of measuring instruments and the registration results are presented on the website of IDG RAS http://idg-comp.chph.ras.ru/~mikhnevo/).

3 Synchronism of Variations of Physical Fields and Meteorological Parameters of Atmospheres

3.1 Synchronous Variations in Electric and Magnetic Fields, Acoustic Vibrations and Meteorological Parameters

The data analysis showed that main time-limited local variations in the electric field are related to the disturbance of meteorological parameters of the atmosphere (passage of atmospheric, in particular, storm fronts, formation of a dense cloud cover, etc.). The example of this type of synchronous variations of all measured values is shown in Fig. 1.

From Fig. 1 it is clear that the passage of powerful cold atmospheric front from 9:45 to 10:20 UT, accompanied by an almost intermittent change of air temperature *T* and atmospheric pressure *P*, and also by increasing air humidity *W* and by bay-type reducing of solar radiation *S*, causes variations not only in the amplitude of atmospheric pressure micropulsations *A*, but also significant variations in the vertical component of the electric field strength *E*. A weaker cold atmospheric front (~ 14:30 UT), accompanied by formation of a dense cumulus cloud cover without thunderstorm activities, also causes, though less feebly marked, but nearly synchronous variations of all researched values.

Here it should be noted that it is the cold front, forming a current of air masses along the Earth's surface and, as a consequence, aeroelectric effects related, in particular, to the turbulent transport of a bulk electrical charge, that causes variations in the electric field in the near-surface atmosphere (warm atmospheric fronts, as a rule, are not accompanied by significant variations in the electric field and by the barometric micropulsations) [3, 4].

The empirical distribution of variations in the electric field *E* in amplitude has a multimodal character (Fig. 2). In doing so we distinguish three specific areas ***I–III***, mainly accumulating events with a variation amplitude respectively in the ranges of 50–100, 300–600 and 1000–3000 V/m. Drawing on the results of meteorological observations, it can be assumed that in area ***I*** there are concentrated mainly events related to field perturbations in powerful atmospheric fronts, characterized by presence of thunderstorm activities (intra-cloud, inter-clouded and cloud-to-ground lightning discharges, strong changes in illumination and air humidity), in area ***II*** – events also related to passage of cold atmospheric fronts, which are accompanied by a dense cloud cover, but no thunderstorm activities, and, finally, in area ***III*** – events in periods of cold atmospheric fronts with a little or no cloud cover. The latter group can also be classified by pulse-type variations of *E*, which presumably can be related to man-made sources (in this case, the recorded field perturbations are not accompanied by variations in atmospheric pressure and air temperature). The boundaries between mentioned groups should naturally be considered conditional, because a clear distinction between the events referred to adjacent groups was not possible to formulate.

The analysis of results of instrumental observations shows that in the vast majority of cases (~ 75%) simultaneously with the variations in the electric field *E* we observed the variations in the barometric micropulsations *A*. Assuming that each of these values characterizes intensity of meteorological disturbances in the near-surface layer of the atmosphere, we can expect a relationship between *E* and *A*. However, as the analysis demonstrates, such a relationship does not exist.

This is clearly seen from Fig. 3, which shows the observation data in *E, A* coordinates (in order to eliminate the influence of random outliers, the amplitudes of variations were determined from smoothed pulses [5, 6]; in the determination of *E* we did not take into account the events, amplitude of which does not fall into dynamic range of recording channel). In view of the carried out statistical estimates we can distinguish groups of data (Fig. 3) related to considered above areas ***I–III***. Within each of these areas in a wide range of changes of microbarometric variation amplitudes A the amplitude of pulse variations *E* remains practically constant.

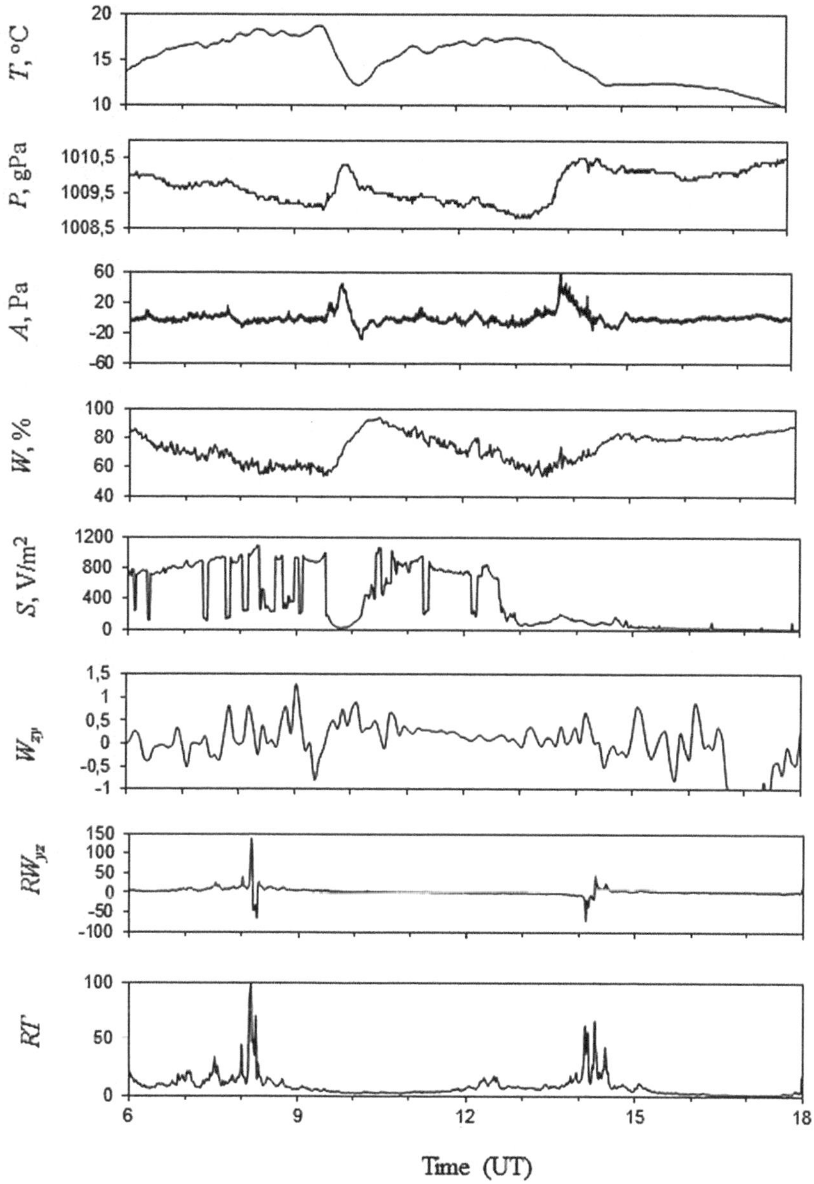

Fig. 1. Examples of synchronous variations of geophysical fields and meteorological parameters of atmosphere at the Geophysical observatory "Mikhnevo" 2014.06.25 in the period of passage of two cold atmospheric fronts at ~9:30 and ~13:30 UT (T and P – temperature and atmospheric pressure, A – barometric micropulsations, W – air humidity, S – solar radiation power, E – variations of vertical component of electric field strength, W_{zy} and RW_{zy} – components of magnetic and "reverse" magnetic tippers, RT – module of "reverse" magnetic tipper)

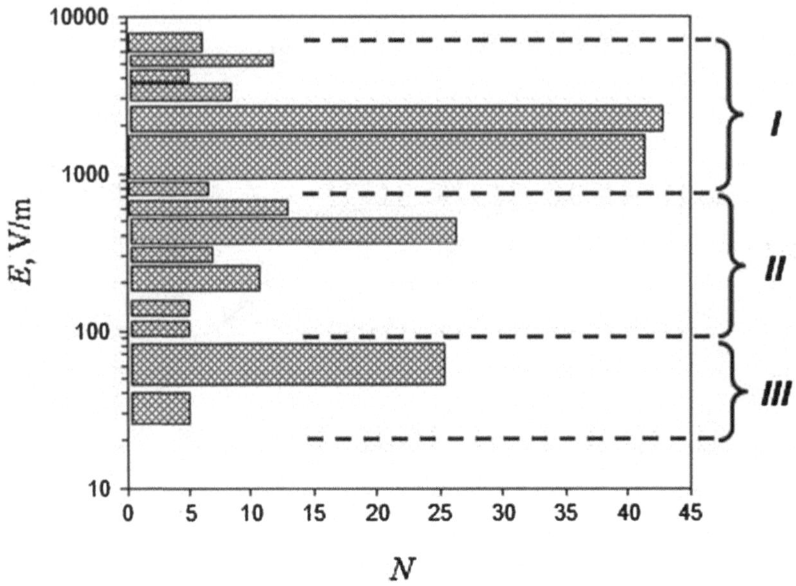

Fig. 2. A histogram of distribution of the event number *N* in variation amplitude *E* (*I* – data obtained during the storm front (mainly with thunderstorm activities); *II* – atmospheric fronts accompanied by formation of a dense cloud cover without thunderstorm activities; *III* – cold atmospheric fronts without forming a dense cloud cover)

In this paper as a characteristic of geomagnetic variations we research a magnetic tipper [7, 8]. The components of the magnetic tipper W_{zx} and W_{zy} are calculated using the ratio:

$$B_z = W_{zx}B_x + W_{zy}B_y, \tag{1}$$

where B_x, B_y and B_z - components of the magnetic induction, presented by the digital series - registration results.

This ratio in the matrix notation has the form:

$$B_z = [W]B_\tau, \tag{2}$$

where $[\boldsymbol{W}] = [W_{zx}, W_{zy}]; B_\tau = [B_x, B_y]^T$.

The tipper magnitude (absolute value) T is given by the equitation:

$$T = \sqrt{|W_{zx}|^2 + |W_{zy}|^2}. \tag{3}$$

To evaluate the magnetic tipper, we used the Larsen technique with signal-to-noise division, which allows, using the surface measurements of the magnetic field, not only to produce calculation of the magnetic tipper, but also to perform interference reduction related to noises of various origins [9, 10].

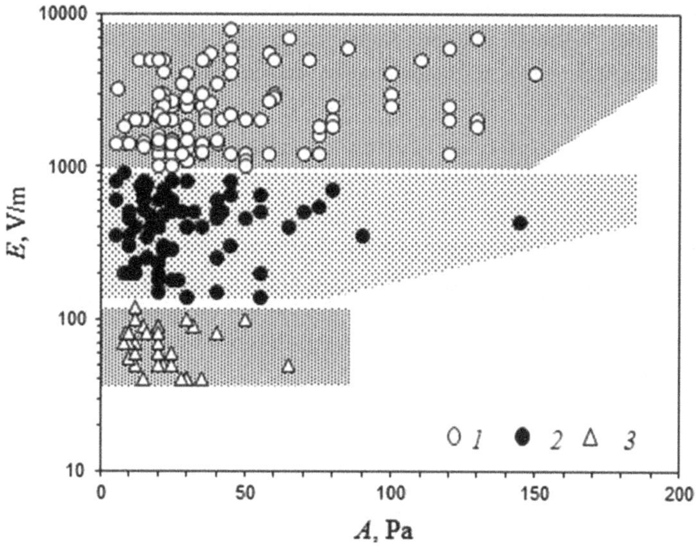

Fig. 3. Data of instrumental observations concerning the amplitudes of synchronous variations in the vertical component of electric field strength E and barometric micropulsations A at the Geophysical observatory "Mikhnevo" from 2011 to 2014 (*1* – atmospheric fronts mainly with thunderstorm activities, *2* – cold atmospheric fronts without thunderstorm activities, *3* – cold fronts without formation of a dense cloud cover)

The detailed analysis indicates the presence of strongly marked variations in a transfer function of the magnetic field as the magnetic tipper [W] in the time periods that are close to the variations E (Fig. 1). In doing so it should be noted that the variations in the magnetic tipper are observed both synchronously with the variations E and in advance of 1–2 h, what has happened, as a rule, in cases of the undisturbed atmosphere before the storm front (the front is preceded by clear weather conditions). This effect is related to generation of acoustic gravity waves by a storm cell, which can be registered at the observation point well (1–6 h) before arrival of storm convective cells [11, 12].

Simultaneously with the magnetic tipper the matrix [$\boldsymbol{RW}$] is calculated as a result of the reverse division of matrices B_z and $\boldsymbol{B}_\tau$ (procedure of division on matrix-multiplicand ("left" matrix division) – reverse division [13]. By analogy with the matrix [$\boldsymbol{W}$] the matrix [$\boldsymbol{RW}$] = $B_z \backslash \boldsymbol{B}_\tau$ may be called a "reverse" magnetic tipper (as opposed to the Nakamura parameter used in seismology [14, 15] the introduced parameter has two components). The data analysis demonstrates that the variations of components of the "reverse" magnetic tipper during periods of the variations of E manifested more clearly as compared with the variations of components of the magnetic tipper. As an example, Fig. 1 shows variations of the magnetic tipper component W_{zy} and the "reverse" magnetic tipper component RW_{zy}, as well as a module of the "reverse" magnetic tipper RT calculated by the equitation:

$$RT = \sqrt{|RW_{zx}|^2 + |RW_{zy}|^2}. \tag{4}$$

As it is demonstrated in Fig. 1, higher variations of the magnetic tipper are recorded approximately 2 h before the passage of the first atmospheric front (9:45–10:20) and ~20 min before the weaker second front (~14:30 UT). The maximum of variations in the "reverse" magnetic tipper is observed 1.5 h before the passage of the first atmospheric front and almost simultaneously with maximum of the E variations during the passage of the second atmospheric front.

3.2 Quantitative Relationships

Of specific interest are the relationships between the quantities which characterize the intensity of the perturbation influence on the atmosphere. We can assume that, when considering cold atmospheric fronts, the amplitudes of change in temperature and atmospheric pressure, variations in the electric field strength and, finally, the magnetic tipper – a parameter characterizing variations in the geomagnetic field – serve as such parameters.

The data analysis demonstrates that there exists a quite definite relationship between the amplitude of the variations in the electric field and the perturbed meteorological parameters of the atmosphere. The data given in Fig. 4 show that as the amplitudes of change in the atmospheric pressure and air temperature increase in the cold atmospheric front the range of variations of the amplitude of the electric field strength is significantly narrowed.

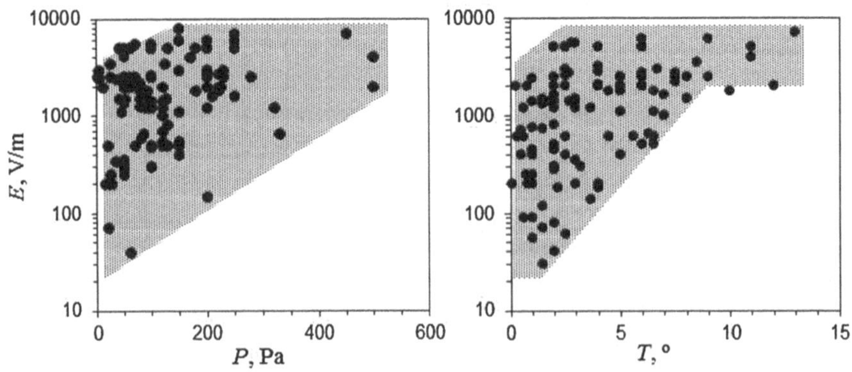

Fig. 4. Experimental values of an amplitude of variations of the vertical component of electric field strength in the near-surface layer of atmosphere E depending on the growth of atmospheric pressure (left) and the fall of air temperature (right) in the cold atmospheric front

If both small-size and large-size variations in the electric field, which are determined by not only heat (convective) turbulence of the air-flow (turbulent transport of electric charges), but also by the presence of electrified clouds, are possible in small variations in the temperature and atmospheric pressure, the powerful atmospheric fronts, as a rule, accompanied by thunderstorm activity, cause only significant in amplitude variations of the electric field strength.

4 Conclusion

Thus, the given results of instrumental observations demonstrate the synchronism of variations in the electric field, magnetic tipper and barometric micropulsations during periods of meteorological disturbances in the Earth's near-surface atmosphere. It is shown that the introduced value – the "reverse" magnetic tipper – is significantly more sensitive to variations of meteorological parameters of the near-surface atmosphere and the electric field in comparison with the magnetic tipper.

References

1. Adushkin, V.V., Spivak, A.A.: Near-surface geophysics: complex investigations of the lithosphere-atmosphere interactions. Izv. Phys. Solid Earth **48**(3), 181–198 (2012)
2. Adushkin, V.V., Ovtchinnikov, V.M., Sanina, I.A., Riznichenko, O.Y.: Mikhnevo: from seismic station N 1 to a modern geophysical observatory. Izv. Phys. Solid Earth **52**(1), 105–116 (2016)
3. Anisimov, S.V., Mareev, E.A., Shikhova, N.M., Shatalina, M.V., Galichenko, S.V., Zilitinkevich, S.S.: Aeroelectric structures and turbulence in the atmospheric boundary layer. Nonlin. Process. Geophys. **20**, 819–824 (2013)
4. Tuomi, T.J.: Atmospheric electrode affect: approximate theory and wintertime observations. Pure. Appl. Geophys. **119**, 31–45 (1981)
5. Gvishiani, A.D., Agayan, S.M., Bogoutdinov, S.R.: Fuzzy recognition of anomalies in time series. Dokl. Earth Sci. **421**(1), 838–842 (2008)
6. Zelinskiy, N.R., Kleimenova, N.G., Kozyreva, O.V., Agayan, S.M., Bogoutdinov, S.R., Soloviev, A.A.: Algorithm for recognizing Pc3 geomagnetic pulsations in 1-s data from INTERMAGNET equatorial observatories. Izv. Phys. Solid Earth **50**(2), 240–248 (2014)
7. Berdichevsky, M.N., Dmitriev, V.I., Golubtsova, N.S., Mershchikova, N.A., Pushkarev, P. Y.: Magnetovariational sounding: new possibilities. Izv. Phys. Solid Earth **39**(9), 701–727 (2003)
8. Vozoff, K.: The magnetotelluric method in the exploration of sedimentary basins. Geophysics **37**(1), 98–141 (1972)
9. Larsen, J.C.: Transfer functions: smooth robust estimates by least-squares and remote reference methods. Geophys. J. Int. **99**(3), 645–663 (1989)
10. Larsen, J.C., Mackie, R.L., Manzella, A., Fiordelisi, A., Rieven, S.: Robust smooth magnetotelluric transfer functions. Geophys. J. Int. **124**(3), 801–819 (1996)
11. Balachandran, N.K.: Gravity waves from thunderstorms. Mon. Weather Rev. **108**(6), 804–816 (1980)
12. Danilov, S.D., Svertilov, A.I.: Internal gravity waves generated by a passage of thunderstorm. Izv. Atmos. Oceanic Phys. **27**(3), 234–242 (1991)

13. Lancaster, P.: Theory of Matrices. Academic Press, New York (1969)
14. Nakamura, Y.: A method for dynamic characteristic estimation of subsurface using microtremor on the ground surface. Q. Report Railway Tech. Res. Inst. **30**(1), 25–33 (1989)
15. Steimen, S., Fah, D., Kind, F., Schmid, Ch., Giardini, D.: Identifying 2D resonance in microtremor wave fields. Bull. Seismol. Soc. Am. **93**(2), 583–599 (2003)

Vertical Mass Transport by Weakly Nonlinear Inertia-Gravity Internal Waves

A. A. Slepyshev[1(✉)] and D. I. Vorotnikov[2]

[1] Marine Hydrophysical Institute, Russian Academy of Sciences, Sevastopol 299011, Russia
slep55@mail.ru
[2] Moscow State University, Moscow 119991, Russia
infsup@yandex.ru

Abstract. In the Boussinesq approximation, free inertia-gravity internal waves are considered in a two-dimensional vertically non-uniform flow. In the linear approximation was find vertical distribution of the amplitude of the vertical velocity and the dispersion relation. The boundary-value problem for internal waves has complex coefficients when the flow velocity component, transverse to the wave propagation direction depends on the vertical coordinate. Therefore, the eigenfunction and frequency of the wave are complex (it is shown that there is a weak damping of the wave). Vertical wave mass fluxes are nonzero. The vertical component of the Stokes drift velocity also differs from zero and contributes to the wave transport. A non-oscillating on a time scale of the wave correction to the average density, which is interpreted as an irreversible vertical fine structure generated by a wave, is determined on the second order of amplitude.

Keywords: Inertia-gravity internal waves · Stokes drift
Wave fluxes of mass · Critical layers

1 Introduction

Vertical exchange in the marine environment is of key importance in the functioning of the ecosystem. The supply of oxygen to the deep layers of the sea is due to metabolic processes. Traditionally, vertical transfer is associated with small-scale turbulence, which is intermittent in the stratified layers of the sea. Therefore, one can speak of an "effective" turbulent exchange coefficient. When the turbulent viscosity and diffusion are taken into account, the internal waves decay [1–3]. In the pycnocline the processes of water mixing are strongly suppressed, and therefore it seems relevant to study the contribution of internal waves to vertical exchange. Nonlinear effects at the propagation of packets of internal waves are manifested in the generation of average wave-scale currents and corrections to the density, which is treated as a vertical fine structure generated by a wave [4, 5].

V. Karev et al. (Eds.): PMMEEP 2017, SPRINGERGEOL, pp. 99–111, 2018.
https://doi.org/10.1007/978-3-319-77788-7_12

This correction is proportional to the square of the current wave amplitude and after the wave packet passes the unperturbed profile the density field is restored.

Vertical wave fluxes of heat, salt and momentum, when turbulent viscosity and diffusion are taken into account, are nonzero [6,7], but for inertia-gravity internal waves (with allowance for the rotation of the Earth) vertical fluxes are non-zero and when turbulent viscosity and diffusion are ignored for vertically non-uniform two-dimensional flow. It is of interest to find vertical wave fluxes of mass and to investigate their contribution to the formation of a vertical fine structure.

2 Problem Definition

In Boussinesq approximation, free internal waves are considered in a stratified two-dimensional flow with rotation. Two components of the average flow velocity depend on the vertical coordinate. In the linear approximation, the boundary value problem for the amplitude of the vertical velocity has complex coefficients, so its solution is a complex function, the wave frequency is also a complex quantity (it is shown that there is a weak damping of the wave).

The system of equations of hydrodynamics for wave perturbations in the Boussinesq approximation has the form

$$\frac{Du}{Dt} - fv + w\frac{dU_0}{dz} = -\frac{1}{\rho_0(0)}\frac{\partial P}{\partial x} \tag{1}$$

$$\frac{Dv}{Dt} + fu + w\frac{dV_0}{dz} = -\frac{1}{\rho_0(0)}\frac{\partial P}{\partial y} \tag{2}$$

$$\frac{Dw}{Dt} = -\frac{1}{\rho_0(0)}\frac{\partial P}{\partial z} - \frac{g\rho}{\rho_0(0)} \tag{3}$$

$$\frac{D\rho}{Dt} = -w\frac{d\rho_0}{dz} \tag{4}$$

$$\frac{\partial u}{\partial x} + \frac{\partial v}{\partial y} + \frac{\partial w}{\partial z} = 0 \tag{5}$$

where x, y, z are two horizontal and vertical coordinates, the z-axis is directed vertically upwards, u, v, w are, respectively, two horizontal and vertical components of the wave flow velocity, ρ and P are wave perturbations of density and pressure, H is the depth of the sea, $\rho_0(z)$ is the average density profile, f is the Coriolis parameter, $U_0(z), V_0(z)$ are two components of the mean velocity flow, g is acceleration of free fall, the action of the operator $\frac{D}{Dt}$ is determined by the formula $\frac{D}{Dt} = \frac{\partial}{\partial t} + (u + U_0)\frac{\partial}{\partial x} + (v + V_0)\frac{\partial}{\partial y} + w\frac{\partial}{\partial z}$.

Boundary condition on the sea surface ($z = 0$) the condition of a “hard cover”, which filters out internal waves from surface waves [8]:

$$w(0) = 0 \tag{6}$$

The boundary condition at the bottom is the condition of “non-flow”:

$$w(-H) = 0 \tag{7}$$

3 Linear Approximation

We seek solutions of the linear approximation in the form:

$$u_1 = u_{10}(z)Ae^{i\theta} + c.c.,\ v_1 = v_{10}(z)Ae^{i\theta} + c.c.,\ w_1 = w_{10}(z)Ae^{i\theta} + c.c.$$
$$P_1 = P_{10}(z)Ae^{i\theta} + c.c.,\ \rho_1 = \rho_{10}(z)Ae^{i\theta} + c.c. \tag{8}$$

where $c.c.$ is complex conjugate term, A is the amplitude factor, θ is phase of the wave; $\frac{\partial\theta}{\partial x} = k$, $\frac{\partial\theta}{\partial t} = -\omega$, k is horizontal wave number, ω is wave frequency.

It is assumed that the wave propagates along the x-axis.

After substituting (8) into system (1)–(5), follows coupling of the amplitude functions $u_{10}, v_{10}, \rho_{10}, P_{10}$ with w_{10}

$$u_{10} = \frac{i}{k}\frac{dw_{10}}{dz}, \qquad \Omega = \omega - k \cdot U_0 \tag{9}$$

$$\frac{P_{10}}{\rho_0(0)} = \frac{i}{k}\left[\frac{\Omega}{k}\frac{dw_{10}}{dz} + \frac{dU_0}{dz}w_{10} + \frac{f}{\Omega}\left(i\frac{dV_0}{dz}w_{10} - \frac{f}{k}\frac{dw_{10}}{dz}\right)\right] \tag{10}$$

$$\rho_{10} = -\frac{i}{\Omega}w_{10}\frac{d\rho_0}{dz}, \qquad v_{10} = \frac{1}{\Omega}\left(\frac{f}{k}\frac{dw_{10}}{dz} - iw_{10}\frac{dV_0}{dz}\right) \tag{11}$$

Function w_{10} satisfies the equation

$$\frac{d^2w_{10}}{dz^2} + k\left[\frac{if\frac{dV_0}{dz}}{\Omega^2 - f^2} - \frac{f^2\frac{dU_0}{dz}}{\Omega(\Omega^2 - f^2)}\right]\frac{dw_{10}}{dz} +$$
$$+kw_{10}\left[\frac{k(N^2 - \Omega^2) + \Omega\frac{d^2U_0}{dz^2} + if\frac{d^2V_0}{dz^2}}{\Omega^2 - f^2} + \frac{ifk\frac{dU_0}{dz}\frac{dV_0}{dz}}{\Omega(\Omega^2 - f^2)}\right] = 0 \tag{12}$$

where $N^2 = -\frac{g}{\rho_0(0)}\frac{d\rho_0}{dz}$ is the square of Brunt-Vaisala frequency.

Boundary conditions for w_{10}:

$$z = 0, \qquad w_{10} = 0 \tag{13}$$
$$z = -H, \qquad w_{10} = 0 \tag{14}$$

Equation (12) has complex coefficients, the imaginary part of which is small, so let us turn to dimensionless variables (the dashed line denotes dimensionless physical quantities)

$$z = Hz',\ t = t'/\omega_*,\ w_{10} = w'_{10}V_{0*},\ V_0 = V'_0V_{0*},\ U_0 = U'_0V_{0*},$$
$$k = k'/H,\ f = f'\omega_*,\ \omega = \omega'\omega_*,\ \Omega = \Omega'\omega_*, \tag{15}$$

where ω_* is characteristic wave frequency, V_{0*} is a characteristic value of the flow velocity which is transverse to the wave propagation direction.

Equation (12) then takes the form:

$$\frac{d^2 w'_{10}}{dz'^2} + k' \left[\frac{i\varepsilon f' \frac{dV'_0}{dz'}}{\Omega'^2 - f'^2} - \frac{\varepsilon f'^2 \frac{dU'_0}{dz'}}{\Omega'(\Omega'^2 - f'^2)} \right] \frac{dw'_{10}}{dz'} +$$

$$+ k' w'_{10} \left[\frac{k'(N'^2 - \Omega'^2) + \varepsilon \Omega' \frac{d^2 U'_0}{dz'^2} + i\varepsilon f \frac{d^2 V'_0}{dz'^2}}{\Omega'^2 - f'^2} + \frac{i\varepsilon^2 f' k' \frac{dU'_0}{dz'} \frac{dV'_0}{dz'}}{\Omega'(\Omega'^2 - f'^2)} \right] = 0 \quad (16)$$

$\varepsilon = V_{0*}/H\omega_*$ is a small parameter. The imaginary part of the coefficients in Eq. (16) is of the order of ε, therefore the imaginary part of the solution w_{10} is also proportional ε i.e. the solution of Eq. (16) is represented in the form:

$$w'_{10} = w'_0(z') + \varepsilon i w'_1(z') \quad (17)$$

where $w'_0(z')$ and $w'_1(z')$ are real functions. The frequency is also expressed as a parameter expansion ε

$$\omega' = \omega'_0 + \varepsilon \sigma'_1 + \ldots \quad (18)$$

then $\Omega' = \Omega'_0 + \varepsilon \sigma'_1 + \ldots$. After substituting (17), (18) into (12), we obtain boundary value problems for $w'_0(z')$ and $w'_1(z')$. Function $w'_0(z')$ satisfies the equation (up to terms $\sim \varepsilon$):

$$\frac{d^2 w'_0}{dz'^2} - \epsilon k' \frac{dw'_0}{dz'} \frac{dU'_0}{dz'} \frac{f'^2}{\Omega'_0(\Omega'^2_0 - f'^2)} +$$

$$+ \frac{k' w'_0}{(\Omega'^2_0 - f'^2)} \left[k'(N'^2 - \Omega'^2_0) + \varepsilon \Omega'_0 \frac{d^2 U'_0}{dz'^2} \right] = 0 \quad (19)$$

The boundary conditions for w'_0

$$w'_0(0) = 0, \qquad w'_0(-1) = 0 \quad (20)$$

Function $w'_1(z')$ satisfies the equation (up to terms $\sim \varepsilon$):

$$\frac{d^2 w'_1}{dz'^2} - \varepsilon k' \frac{dw'_1}{dz'} \frac{dU'_0}{dz'} \frac{f'^2}{\Omega'_0(\Omega'^2_0 - f'^2)} +$$

$$+ \frac{k' w'_1}{(\Omega'^2_0 - f'^2)} \left[k'(N'^2 - \Omega'^2_0) + \varepsilon \Omega'_0 \frac{d^2 U'_0}{dz'^2} \right] = F'(z') \quad (21)$$

where

$$F'(z') = -k' \frac{dw'_0}{dz'} \frac{dV'_0}{dz'} \frac{f'}{(\Omega'^2_0 - f'^2)} + ik' \frac{dw'_0}{dz'} \frac{dU'_0}{dz'} \frac{\sigma'_1 f'^2 (3\Omega'^2_0 - f'^2)}{\Omega'^2_0 (\Omega'^2_0 - f'^2)^2} -$$

$$- \frac{k' w'_0}{(\Omega'^2_0 - f'^2)} \left[k' \frac{2i\Omega'_0 \sigma'_1 (N'^2 - f'^2)}{(\Omega'^2_0 - f'^2)} + \varepsilon \frac{d^2 U'_0}{dz'^2} \frac{i\sigma'_1 (\Omega'^2_0 + f'^2)}{(\Omega'^2_0 - f'^2)} + \right.$$

$$\left. + f' \frac{d^2 V'_0}{dz'^2} + \varepsilon \frac{f' k'}{\Omega'_0} \frac{dU'_0}{dz'} \frac{dV'_0}{dz'} \right]$$

The boundary conditions for w_1'

$$w_1'(0) = 0, \qquad w_1'(-1) = 0 \tag{22}$$

After the transition to dimensional variables, Eq. (19) takes the form:

$$\frac{d^2 w_0}{dz^2} - k\frac{dw_0}{dz}\frac{dU_0}{dz}\frac{f^2}{\Omega_0(\Omega_0^2 - f^2)} + \\ + \frac{kw_0}{(\Omega_0^2 - f^2)}\left[k(N^2 - \Omega_0^2) + \Omega_0\frac{d^2 U_0}{dz^2}\right] = 0 \tag{23}$$

where $\Omega_0 = \omega - kU_0$ is wave frequency with Doppler shift.

Equation (23) should be supplemented by boundary conditions:

$$w_0(0) = 0, \qquad w_0(-H) = 0 \tag{24}$$

The boundary-value problem (23), (24) in the absence of flow for $U_0 = 0$ has a countable set of eigenfunctions, a set of modes. Moreover, to each value of the wavenumber k corresponds to a certain frequency value $\omega_0 < max(N)$ corresponding to the given mode. When $U_0 \neq 0$ the discrete spectrum of real eigenfrequencies may not exist [9]. This is connected with the singularities in Eq. (23) with $\Omega_0 = 0$ and $\Omega_0 = \pm f$ (hydrodynamically stable flows are considered). In the presence of singular $\Omega_0 = 0$, there is a critical layer [10], where the phase velocity of the wave is equal to the flow velocity. With allowance for the rotation of the Earth, the singularity shifts to a level where $\Omega_0 = f$ [11]. The effect of this singularity on the dispersion curves is illustrated by the calculations given below.

Let $a(z) = -\frac{f^2 k}{\Omega_0(\Omega_0^2 - f^2)}\frac{dU_0}{dz}$, $b(z) = \frac{k}{(\Omega_0^2 - f^2)}\left[k(N^2 - \Omega_0^2) + \Omega_0\frac{d^2 U_0}{dz^2}\right]$, then Eq. (23) can be written in the form:

$$\frac{d^2 w_0}{dz^2} + a(z)\frac{dw_0}{dz} + b(z)w_0 = 0 \tag{25}$$

Equation (25) leads to a self-adjoint form, multiplying both sides of the equation by $p(z) = \exp(\int a(z)\,dz)$:

$$\frac{d}{dz}\left(p(z)\frac{dw_0}{dz}\right) - q(z)w_0 = 0 \tag{26}$$

here $q(z) = -b(z)p(z)$.

After the transition to dimensional variables, Eq. (21) is transformed to the form

$$\frac{d^2 w_1}{dz^2} + a(z)\frac{dw_1}{dz} + b(z)w_1 = F_1(z) \tag{27}$$

where

$$F(z) = -k\frac{dw_0}{dz}\frac{dV_0}{dz}\frac{f}{(\Omega_0^2 - f^2)} + ik\frac{dw_0}{dz}\frac{dU_0}{dz}\frac{\sigma_1 f^2(3\Omega_0^2 - f^2)}{\Omega_0^2(\Omega_0^2 - f^2)^2} - \\ -\frac{kw_0}{(\Omega_0^2 - f^2)}\left[k\frac{2i\Omega_0\sigma_1(N^2 - f^2)}{(\Omega_0^2 - f^2)} + i\frac{d^2U_0}{dz^2}\frac{\sigma_1(\Omega_0^2 + f^2)}{(\Omega_0^2 - f^2)} + \right. \\ \left. + f\frac{d^2V_0}{dz^2} + \frac{fk}{\Omega_0}\frac{dU_0}{dz}\frac{dV_0}{dz}\right]$$

The boundary conditions for the function w_1:

$$w_1(0) = 0, \qquad w_1(-H) = 0 \tag{28}$$

We multiply both sides of the linear inhomogeneous Eq. (27) by the function $p(z)$ we obtain on the left-hand side a self-adjoint operator, the same as in the linear homogeneous Eq. (26):

$$\frac{d}{dz}\left(p(z)\frac{dw_1}{dz}\right) - q(z)w_1 = F_1(z) \tag{29}$$

where $F_1(z) = p(z)F(z)$.

The solvability condition for the boundary value problem (28), (29) [12]:

$$\int_{-H}^{0} F_1 w_0 \, dz = 0 \tag{30}$$

Hence the expression for σ_1:

$$\sigma_1 = \frac{a}{b}$$

where

$$a = ifk\int_{-H}^{0}\frac{pw_0}{(\Omega_0^2 - f^2)}\left(\frac{dw_0\frac{dV_0}{dz}}{dz} + w_0\frac{k}{\Omega_0}\frac{dU_0}{dz}\frac{dV_0}{dz}\right)dz,$$

$$b = \int_{-H}^{0}\frac{pkw_0}{(\Omega_0^2 - f^2)^2}\left[w_0\left(2k\Omega_0(N^2 - f^2) + \frac{d^2U_0}{dz^2}(\Omega_0^2 + f^2)\right) - \right. \\ \left. - f^2\frac{dw_0}{dz}\frac{dU_0}{dz}\frac{(3\Omega_0^2 - f^2)}{\Omega_0^2}\right]dz \tag{31}$$

4 Nonlinear Effects

The velocity of the Stokes drift of the liquid particles is determined by the formula [13]:

$$\mathbf{u}_s = \overline{\int_0^t \mathbf{u}d\tau\nabla\mathbf{u}}, \tag{32}$$

where **u** is the field of wave Euler velocities, the bar above denotes averaging over the wave period.

The vertical component of the Stokes drift velocity is determined by the formula:

$$w_s = iA_1A_1^* \left(\frac{1}{\omega} - \frac{1}{\omega^*}\right) \frac{d}{dz}(w_{10}w_{10}^*) \tag{33}$$

where $A_1 = A\exp(\delta\omega \cdot t)$, $\delta\omega = \sigma/i$ is the damping decrement of the wave, the value of σ_1 is purely imaginary, $A_1 = A$ at the initial time at $t = 0$.

In the presence of an average flow, in which the velocity component transverse to the direction of propagation of the wave V_0 depends on the vertical coordinate, the value of w_s is distinct from zero.

The vertical wave mass flux is determined by the formula:

$$\overline{\rho w}/\left|A_1\right|^2 = -w_{10}w_{10}^* \left(\frac{i}{\Omega} - \frac{i}{\Omega^*}\right) \frac{d\rho_0}{dz} \tag{34}$$

The presence of a vertical wave mass flux leads to an irreversible deformation of the density field, which can be regarded as a vertical fine structure generated by a wave. The equations for the nonoscillating on a time scale, the correction to the mean density $\overline{\rho}$:

$$\frac{D\overline{\rho}}{Dt} + \frac{\partial\overline{\rho u}}{\partial x} + \frac{\partial\overline{\rho v}}{\partial y} + \frac{\partial\overline{\rho w}}{\partial z} + w_s\frac{d\rho_0}{dz} = 0$$

from here we have:

$$\frac{\partial\overline{\rho}}{\partial t} + \frac{\partial\overline{\rho w}}{\partial z} + w_s\frac{d\rho_0}{dz} = 0 \tag{35}$$

Integrate Eq. (35) in time

$$\overline{\rho(t)} = \overline{\rho(0)} - \int_0^t \left(\frac{\partial\overline{\rho w}}{\partial z} + w_s\frac{d\rho_0}{dz}\right) dt' \tag{36}$$

Substituting $\overline{\rho w}$ (34) and the vertical component of the Stokes drift velocity w_s (33), we obtain in (36) after integration

$$\overline{\Delta\rho} = \overline{\rho(t)} - \rho(0) = \left[\frac{\partial\overline{\rho w_0}}{\partial z} + w_{s0}\frac{d\rho_0}{dz}\right] \cdot \frac{1}{2\delta\omega}\left(1 - \exp^{2\delta\omega\cdot t}\right), \tag{37}$$

where

$$\overline{\rho w_0} = i\left|A\right|^2 w_{10}w_{10}^*\left(\frac{1}{\Omega^*} - \frac{1}{\Omega}\right)\frac{d\rho_0}{dz}, \qquad w_{s0} = i\left|A\right|^2\left(\frac{1}{\omega} - \frac{1}{\omega^*}\right)\frac{d}{dz}(w_{10}w_{10}^*)$$

Passing to the limit in (37) for $t \to \infty$ taking into account that $\delta\omega < 0$ we find $\overline{\Delta\rho}$

$$\overline{\Delta\rho} = \left[\frac{\partial\overline{\rho w_0}}{\partial z} + w_{s0}\frac{d\rho_0}{dz}\right] \cdot \frac{1}{2\delta\omega} \tag{38}$$

5 Results of Calculations

Wave mass fluxes are calculated for internal waves that were observed during the full-scale experiment in the third stage of the 44th voyage of the research vessel "Mikhail Lomonosov" on the northwestern shelf of the Black Sea.

At the Fig. 1 presents four realizations of elevations of temperature isolines obtained from GRAD instruments (gradient-distributed temperature sensors) [14].

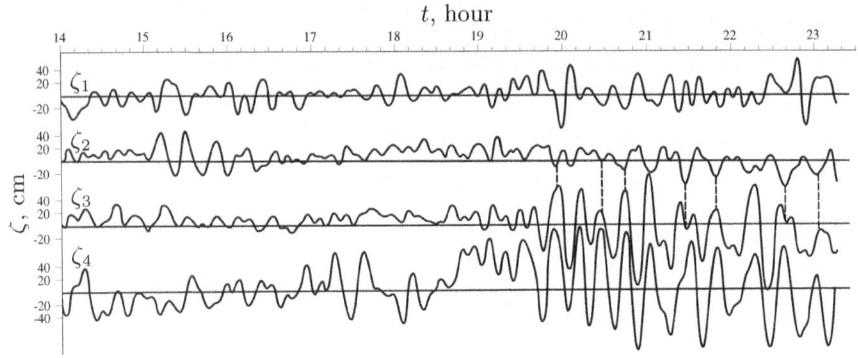

Fig. 1. Time course of vertical displacements of temperature isolines

The first device was located in the 5–15 m layer, the second in the layer 15–25 m, the third in the layer 25–35 m, the fourth in the layer 35–60 m. It is easy to see that powerful oscillations with a period of 15 min in the 25–60 m layer are in antiphase with oscillations in the 15–25 m layer, which indicates the presence oscillation of the second mode. The vertical profiles of the two components of the flow velocity are shown in Fig. 2, the Brent-Vaisala frequency is shown in Fig. 3a. The boundary value problem (23), (24) for internal waves is solved numerically according to the implicit Adams scheme of the third order of accuracy. The wavenumber is found by the method of adjusting from the necessity of performing boundary conditions (24). The eigenfunction of the 15-min internal waves of the second mode is shown in Fig. 3b.

The wavenumber is 0.032 rad/m. The normalizing factor A_1 we find from the known value the maximum amplitude of the vertical displacements. To do this, we express the vertical displacement ζ using the relation $\frac{d\zeta}{dt} = w$:

$$\zeta = \frac{iw_0}{\Omega_0} A_1 \exp(ikx - i\omega_0 t) + c.c.$$

This implies:

$$A_1 = \frac{\max \zeta}{2 \max |w_0/\Omega_0|}$$

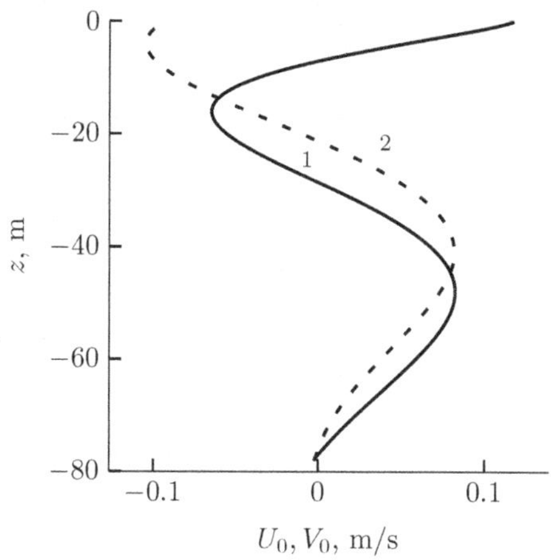

Fig. 2. Vertical profiles of flow velocity components U_0 (1), V_0 (2)

Thus, the amplitude of vertical displacements is proportional to w_0. Extremums of the function w_0 correspond to the maximum vertical displacements from the experimental data (Figs. 1 and 3b) i.e. in the experiment, the second mode was observed. The wavelength of fifteen-minute internal waves of the second mode is 196 m. Dispersion curves of the first two modes are shown in Fig. 4. If the flow were not taken into account, the dispersion curves in the low-frequency area would begin with a minimum frequency equal to the inertial one. When the flow is taken into account, because of the effect of the singularity $\Omega_0 = f$ the dispersion curves are cut off at low frequencies. The minimum frequency of the first mode corresponds to $1.13 \cdot 10^{-4}$ rad/s, for the second mode $3.49 \cdot 10^{-4}$ rad/s (for comparison we indicate that the Coriolis frequency is

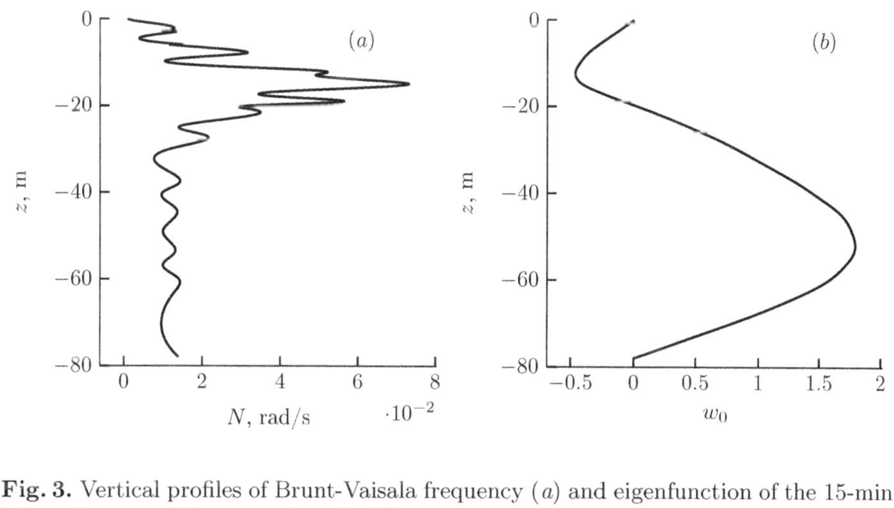

Fig. 3. Vertical profiles of Brunt-Vaisala frequency (*a*) and eigenfunction of the 15-min internal waves of the second mode (*b*)

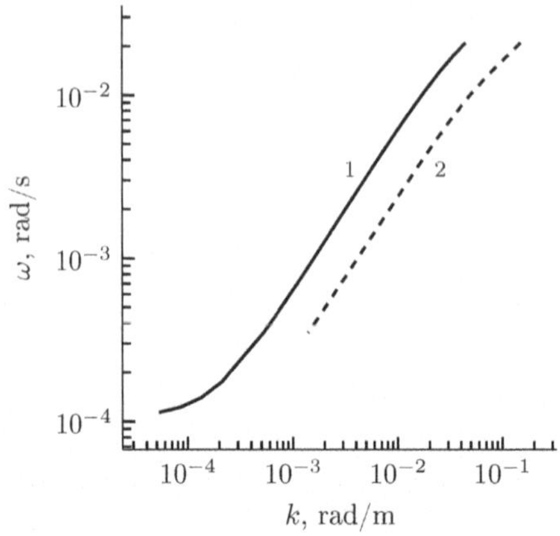

Fig. 4. Dispersion curves of the first (1) and second (2) modes

$1.048 \cdot 10^{-4}$ rad/s). Cutoff dispersion curves occurs due to the influence of critical layers, where the frequency of the wave with the Doppler shift is equal to the inertial one.

A boundary value problem for the definition of a function w_1 (28), (29) solve numerically according to the implicit Adams scheme of the third order of accuracy, we find the unique solution orthogonal w_0 and wave damping decrement $\delta\omega$. For the 15-min internal waves of the second mode, the attenuation decrement is $\delta\omega = -1.15 \cdot 10^{-5}$ rad/s. Attenuation decrement is two orders of magnitude smaller than the wave frequency. The dependence of the attenuation decrement on the wavenumber for the first two modes is shown in Fig. 5. Difference in behavior $\delta\omega$ in the low-frequency area is due to the cutoff of the dispersion curves in the vicinity of the inertial frequency.

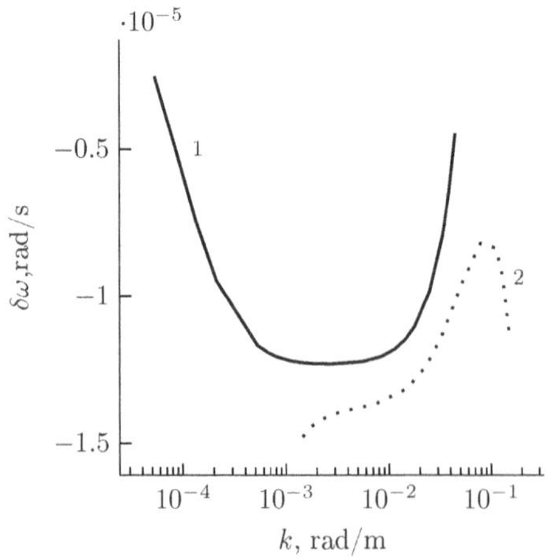

Fig. 5. Dependence of the wave attenuation decrement on the wavenumber for the first (1) and second (2) modes

Wave mass flux $\overline{\rho w}$ (34) are compared for the first two modes at the same wave amplitude in Fig. 6. The wave flux of the first mode dominates in the upper 30-m layer. Deeper these flows are comparable in magnitude.

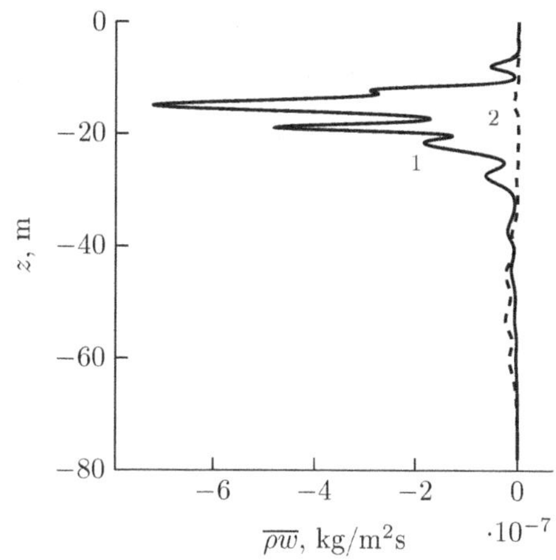

Fig. 6. Vertical wave mass fluxes $\overline{\rho w}$ for the first (1) and second (2) modes

The total vertical wave mass flux is added from the stream $\overline{\rho w}$ (34) and the flux due to the vertical component of the Stokes drift velocity $J_{\rho s} = \rho_0(z) w_s$. A comparison of the total fluxes for the first two modes with the corresponding turbulent flux $\overline{\rho' w'}$ is shown in Fig. 7. The turbulent flux is determined by the formula $\overline{\rho' w'} = -M_z \frac{d\rho_0}{dz}$. The coefficient of vertical turbulent diffusion is estimated

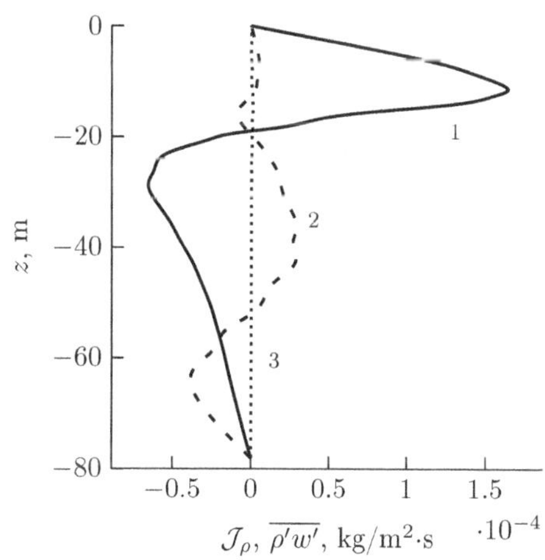

Fig. 7. A comparison of the total wave and turbulent (3) mass fluxes for the first (1) and second (2) modes

from formula $M_z \cong 0.93 \cdot 10^{-4} N_c^{-1} m^2/s$, N_c corresponds to the BruntVaisala frequency in the cycle/hour [15]. The first mode dominates in the upper 40-m layer, deeper these fluxes are comparable in magnitude (Fig. 7). The wave fluxes exceed the turbulent flux in absolute value.

The vertical density profile is shown in Fig. 8a. A nonoscillating wave on a time scale, the corrections to the mean density $\overline{\Delta\rho}$ (37) contains the quantity $|A^2|$ which is exactly equal to $|A_1^2|$ at the initial time. The nonoscillating correction to the average density is shown in Fig. 8b and is a vertical fine structure generated by a wave which is irreversible, with no inversions in the field of average density.

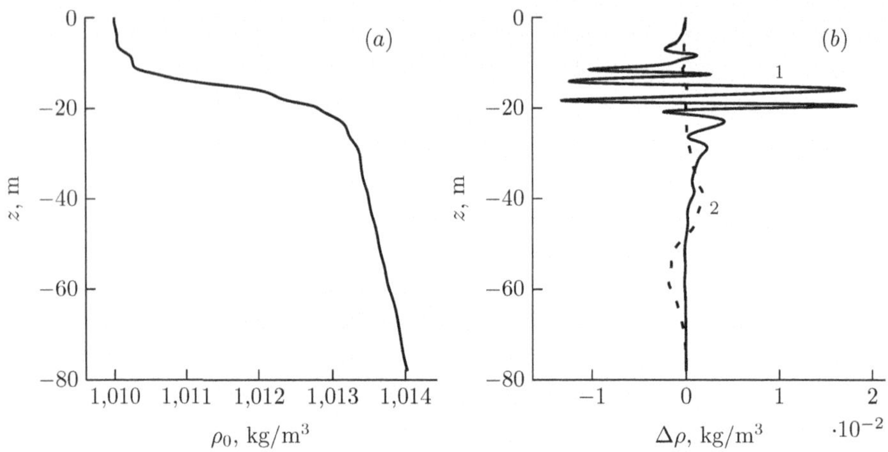

Fig. 8. Vertical profiles of medium density (*a*) and nonoscillating wave-scale corrections to the density (*b*) for the first (1) and second (2) modes

6 Conclusions

- The vertical component of the Stokes drift velocity of internal waves is different from zero and makes a decisive contribution to the wave transport of the mass.
- The vertical wave mass flux leads to an irreversible deformation of the average density profile-the fine structure generated by the wave.
- The wave flux of the first mode in the upper 40-meter layer exceeds the flux of the second mode. Wave fluxes dominate over turbulent fluxes.

Notes and Comments. The work was carried out within the framework of the state task on the topic № 0827-2014-0010 "Complex interdisciplinary studies of oceanological processes determining the functioning and evolution of the systems of the Black and Azov Seas on the basis of modern methods for monitoring the state of the marine environment and grid technologies".

References

1. Le Blond, P.H., Mysak, L.: Waves in the Ocean. Elsevier, Amsterdam (1977); Part 2, 363 p. Mir, Moscow (1981)
2. Le Blond, P.H.: On damping of internal gravity waves in a continuously stratified ocean. J. Fluid Mech. **25**(1), 121–142 (1966). https://doi.org/10.1017/S0022112066000089
3. Ostrovskii, L.A., Soustova, I.A.: The upper mixing layer as a sink of internal wave energy. Okeanologiya **19**(6), 973–981 (1979)
4. Borisenko, Y.D., Voronovich, A.G., Leonov, A.I., Miropolsky, Y.Z.: On the theory of nonstationary weak-nonlinear internal waves in a stratified fluid. Izv. Akad. Nauk SSSR: Fiz. Atmos. Okeana **12**(3), 293–301 (1976)
5. Voronovich, A.G., Leonov, A.I., Miropolsky, Y.Z.: On the theory of the formation of a fine structure of hydrophysical fields in the ocean. Okeanologiya **11**(5), 490–497 (1976)
6. Slepyshev, A.A.: Transport processes induced by weakly nonlinear internal waves in the presence of turbulence. Izv. Atmos. Ocean. Phys. **33**(4), 536–548 (1997)
7. Nosova, A.V., Slepyshev, A.A.: Vertical flows induced weakly nonlinear internal waves on the shelf. Fluid Dyn. **50**(1), 12–21 (2015). https://doi.org/10.1134/S0015462815010020
8. Miropolsky, Y.Z.: Dynamics of Internal Gravitational Waves in the Ocean, p. 30. Gidrometeoizdat, Leningrad (1981)
9. Banks, W.H., Drazin, P.G., Zaturska, M.B.: On the normal modes of parallel flow of inviscid stratified fluid. J. Fluid Mech. **75**(1), 149–171 (1976). https://doi.org/10.1017/S0022112076000153
10. Booker, J.B., Brethertone, F.P.: The critical layer for internal gravity waves in a shear flow. J. Fluid Mech. **27**(4), 513–539 (1967). https://doi.org/10.1017/S0022112067000515
11. Jones, W.L.: Propagation of internal waves in fluids with shear flow and rotation. J. Fluid Mech. **30**(3), 439–448 (1967). https://doi.org/10.1017/S0022112067001521
12. Kamke, E.: Differentialgleichungen. Lsungsmethoden und Lsungen, p. 244. Akademische Verlagsgesellschaft, Leipzig (1959)
13. Longuet-Higgins, M.S.: On the transport of mass by time varying ocean current. Deep-Sea Res. **16**(5), 431–447 (1969)
14. Report on the work on the 44th voyage (3rd stage) of the NIS Mikhail Lomonosov August 7–September 15, 1985. MHI Academy of Sciences of the Ukrainian SSR, Sevastopol, vol. 1, p. 135 (1985)
15. Ivanov, V.A., Samodurov, A.S., Chukharev, A.M., Nosova, A.V.: Intensification of vertical turbulent exchange in areas of pairing of the shelf and the continental slope in the Black Sea. Dop. Nats. Akad. Nauk Ukr. **6**, 108–112 (2008)

Check for updates

Field Investigation and Numerical Simulation of Wind-Wave Interaction at the Middle-Sized Inland Reservoirs

G. A. Baydakov(✉), A. M. Kuznetsova, V. V. Papko, A. A. Kandaurov, M. I. Vdovin, D. A. Sergeev, and Yu. I. Troitskaya

Institute of Applied Physics, Russian Academy of Sciences, 46 Ul'yanov Street, 603950 Nizhny Novgorod, Russia
baydakov@ipfran.ru

Abstract. An attempt is made to apply the modern methods of surface wave simulation developed for oceanic conditions to the modeling of waves in medium-size inland reservoirs (10–100 km). The results of field measurements of wind speed and waves are described, and on their basis the parameterization $C_D(U_{10})$ is proposed. WAVEWATCH III spectral wave model was adapted to the conditions of a medium-size inland reservoir. The simulated data are compared with the field data. The use of the new parameterization $C_D(U_{10})$ allowed reducing the values of the wind wave growth rate that improved consistency in data from the field experiment and numerical modeling concerning the height of significant waves. Further steps towards improving the quality of prediction of the adapted WAVEWATCH III model are discussed.

Keywords: Field experiment · Numerical simulation · Wind-wave interaction WAVEWATCH III

1 Introduction

Prediction of surface wind waves on the inland water bodies is recognized as an important problem involving many environmental applications, such as safety of the inland navigation and protection from the banks erosion. Lake waves also strongly affect the processes of exchange of momentum, heat and moisture in the low atmosphere and thus determine microclimate of the adjacent areas, which should be taken into account in planning structure of recreation zones.

The major physical problem of numerical wave modeling in inland water bodies is associated with short fetches, when parameters of wave excitation and development are strongly different from long-fetch condition typical for the open ocean [1]. Typically, in these conditions, the numerical description of waves in lakes and reservoirs are based on empirical models (see e.g. [2, 3]). But the empirical relationships are based on the averaged characteristics that cannot predict the extremes important for many tasks of operational meteorology (storm conditions such as Great Lakes storm, discussed in [4]), and numerical wave models are required. Now there is a number of examples of

V. Karev et al. (Eds.): PMMEEP 2017, SPRINGERGEOL, pp. 112–124, 2018.
https://doi.org/10.1007/978-3-319-77788-7_13

application of third generation models for waves forecast in large lakes. So, WAVEWATCH III [1] is used successfully for the wave forecasts on the Great Lakes in the USA [5, 6]. The data for a current wave situation is presented on the open web-site and is updated every three hours [7]. Furthermore, WAVEWATCH III and SWAN [8] are applied to Caspian Sea and Ladoga Lake to analyze the wind and waves climate hindcasting [9]. Nevertheless, lakes and reservoirs of smaller sizes (less than 100 km linear size, so called middle-sized reservoirs) also have examples of hurricane-force wind and severe surface wave states conditions. The first attempt of application of a global wave model WAM [10] for the wave forecasting on a middle-sized reservoir was reported recently in [11].

Among the peculiarities of the low-fetch waves at the middle-sized reservoirs is the stronger wind input, which is proportional to the ratio of wind friction velocity (or 10-m wind speed) to the wave phase velocity [1]. Another feature is the enhanced non-linearity caused by higher steepness of the waves. Then the tuning of the ocean wave model to the inland water conditions should be two-fold: adjusting of the wind source and "collision integral". Dissipation due to wave breaking can be expected to be similar to wave conditions due to their universal nature.

One more problem of tuning of numerical models to the conditions of middle-sized reservoirs and lakes is a small amount of experimental data that can be used for its verification. Rare examples of such experiments are studies [12, 13], which show the specificity of wind-wave interaction in the indicated circumstances. In this paper we present a tuning of the wind input term in WAVEWATCH III model to the conditions of the middle-sized reservoir on an example of Gorky Reservoir belonging to the Volga Cascade. The tuning is based of the data of the field experiment held by our group. The methods of the experiment is different from that used in [12, 13], and it focuses on the study of airflow in the close proximity to the water surface. The comparison of results of the numerical experiments with the results of the field experiments on Gorky Reservoir is presented.

2 Parameterizations of Wind-Wave Interaction

The software system of the WAVEWATCH III model is based on the numerical solution of the Hasselman's equation for the wave action spectral density $N(k, \theta, x, t)$ [1] which has the following form for the case of deep water:

$$\frac{\partial N}{\partial t} + \nabla_x \dot{x} N + \frac{\partial}{\partial k}\dot{k}N + \frac{\partial}{\partial \theta}\dot{\theta}N = \frac{1}{\sigma}(S_{in} + S_{dis} + S_{nl}). \quad (1)$$

The left part of the equation described the wave kinematics (where $\dot{x}$ is the group velocity; k is the wave number; θ is the angular direction). The right part includes dynamic summands: S_{nl} describes the four-wave interaction, S_{in} describes wind-induced wave growth, S_{dis} describes dissipation mainly caused by the wave collapse; σ is circular frequency.

In medium- and small-size reservoirs characterized by small fetches, wind effects become considerable and need more accurate description. In general case, wind effects S_{in} are specified according to the Miles model of wind-induced wave growth [14]:

$$S_{in} = \beta N \sigma. \tag{2}$$

There β is the dimensionless coefficient through which the rate of wind-induced wave height growth is expressed [14]:

$$\operatorname{Im} \sigma = \frac{1}{2} \sigma \frac{u_*^2}{c^2} \beta, \tag{3}$$

where c is the phase velocity of the wave; the coefficient β depends on wind friction velocity u_* determined through the turbulent momentum flux:

$$\tau_{turb} = \rho_a \langle u_x' u_z' \rangle = \rho_a u_*^2, \tag{4}$$

(ρ_a is air density; u_x' and u_z' are the pulsation components of wind speed).

The experimental determination of the value of the turbulent momentum flux is a complex problem. The most widespread methods are the profiling, pulsation, and dissipation methods. In the pulsation method the momentum flux is retrieved by the direct measurement of eddy fluxes [15]. The dissipation method consists in the analysis of distribution of spectral density turbulence and is based on the assumption that there is balance between the generation and decay of turbulence. The profiling method uses the logarithmic law based on the Prandtl-Karman boundary layer theory for the flat plate: under conditions of neutral stratification the wind speed profile in the constant flow layer (where the turbulent momentum flux does not depend on height) is close to logarithmic [16]:

$$U(z) = \frac{u_*}{\kappa} \ln \frac{z}{z_0}. \tag{5}$$

where $\kappa = 0,4$ is the Karman constant; z_0 is the surface roughness height. By analogy with the flat plate resistance, the aerodynamic drag coefficient of water surface is introduced, it connects the measured wind speed and the turbulent momentum flux (wind friction velocity):

$$C_D = \frac{\tau_{turb}}{\rho_a U_{10}^2} = \frac{u_*^2}{U_{10}^2}, \tag{6}$$

where U_{10} is the wind speed reduced to the height of 10 m.

There is a number of empirical models that describe the coefficient β of interaction between wind and waves. They differ much for long waves typical of the oceans but are similar in the frequency range typical of the conditions under study [17]; therefore the type of parameterization of the wind-wave interaction coefficient under conditions of medium-size inland reservoirs is not essential. Thus, to provide the more exact specification of wind effects, the parameterization of coefficient C_D is needed which defines transition from the measured speed U_{10} to the wind friction velocity u_* included into the parameterization.

Calculations used the WAM 3 Snyder's parameterization [18, 19], the most convenient of the WAVEWATCH III model parameterizations from the point of view of modification. The WAM 3 model [18–20] is specified by two empirical formulae. The first formula estimates the wind-induced wave growth rate:

$$\operatorname{Im}\sigma = C_{in}\frac{\rho_a}{\rho_w}\max\left[0, \left(\frac{28u_*}{c}\cos(\theta-\theta_w)-1\right)\right]\sigma \tag{7}$$

where $C_{in} = 0.25$ is the constant; ρ_a/ρ_w is the ratio of air density to water density; θ_w is the main wind direction. The second formula represents the parameterization of aerodynamic drag coefficient of water surface C_D and was proposed in [20]

$$C_D = 0,001 \times (0,8 + 0,65U_{10}) \tag{8}$$

This parameterization provides the relation between the wind speed U_{10} and wind friction velocity $u_* = U_{10}\sqrt{C_D}$.

The parameterization of the dependence $C_D(U_{10})$ used to modify the WAM 3 model was proposed as a result of the series of field experiments in the Gorky Reservoir area.

3 Instruments and Methods of Field Experiments

The measurements were carried out from May to October in 2014–2016 in the Gorky Reservoir area (the depth of the reservoir is 4–20 m and in the area of measurements the depth is 9–12 m). The prolate shape of the reservoir (see Fig. 1a, b) enables studying wind waves for different values of fetch depending on the wind direction.

The measuring instruments were installed at the buoy station that was originally worked out on the base of the oceanographic Froude buoy. The buoy represented a partially submerged mast that is held in the vertical position using a float near the surface and using a weight under water (Fig. 2). The total length of the buoy is 12 m and the length of its above-water part is 5.3 m. The resonance frequency of vertical fluctuations is 0.25 Hz that corresponds to the wavelength of 25 m. Four WindSonic wind speed sensors (made by Gill Instruments) were installed at the buoy mast at the height of 0.85, 1.3, 2.27, and 5.26 m (sensors 2–5). The fifth sensor (1) was located on the float tracking the wave to measure wind speed in the immediate proximity to the water surface. The distance from the float to the buoy mast was equal to about 1 m, and the height of the wind speed measurement zone from the water surface was equal to 10 cm. The buoy was equipped with the sensors of air temperature (at the height of 0.1 (a float), 0.85, and 1.3 m) and water temperature and with the three-channel string wave recorder which enables retrieving spatiotemporal profiles of waves.

WindSonic is an ultrasonic two-component wind speed sensor (the measurement error is 4% and the speed resolution is 0.01 m/s). The range of measured values of wind speed

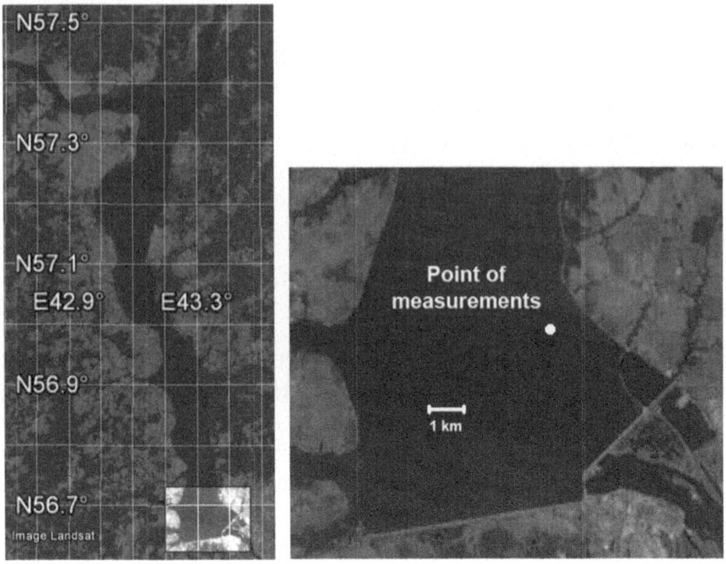

Fig. 1. Left: Gorky Reservoir (Google Earth data); right: zoom view of the measurements area.

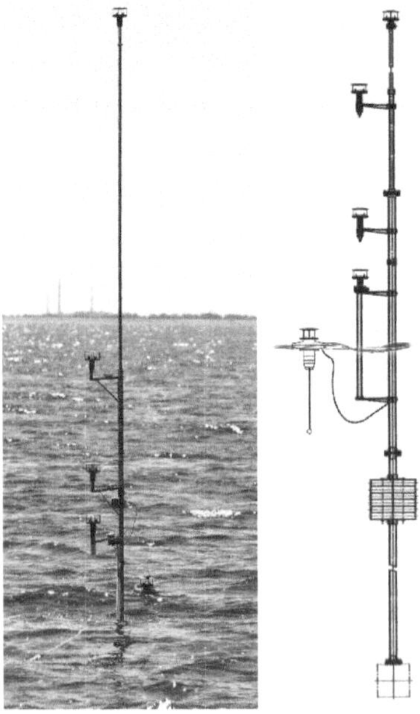

Fig. 2. Froude buoy: (a) real view of the operating state, (b) scheme

(0–60 m/s) includes the values typical of calm conditions. Resistive temperature sensors measure the temperature of the environment with the error of 3% and resolution of 0.01 °C. The string wave recorder consists of three pairs of string resistive sensors located at the vertices of the equilateral triangle with the side equal to 62 mm; the sampling frequency is 100 Hz. The system allows estimating the parameters of waves whose length exceeds the double distance between the sensors ($k_{max} \approx 0.5\ \mathrm{cm}^{-1}$). The algorithm of the processing of signals received from the instrument uses the Fourier transform and is described in detail in [21] (paper [22] presents a similar algorithm using the wavelet transform).

The location of wind speed sensors corresponds to the structure of the airflow. In the presence of waves on the water surface, the stream function in the air can be represented in the form of the sum of mean and wave components [23]

$$\Phi = \int_0^z U(\eta)d\eta + \varphi \tag{9}$$

where z is the vertical coordinate; φ is the wave perturbation of the stream function. In case of a traveling monochromatic wave where the elevation of the surface $z = \varsigma(x,t) = A\,\mathrm{Re}\,e^{-ik(ct-x)}$, φ can be calculated from the following equation:

$$(U - c)\left(\varphi'' - k^2\varphi\right) - U''\varphi = 0 \tag{10}$$

If the value of $U''/k^2(U - c)$ is much above or much below 1, the following function represents the approximate solution of the equation:

$$\varphi = A(U - c)e^{-kz} \tag{11}$$

where A is the wave amplitude. In the case of the logarithmic profile of speed (5) this condition takes the form of $u_*/\kappa(kz)^2|U - c| \ll 1$ (or $u_*/\kappa(kz)^2|U - c| \gg 1$) and is well met at the height z that is about the wave amplitude and higher. Thus, the basic disturbance contributed by waves to the air flux (wind bending along the surface) exponentially decreases with height. Hence, to provide the immobility of the sensor relative to the mean stream lines, the wave velocity at the distance from the water surface should be measured at the fixed level. The measurements near the surface should be carried out from the float using the sensor tracking the wave shape. It is important that the lower sensor is not located in the wave boundary layer whose value ε can be estimated in accordance with [24]:

$$k\left(\frac{u_*}{\kappa}\ln\left(\frac{z}{z_0}\right) - c\right)\Bigg|_{z=\varepsilon} = \frac{\kappa u_* z}{\varepsilon^2}\Bigg|_{z=\varepsilon} \tag{12}$$

Under conditions of the Gorky Reservoir ($k = (2 \div 3)\ \mathrm{m}^{-1}$, $u_* = (0.1 \div 0.4)\ \mathrm{m/s}$), $\varepsilon \sim 1$ mm that is much below the measurement height of the lower wind speed sensor.

The study of wind flow parameters was carried out by the profiling method. The general record of wind speed with the duration up to 5 h was divided into the periods of 5 min (300 measurement points) overlapping by 50%. As a result of the averaging, five values of wind speed corresponding to five measurement levels were obtained for each time period. The obtained mean profile was approximated by function (5) with the approximation parameters u_* and z_0. The values of wind speed U_{10} at the height of 10 m and the aerodynamic drag coefficient C_D were retrieved from the obtained approximation.

4 Results of the Field Experiment

The effects of the data from separate horizons on the result of wind speed profile approximation were analyzed. Figure 3a presents the comparison of retrieved dependences $C_D(U_{10})$ for two combinations of wind speed sensors: with and without the lower sensor as well as the results presented in [12, 13] and the oceanic parameterization [25]. It is clear that the values of $C_D(U_{10})$ obtained without data from the lower sensor are higher and closer to the results presented in [12, 13, 25]. In the case of using the data from the lower sensor, the values of the aerodynamic drag coefficient are lower. Figure 3b presents the comparison of retrieved dependences $C_D(U_{10})$ using only the data of two lower sensors and the data of five sensors of wind speed. In case of using two sensors only, significant differences are observed in retrieving wind parameters in the range of small values of wind speed.

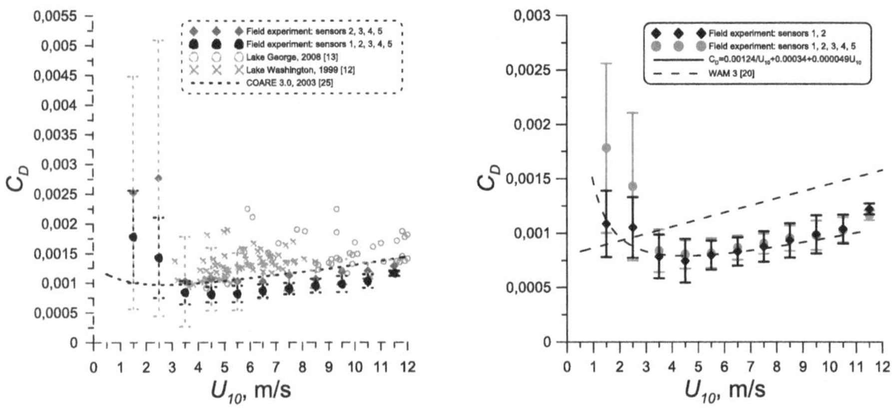

Fig. 3. The comparison of dependences $C_D(U_{10})$ retrieved using different combinations of sensors. Left: with and without the lowest sensor; right: with and without upper sensors.

These results can be explained by the distinction of wind speed profile shape from the logarithmic one. This distinction is probably caused by the stratification of the atmospheric surface layer and by the non-stationary nature of wind because the lower part of the profile is adapted to varying wave conditions more quickly. The airflow parameters on the water-air interface define the momentum transfer from wind to waves.

Thus, the use of the lower sensor (in the case under consideration, the use of two lower sensors only) affects the measurement result considerably. To determine the correctness of the measured dependence $C_D(U_{10})$, this dependence was used in the numerical modeling of wind waves in the WAVEWATCH III model. For this purpose the experimental data were approximated (see Fig. 3b) by the following function:

$$C_D = 0.00124 \cdot U_{10}^{-1} + 0.00034 + 0.000049 \cdot U_{10} \quad (13)$$

5 Numerical Experiment

The WAVEWATCH III model was adapted to inland reservoir conditions. For this purpose, the minimum value of significant wave height (H_S) was changed in the open software code. To describe the reservoir, the topographic grid of the Gorky Reservoir with the size of 72 × 108 and grid spacing of 0.00833° was used. The grid was taken from the NOAA Global Land One-kilometer Base Elevation (GLOBE) data. In view of the absence of open trustworthy bathymetric data for the Gorky Reservoir and taking into account that the navigation maps indicate the rather large depth of the reservoir, deep water approximation was chosen. Besides, the waves with the length of more than 4.5 m were not observed in field experiments. In view of this, the bottom topography was not taken into account for calculations, and the depth was selected to be equal to 9 m. The frequency range was changed in accordance with that observed in the experiment, from 0.2 to about 4 Hz. For the modeling it was divided into 31 frequencies and was specified by the logarithmic formula for the frequency increase $\sigma_N = (\delta)^{N-1}\sigma_1$ where the increment $\delta = 1.1$ was chosen in accordance with recommendations [1]; 31 angular directions were considered. The waves in the reservoir were simulated using the prescribed topographic data and data on the speed and direction of wind and on difference in the values of temperature at the water-air interface and at the prescribed Gaussian initial perturbation for different parameterizations of wind effects.

The range was considered of the moderate speed of wind (1–9 m/s) of different directions with the constant values over the whole Gorky Reservoir surface. In practice, reanalysis data are commonly used to specify wind effects for modeling wind waves on the sea and ocean surface. This approach is unsuitable for the water areas of medium-size inland reservoirs due to the too low spatial resolution (2.5°). Besides, only two weather stations (Yur'evets and the Volga River hydrometeorological observatory) located on the shore are situated in the area under consideration. It was found that the values of wind speed in the coastal part of the reservoir and over its water area differ significantly. In view of this, calculation was carried out using the input data measured in the field experiment and updated every 15 min: the speed and direction of wind at the height of 10 m and difference in the values of temperature at the water-air interface. If the speed and direction of wind are forcedly prescribed to be the same over the reservoir, this may lead to errors in the numerical experiment because the prolate shape of the reservoir and its high shores can be a reason for considerable spatial variability with the scale of about or below 1 km.

The comparison was carried out for the following output data: one-dimensional elevation spectra, significant wave height H_S, and weighted mean wave period T_m. The value of H_S in the model and in the experiment was calculated from the formula

$$H_S = 4\sqrt{E}, \tag{14}$$

where $E = \int_{f_{\min}}^{f_r} E(f)df$ is full energy, $E(f)$ is the spectral density of wave force.

The weighted mean wave period T_m was calculated from the following formula:

$$T_m = T_{m0,-1} = \left(\int_{f_{\min}}^{f_r} E(f)df \right)^{-1} \int_{f_{\min}}^{f_r} E(f)f^{-1}df \tag{15}$$

All model data were obtained at the point corresponding to the observational point and were averaged for 15 min to agree with the field experiment data averaged in a similar way.

The calculations were carried out in two ways: in the framework of the WAM 3 oceanic parameterization using the linear dependence of C_D on U_{10} [20]; using the parameterization of C_D proposed by the authors and the wind-induced wave growth rate from WAM 3. Difference in parameterization is demonstrated in Fig. 3b. It is clear that if the wind speed is below 2.5 m/s, the values of C_D obtained as a result of the field experiment are higher than those obtained from the oceanic parameterization; if wind speed is above 3 m/s, the opposite picture is observed.

6 Comparison of the Results of the Numerical and Field Experiments

One-dimensional elevation spectra at the measurement point obtained in the field experiment were compared with those of the numerical experiment using different parameterizations of wind effects. It is clear from Fig. 4 that the values are overestimated too much in case of using oceanic wind effects, whereas the use of the new parameterization improves the agreement between the results of the numerical and field experiments.

The comparison of integral characteristics of the spectra (significant wave heights and weighted mean period of the spectrum) was carried out for all experiments. In Fig. 5, the black dash line is a bisectrix of the angle corresponding to the equality of the field experiment characteristics. We demonstrate data computed using the oceanic wind effects from WAM 3 and using the modified parameterization of WAM 3 with the new parameterization of C_D.

At using the oceanic model, the systematic overestimation of significant wave height is observed as well as the underestimation of weighted mean wave period. The standard deviation of the calculated values of H_S for WAM 3 is 52%. The use of the new parameterization of C_D reduces the standard deviation for WAM 3 from 52 to

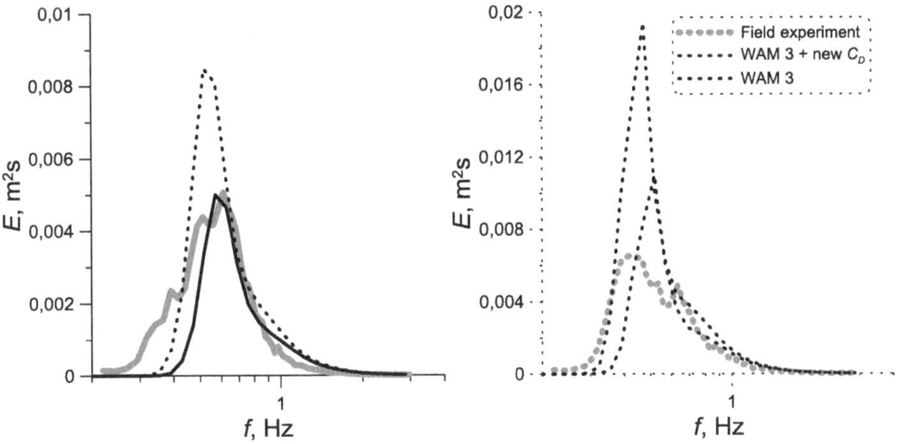

Fig. 4. 1D wave spectra. The experimental spectrum is indicated by light-gray bold solid wideline, the simulated spectrum with the parameterization WAM 3 – dashed line, WAM 3 with the new C_D – black solid line

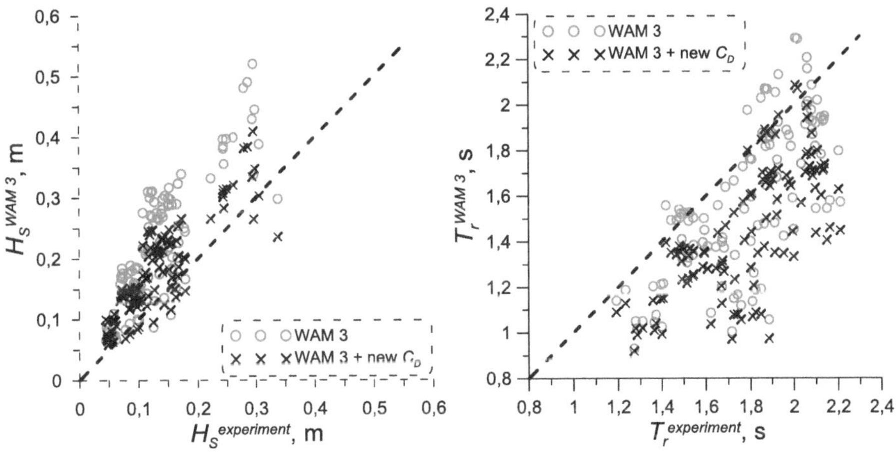

Fig. 5. H_S (left) and T_m (right) in comparison with the data of field experiment for parameterization WAM 3 (gray circle) and WAM 3 with the new C_D (black crosses),

39%. This is an expected result because in the numerical experiment the rate of wind-induced wave growth with the proposed parameterization of C_D is specified more accurately, i.e., the amount of energy coming to the system is simulated more accurately.

However, it is clear from Fig. 5 that the prediction of weighted mean wave periods has a significant error and the correction in the specifying of the rate of wind-induced wave growth did not result in considerable changes. Perhaps, this is associated with the fact that the WAVEWATCH III model is adapted to marine conditions. This is manifested not only in the wind effect function but also in the features of parametric accounting of nonlinearity which causes the spectral redistribution of received energy. The model is intended to describe waves typical of marine and oceanic conditions that have a smaller steepness ratio as compared with the waves in a medium-size inland reservoir. Proportionality coefficients in the DIA scheme [26, 27] are adapted to marine conditions. To describe steeper waves in a medium-size inland reservoir, other adjustment parameters can be required. These parameters should correspond to the situation with more significant nonlinearity that will quicken the frequency shift towards the low-frequency range. Hence, weighted mean wave periods should also be smaller. Probably, such adaption of the scheme of nonlinearity will not affect the quality of prediction of the value of H_S characterizing the amount of energy coming to the system and will increase the accuracy of prediction of weighted mean wave periods. It is planned to test this hypothesis in future numerical experiments.

7 Conclusions

The possibility is considered of the WAVEWATCH III model adaptation to a medium-size inland reservoir by an example of the Gorky Reservoir which was specified in the model using the NOAA GLOBE real topographic grid. To carry out the calculations, the original values of model parameters were changed according to the data of field experiments in the reservoir. In particular, the minimum significant wave height was changed, the frequency range is from 0.2 to about 4 Hz. The waves developed under the influence of homogeneous non-stationary wind (specified as a result of the data of the field experiment) were calculated using both parameterizations of wind effects adapted to the open ocean conditions and parameterizations with the modified specification of $C_D(U_{10})$ which was obtained from field experiments. The data of field experiments in the Gorky Reservoir demonstrated that the value of the aerodynamic drag coefficient of the surface C_D in the area of moderate and strong wind is by about 50% smaller than the values typical of oceanic conditions. The results of the numerical experiment were compared with the results of the field experiment in the Gorky Reservoir. The use of the original parameterization demonstrated the considerable overestimation of the calculated data on H_S as compared with the experimental data. The authors explained this by the considerable overestimation of turbulent wind stress (values of wind friction velocity w,) and, hence, of wind effects. The use of the new parameterization of $C_D(U_{10})$ obtained from the measurement data lead to decrease in the values of w and, consequently, in the rate of wind-induced wave growth that improved the consistency in H_S data between the field experiment and numerical modeling. The comparison of the results of calculation in the framework of original oceanic models of wind effects also demonstrated the overestimated values of weighted mean wave period T_m. At the same time, the variations of wind effects did not affect considerably agreement in the values of T_m between the results of numerical simulation and the field experiment. This is probably associated with the fact that the scheme of nonlinearity is also adapted to marine and oceanic conditions. In the future it is necessary to adapt the parameters of DIA nonlinearity scheme to the conditions of a medium-size inland reservoir.

Besides, the use of deep water approximation for the calculations can be a possible source of differences. The accounting of the real bathymetry of the Gorky Reservoir as well as the use of shallow-water-related parameterizations in WAVEWATCH III or the nesting of the SWN model for the coastal zone can essentially improve the results.

One more source of the possible errors of the numerical experiment should also be noted. Due to the absence of sufficient experimental data, wind speed was assumed to be uniform over the whole water area of the reservoir taking into account the temporal variability specified as a result of the experiments. In reality the non-uniform distribution of wind speed and wind direction can be expected because such factors as the prolate shape of the reservoir and high shores may result in the considerable spatial variability with the scales of about or below 1 km. It is also impossible to specify wind speed from the reanalysis data due to the too low spatial resolution (2.5°). The accounting of high spatial variability is a complex problem, for its solution it is planned to use high- and very-high-resolution atmospheric models (for example, WRF (Weather Research and Forecasting) with the LES (Large Eddy Simulation) block).

Acknowledgments. The research was supported by the Russian Foundation for Basic Research (grants 17-05-41117 and 15-45-02580). The field experiments were supported by the Russian Scientific Foundation (grant 15-17-20009).

References

1. Tolman H.L., WAVEWATCH III Development Group: User Manual and System Documentation of WAVEWATCH III Version 4.18. Environmental Modeling Center, Marine Modeling and Analysis Branch (2014)
2. Poddubnyi, S.A., Sukhova, E.V.: Modeling the Effects of Hydrodynamic and Anthropogenic Factors on the Distribution of Hydrobionts in Reservoirs. User's Manual. Rybinskii Dom Pechati, Rybinsk (2002). (in Russian)
3. Sutyrina, E.N.: Determination of Wave Characteristics in the Bratsk Reservoir. Izvestiya Irkutskogo Gosudarstvennogo Universiteta, vol. 2, no. 4 (2011). (in Russian)
4. Newton Matza, M.: Disasters and Tragic Events: An Encyclopedia of Catastrophes in American History. ABC-CLIO, Santa Barbara (2014)
5. Alves, J.-H.G.M., Chawla, A., Tolman, H.L., et al.: The great lakes wave model at NOAA/NCEP: challenges and future developments. In: 12th International Workshop on Wave Hindcasting and Forecasting, Kohala Coast, Hawaii (2011)
6. Alves, J.-H.G.M., Chawla, A., Tolman, H.L., et al.: The operational implementation of a great lakes wave forecasting system at NOAA/NCEP. Weather Forecast. **29**, 1473–1497 (2014)
7. NWW3 Product Viewer. http://polar.ncep.noaa.gov/waves/viewer.shtml7-glw-latest-hs-grl. Accessed 27 Nov 2017
8. SWAN Team: SWAN - User Manual. Delft University of Technology, Environmental Fluid Mechanics Section (2006)
9. Lopatoukhin, L.J., Boukhanovsky, A.V., Chernyshova, E.S., Ivanov, S.V.: Hindcasting of wind and wave climate of seas around Russia. In: Proceedings of the 8th International Workshop on Waves Hindcasting and Forecasting, North Shore, Oahu, Hawaii (2004)
10. Gunter, H., Hasselmann, S., Janssen, P.A.E.M.: The WAM model cycle 4. Technical report no. 4. DKRZ WAM4 Model Documentation, Hamburg (1992)

11. Hesser, T.J., Cialone, M.A., Anderson, M.E.: Lake St. Clair: Storm Wave and Water Level Modeling. The US Army Research and Development Center (ERDC) (2013)
12. Atakturk, S.S., Katsaros, K.B.: Wind stress and surface waves observed on lake Washington. J. Phys. Oceanogr. **29**, 633–650 (1999)
13. Babanin, A.V., Makin, V.K.: Effects of wind trend and gustiness on the sea drag: lake George Study. J. Geophys. Res. **113**, C02015 (2008)
14. Miles, J.W.: On the generation of surface waves by shear flows. J. Fluid Mech. **3**(2), 185–204 (1957)
15. Weber, R.O.: Remarks on the definition and estimation of friction velocity. Bound. Layer Meteorol. **93**, 197–209 (1999)
16. Setton, O.G.: Micrometeorology. Gidrometeoizdat, Leningrad (1958). (Transl. from Engl.)
17. Zakharov, V.E.: On the domination of nonlinear wave interaction in the energy balance of wind-driven sea. In: Proceedings of 11th Wave Workshop, Halifax, Canada (2009)
18. Komen, G.L., Hasselmann, S., Hasselmann, K.: On the existence of a fully developed wind-sea spectrum. J. Phys. Oceanogr. **8**(14), 1271–1285 (1984)
19. Snyder, R.L., Dobson, F.W., Elliott, J.A., Long, R.B.: Array measurements of atmospheric pressure fluctuations above surface gravity waves. J. Fluid Mech. **102**, 1–59 (1981)
20. Wu, J.: Wind-stress coefficients over sea surface from breeze to hurricane. J. Geophys. Res. **87**(C12), 9704–9706 (1982)
21. Troitskaya, Y.I., Sergeev, D.A., Kandaurov, A.A., et al.: Laboratory and theoretical modeling of air-sea momentum transfer under severe wind conditions. J. Geophys. Res. **117** (C11), C00J21 (2012)
22. Donelan, M.A., Drennan, W.M., Magnusson, A.K.: Nonstationary analysis of the directional properties of propagating waves. J. Phys. Oceanogr. **26**(9), 1901–1914 (1996)
23. Brooke, B.T.: Shearing flow over a wavy boundary. J. Fluid Mech. **6**, 161–205 (1959)
24. Belcher, S.E., Hunt, J.C.R.: Turbulent shear flow over slowly moving waves. J. Fluid Mech. **251**, 109–148 (1993)
25. Fairall, C.W., Bradley, E.F., Hare, J.E., et al.: Bulk parameterization of air-sea fluxes: updates and verification for the COARE algorithm. J. Clim. **4**(16), 571–591 (2003)
26. Hasselmann, S., Hasselmann, K.: Computations and parameterizations of the nonlinear energy transfer in a gravity-wave spectrum. Part I: a new method for efficient computations of the exact nonlinear transfer integral. J. Phys. Oceanogr. **15**, 1369–1377 (1985)
27. Hasselmann, S., Hasselmann, K., Allender, J.H., Barnett, T.P.: Computations and parameterizations of the nonlinear energy transfer in a gravity-wave spectrum. Part II: parameterizations of nonlinear energy transfer for application in wave models. J. Phys. Oceanogr. **15**, 1378–1391 (1985)

Multidecadal Variability of Hydro-Thermodynamic Characteristics and Heat Fluxes in North Atlantic

N. A. Diansky[1,2,3(✉)] and P. A. Sukhonos[4]

[1] Faculty of Physics, M.V. Lomonosov Moscow State University, GSP-1, Leninskie Gory, Moscow 119991, Russia
nikolay.diansky@gmail.com
[2] Institute of Numerical Mathematics of the RAS, Gubkina street., 8, Moscow 119333, Russia
[3] N.N. Zubov State Oceanographic Institute, Kropotkinsky Lane, 6, Moscow 119034, Russia
[4] Institute of Natural and Technical Systems, Lenina street, 28, Sevastopol 299011, Russia
pasukhonis@mail.ru

Abstract. The article describes a manifestation of Atlantic multidecadal oscillation (AMO) in the variability of hydro-thermodynamic characteristics and ocean surface heat fluxes in North Atlantic. It is shown that multidecadal changes in the upper mixed layer temperature and mixed layer depth, as well as sea surface height (SSH), are statistically significant and show physically consistent changes. Convective mixing of waters at high latitudes is stronger for negative AMO phase with an exception the case of Great Salinity Anomaly. Large-scale fluctuations in SSH are most pronounced in the eastern part of North Atlantic. Multidecadal changes in the net surface heat fluxes and horizontal heat advection are coherent. The increase (decrease) of horizontal heat advection in the negative (positive) AMO phase leads to an increase (decrease) of the ocean heat release to the atmosphere. Multidecadal variability of the horizontal heat advection is due to changes in the dynamics of currents, rather than temperature gradients.

Keywords: Atlantic multidecadal oscillation · Variability · North Atlantic

1 Introduction

Natural long-term sea surface temperature (SST) fluctuations stand out at a significant level in North Atlantic [1–7]. This phenomenon was called the Atlantic multidecadal oscillation (AMO) [8]. This climatic signal in the temperature field of World Ocean is one of the main signals at scales from interannual to multidecadal and this signal is not connected with El Niño-Southern Oscillation [9].

The AMO is characterized by an index, which is determined from the SST in the North Atlantic basin, averaged typically over 0–70 °N [3]. The AMO index is calculated as a detrended 10-year low-pass filtered annual-mean of this averaged SST anomaly. A coherent pattern of variability covering the entire North Atlantic and

V. Karev et al. (Eds.): PMMEEP 2017, SPRINGERGEOL, pp. 125–137, 2018.
https://doi.org/10.1007/978-3-319-77788-7_14

exhibiting a long-period quasi-cyclic character with a period of about 50–70 years was first identified in paper [6]. Later model studies also confirm that the multidecadal climate variability in North Atlantic is a single-mode [10, 11].

Several studies have revealed the influence of AMO on climatic conditions over Northern Hemisphere. The AMO is associated with changes in North American, European and Arctic air temperature, rainfall and river flow [3, 12–15], and Sahel drought [16]. The AMO also affects the low frequency hurricane variability [17] and atmospheric blocking [18] in the North Atlantic. The AMO effectively influences atmospheric heat transports on seasonal scale in the Atlantic-European region and significantly modifies the production rate of North Atlantic waters [19]. A multidecadal mode of variability between the observed concentration of sea ice in the Greenland Sea and sea level pressure at high latitudes being coherent with the observed AMO variability is found in [20].

The AMO is described as a mode of variability with a period of about 50–70 years [3, 21]. At the same time, the results of later studies indicate the existence of another mode of interdecadal variability with a period of about 20–30 years [22, 23].

The authors of [24] showed that there is a significant multidecadal variability of surface heat fluxes associated with the low-frequency mode of the North Atlantic Oscillation (NAO). However, the results of [4] indicate that there is no significant correlation between AMO and NAO. Obviously, this is due to the different nature of these oscillations. Since AMO is mainly determined by the low-frequency variability of oceanic circulation, while the NAO – atmospheric.

The nature of AMO formation has been studied much worse than other known climatic indices, since AMO is less associated with the atmosphere as compared to other indices. For example, the physical mechanisms of Pacific Decadal Oscillation have been studied quite well [25].

The nature of the AMO still requires for clarification. In particular, the relationship between AMO and the SST variability in the Pacific Ocean has not been fully clarified. Some authors [3, 7, 12] indicate the presence of some connection between the AMO and SST anomalies in the Gulf of Alaska and the tropical Pacific. Other authors [4] argue that outside the North Atlantic, there is no significant relationship between AMO and SST anomalies.

At present, the mechanism for generating multidecadal variability in the North Atlantic is a subject of discussion. This is due to the fact that the role of atmosphere in the observed climate variability is not fully understood [26]. A number of authors point to the joint nature of interactions in the ocean-atmosphere system due to the interaction of SST anomalies with the NAO [27]. Others believe that low-frequency oceanic modes are excited by atmospheric noise associated with synoptic weather fluctuations [28]. Still others insist that the observed multidecadal climate variability is an oceanic response to stochastic atmospheric forcing [24, 29]. Some authors claim that there are purely oceanic modes, the source of energy for which is the internal instability of large-scale ocean circulation [30–32]. The advection time of thermohaline anomalies from the Tropical Atlantic to subarctic latitudes determines the phase shift between long-period changes of the heat fluxes at the ocean-atmosphere boundary at the high latitudes of the North Atlantic and the meridional heat transport in the Subtropical Atlantic. This phase shift is one of the possible mechanisms for maintaining the interdecadal variability [21, 33, 34]. The results of the model study showed that the long-period ($\sim$60 years) variability of thermohaline circulation in the North Atlantic affects the thermal state of the ocean surface with a delay of about 10 years [35]. At the same time, in some studies the variability of meridional overturning circulation is not the main cause of multidecadal climate variations in the North Atlantic. This variability is explained by the combined effect of «thermal Rossby modes» in the North Atlantic and «saline Rossby modes» from the Arctic [36, 37].

The thermohaline circulation in the North Atlantic, sea ice and fresh water export from the Arctic and atmospheric dynamics have some memory and generate multidecadal variability due to negative feedback [38]. A similar mechanism was proposed using the analysis of the 500-year integration of the coupled atmosphere-sea-ice-ocean model [39]. These authors found pronounced multidecadal oscillations of the Atlantic meridional overturning circulation and the associated meridional heat transport for the periods of 70–80 years, which significantly correlate with the convective activity in the deep-water formation regions and the freshwater export from the Arctic. However, the results of the study [40] showed that the North Atlantic SST variability – a positive feedback between North Atlantic SST, African dust, and Sahel rainfall on multidecadal time scales. Thus, the AMO is a pronounced climatic signal of a multidecadal scale and manifested in a number of climatic characteristics of the ocean-atmosphere system. However, a single point of view on the causes that induce their long-term changes has not yet been obtained.

The goal of this paper is to analyze the extent and nature of the manifestation of AMO in the hydrothermodynamic characteristics and surface heat fluxes in the North Atlantic.

2 Data and Methods

The monthly data of the ocean temperature, ocean salinity, sea surface height (SSH), zonal, and meridional components of the ocean current velocity vector from ocean reanalyses ORA-S3 for 1959–2011 [41], ORA-S4 for 1958–2014 [42], SODA 2.1.6 for 1958–2008 [43], GFDL for 1961–2015 [44], GECCO2 for 1948–2014 [45] were used.

The monthly data of the mixed layer depth (MLD) were used from ocean reanalyses ORA-S3 and GFDL. In order to calculate the MLD in the ORA-S3 reanalysis, a scheme based on the semi-empirical theory of turbulence is applied. The main idea of the scheme is to calculate the Richardson number in the form by [46]. The MLD is assumed to be equal to the depth at which the Richardson number reaches a critical value. The MLD in the GFDL reanalysis is defined by fixed density criterion; namely as a depth where density increases compared to the density at 10 m depth by 0.03 kg/m^3 [44].

The monthly data of the latent and sensible heat fluxes, balances of shortwave and longwave radiation on the ocean surface with a spatial resolution of 1° × 1° were used from the Coordinated Ocean Research Experiments version 2 (COREv2) (http://data1.gfdl.noaa.gov/nomads/forms/core/COREv2.html) for 1949–2006 [47]. This data set is a set of atmospheric boundary conditions and is designed to simulate the ocean circulation taking into account the performance of heat balance on the ocean surface.

According to the set of heat flux components, the net surface heat fluxes for each month are calculated. Their mean annual values are then calculated.

According to the above-mentioned data, the temperature, salinity, zonal and meridional components of the ocean current velocity vector, zonal and meridional temperature gradients, zonal, meridional and total horizontal heat advection were calculated within upper 0–300 m layer for each month and annual values for the entire period. The area of the North Atlantic is restricted by the following coordinates: 0–60 °N, 10–80 °W.

The values of the AMO index for the study period were taken from the database NOAA Earth System Research Laboratory (http://www.esrl.noaa.gov/psd/data/timeseries/AMO/) for 1948–2010 (Fig. 1) [3].

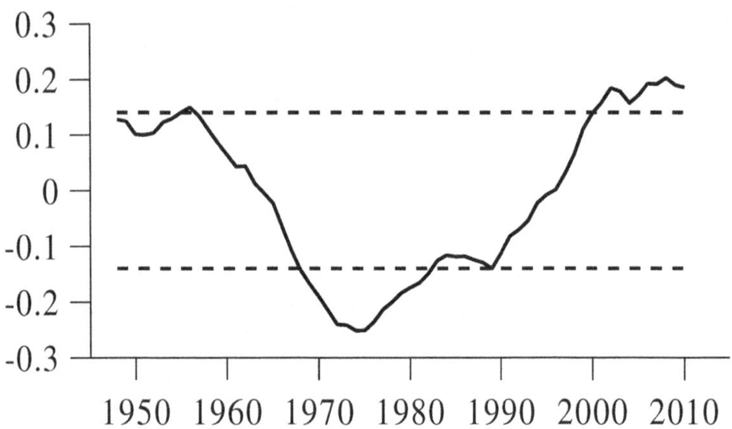

Fig. 1. Time evolution of the average annual values of the smoothed AMO index for the period 1948–2010. The dashed line characterizes the values ±0,14 °C.

The monthly data of the Gulf Stream North Wall index for 1966–2014 were used. This index characterizes the meridional displacement of position of northern boundary of the Gulf Stream off the east coast of North America. This index is calculated as the first Principal Component of the position of the North Wall of the Gulf Stream. Data on the Gulf Stream index are taken from the site (http://www.pml-gulfstream.org.uk/data.htm). According to these data, their annual mean values and linear and polynomial trends were calculated.

For the studies, two methods of processing the initial data were used. First, the time evolution of annual mean upper mixed layer temperature, MLD in January, annual mean SSH from the ORA-S3 reanalysis and net heat flux from the ocean to the atmosphere from COREv2 data were averaged over two latitudinal bands: 15–40 °N and 41–65 °N. Based on these data, linear and polynomial trends were calculated.

Second, the composite analysis is used for this study. According to the time series of the AMO index, anomalous years were chosen in which the value of index exceeds the mean value by ±1 standard deviation (0,14 °C). Thus, two samples of years with values of the index above and below the set limits, corresponding to the positive and negative phases of the AMO, were obtained. The positive phase of AMO is the twelve-year period from 2000 to 2011 and the negative phase – the twelve-year period from 1969 to 1980. Then the linear trend was removed from all the time series. The average value and the standard deviation for the positive and negative AMO phases were found for the each parameter considered. Then we determine the difference between the means by sample for each grid node (the so-called difference composite). Statistical significance of composites was determined by the standard algorithm using the Student's test.

3 Results

3.1 Analysis of Long-Term Trends in the North Atlantic Subtropical and Subpolar Gyres

The time evolution of annual mean upper mixed layer temperature, SSH, net surface heat fluxes and MLD in January averaged over the regions of large-scale subtropical (15–40 °N) and subpolar (41–65 °N) gyres is shown in Fig. 2.

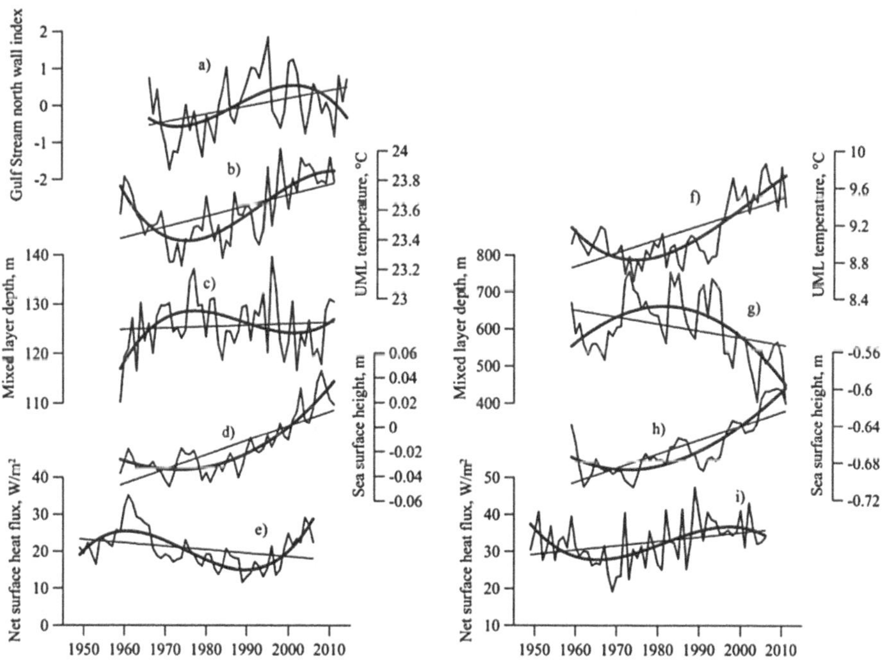

Fig. 2. Time evolution of annual mean Gulf Stream north wall index (a), annual mean upper mixed layer temperature, °C (b, f), MLD in January, m (c, g), annual mean SSH, m (d, h) according to the ORA-S3 data and annual mean net surface heat flux according to the COREv2 data, W/m^2 (e, i) averaged over the 15–40 °N (b, c, d, e) and 41–65 °N (f, g, h, i) in the North Atlantic basin. The thin lines are linear trends. The thick lines are an approximate polynomial of third degree.

Linear trends are distinguished in all the time series used. The upper-ocean active layer is characterized by a long-term tendency to warming. It leads to an increase in the large-scale SSH and the weakening of convective mixing in high latitudes. The MLD and net surface heat fluxes in the middle and high latitudes exhibit oppositely directed trends. The MLD in the subtropical gyre undergoes a long-period increase, and in the subpolar gyre – decrease. Long-term decrease in the heat release by the ocean in the second half of the 20th century occurs in the subtropics, and the increase – in subpolar latitudes.

Multidecadal changes were determined by calculating polynomials of the third order from the time evolution of the characteristics considered. They clearly show a signal of quasi-sixty-year oscillations. Multidecadal SSH changes in the middle and high latitudes are well coordinated with changes in the upper mixed layer temperature and MLD and consist in an increase in these characteristics since the late 1990s. During this period, against the background of intensive interannual variability, the intensity of convective mixing in high latitudes decreases (Fig. 2g). Multidecadal upper mixed layer temperature changes anticorrelate with the MLD. This means the predominance of the dynamic factor in the upper mixed layer heat content changes. At the same time, multidecadal changes in the intensity of net surface heat flux in the middle and high latitudes in the North Atlantic are observed in opposite phase. In the mid-1960s, in the Subtropical Atlantic, there was an intense heat release from the ocean, in the Subpolar Atlantic, on the contrary, the loss of heat by the ocean is minimal due to the Great Salinity Anomaly (GSA) [48]. The increase in heat outflow from the ocean since the late 1990s in the Subtropical Atlantic and the corresponding weakening in the Subpolar Atlantic can be interpreted as the meridional displacement of the Gulf Stream core in a southerly direction into the positive phase of the AMO (Fig. 2a).

3.2 Composite Analysis of Changes in the Warm and Cold Phases of AMO

An investigation of the multidecadal variability of the hydrothermodynamic parameters for the North Atlantic Ocean is carried out on the basis of difference composites. For illustration, the spatial features of difference composites are given from the ORA-S3 oceanic reanalysis and COREv2 data (Fig. 3). Further in the text, only significant results are given that are consistent across all data sets.

The temperature in the 0–300 m layer in the positive AMO phase, relatively negative, is characterized by a statistically significant increase of 0.4 °C at high latitudes and 0.2 °C in the eastern part of the subtropical gyre. Significant negative temperature values are noted in the Gulf Stream area and near 15 °N. However, for most of the North Atlantic basin positive but statistically insignificant temperature values are typical (Fig. 3a).

The salinity in the 0–300 m layer in the positive AMO phase, relatively negative, in all seasons significantly increases in the inner parts of the subtropical and subpolar gyres. Salinity significantly decreases in the positive AMO phase south of 10 °N and in the vicinity of the Gulf Stream (Fig. 3b).

In the region of the subpolar gyre in difference composites for the temperature and salinity, a very strong signal of GSA is detected. It is associated with a decrease in

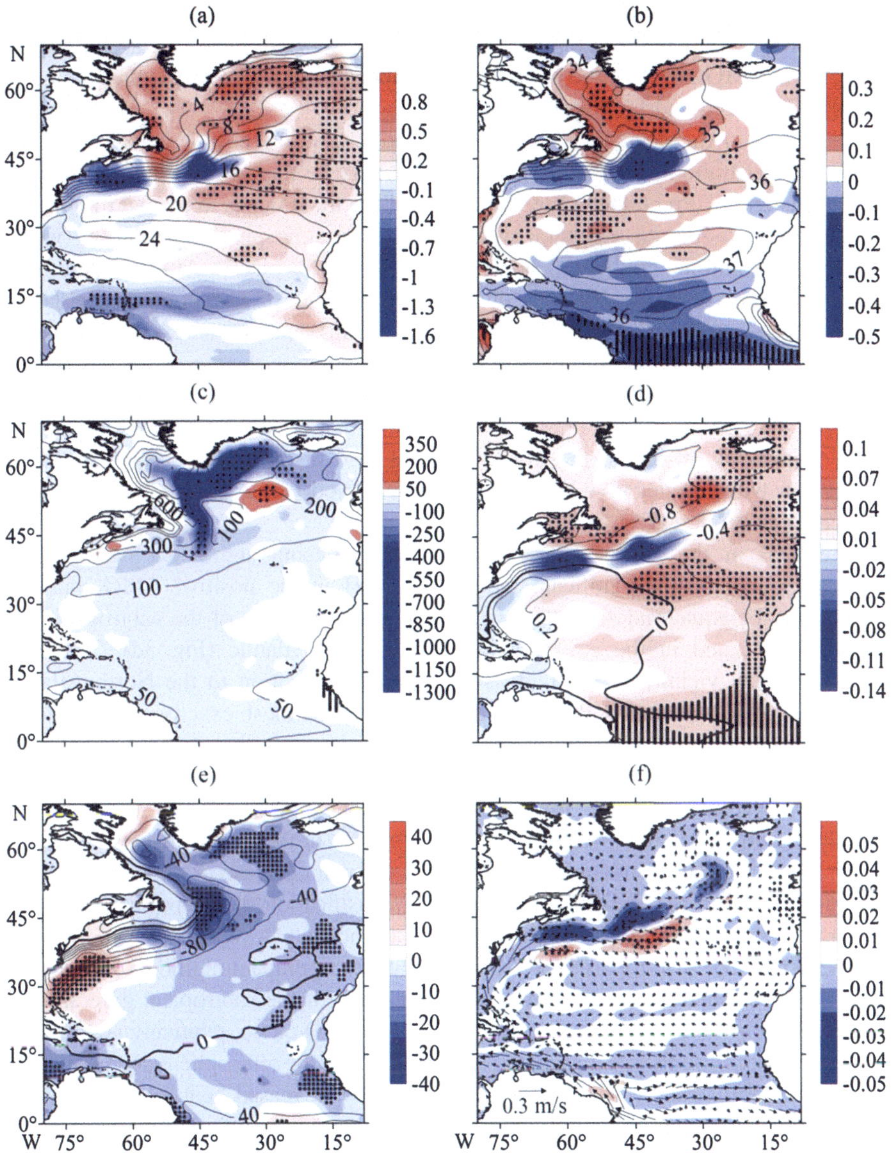

Fig. 3. The difference between the detrended anomalies of annual mean temperature in the 0–300 m layer (°C, a), annual mean salinity in the 0–300 m layer (PSU, b), MLD in January (m, c), annual mean SSH (m, d), annual mean net heat fluxes (W/m^2, e), current velocity modulus in the 0–300 m layer (m/s, f) in the positive and negative AMO phases. The positive AMO phase is a twelve-year period from 2000 to 2011, and a negative one is a twelve-year period from 1969 to 1980. The black dots show the nodes of spatial grid, in which the difference in composites is significant at 95% confidence level. Contours and vectors show average long-term mean fields. Figures a, b, c, d, f are prepared according to the ORA-S3 reanalysis data, and figure e – according to COREv2 data.

salinity here due to the export of more fresh and cold Arctic waters, thereby causing a decrease in the AMO index in the late 1960's and early 1970's. The increase in salinity in the warm AMO phase in the subpolar gyre, apparently, is due to the weakening of the East Greenland and Labrador Currents.

The MLD in the positive AMO phase, relatively negative, is characterized by a significant penetration at the African coast. This may be due to the weakening of the North Equatorial Current due to the weakening of the trade winds in the warm phase of the AMO. This is due to a general weakening in the atmospheric circulation (weakening of the latitudinal gradient of sea level pressure in this phase). At the same time, a significant decrease in the MLD is observed in the area of formation of the North Atlantic deep water in the Labrador Sea (over 300 m), and an increase in the MLD occurs in the center of the subpolar gyre (Fig. 3c). Such a change of the MLD here indicates a general weakening of the intensity of the subpolar gyre during the warm period of the AMO. The obtained result indicates that the convective mixing in the subpolar latitudes of the North Atlantic is weakened in the positive AMO phase in comparison with the negative AMO phase. The period of existence of GSA in this phase is an exception (Fig. 2f).

The SSH in the North Atlantic is also experiencing significant multidecadal changes. A statistically significant increase in SSH in the positive AMO phase is observed in the equatorial zone (0–10 °N), in the central parts of the subtropical and subpolar gyres and in the eastern part of the North Atlantic (Fig. 3d). The SSH decreases in the vicinity of the Gulf Stream and its transition to the North Atlantic Current. It should be noted that the changes in SSH almost exactly correspond to changes in temperature and salinity in the 0–300 m layer by their contribution to the change in sea water density in this layer. The greater contribution to the SSH change is caused by a thermal factor that causes a decrease in sea water density. The East-Greenland and Labrador Currents are exceptions. In the region of these currents, an increase in salinity compensates the factor of thermal decrease in sea water density. General changes in the SSH also indicate the intensification of subtropical gyre and, conversely, the weakening of subpolar gyre in the positive AMO phase.

The net surface heat fluxes experience statistically significant multidecadal changes in the equatorial region, the eastern and western parts of the subtropical gyre and the currents that form the subpolar gyre. In the positive AMO phase, relatively negative, in the western part of subtropical gyre, the heat release into the atmosphere decreases. The rest of the North Atlantic basin is mainly marked by increased heat release from the ocean. Typical values are about 20 W/m^2 (рис. 3e). At the same time, a decrease in the heat release to the atmosphere in the subpolar gyre confirms the weakening of convective processes here, which is evident from the changes in MLD.

Significant multidecadal changes in the current velocity modulus in the 0–300 m layer, caused by the AMO, are not observed throughout the entire North Atlantic Ocean. In the positive AMO phase in the dynamic system «Gulf Stream – North Atlantic Current» there is a significant decrease in current velocities, and to the south – an increase. This clearly indicates the meridional displacement of axis of western boundary current to the south in the positive AMO phase. At the same time, the East-Greenland and Labrador currents are weakening in the warm AMO phase. This agrees with the aforementioned increase in salinity in these currents. These multidecadal changes in the magnitude of current velocity also correspond to the above observed SSH changes associated with the intensification of the subtropical gyre and the weakening of the subpolar gyre and changes in net surface heat fluxes in the warm AMO phase. In addition, multidecadal changes in the dynamics of the North Equatorial Countercurrent are significantly distinguished for the summer season. The intensification of this current is noted in the positive AMO phase.

The structure of the multidecadal changes in horizontal heat advection in the 0–300 m layer, associated with the AMO, is generally consistent with the corresponding changes in the current velocity field. In the positive AMO phase, relatively negative, the weakening of advective heat transports is noted in the dynamic system «Gulf Stream – North Atlantic Current». This allows us to conclude that multidecadal variations of horizontal heat advection are due to changes in the dynamics of currents, rather than temperature gradients.

4 Discussion and Conclusion

We must emphasize that, except for linear trends, we study own multidecadal variability of the ocean-atmosphere system in the North Atlantic. Multidecadal changes in temperature and salinity in the 0–300 m layer, as well as in SSH, are statistically significant and show physically agreed changes. Their values in the subtropical and subpolar latitudes increase in the positive AMO phase. The dynamic system «Gulf Stream – North Atlantic Current» is exception. The MLD, reflecting the intensity of convective mixing at high latitudes, is higher in the negative AMO phase. In the positive AMO phase large-scale SSH fluctuations are characterized by an increase in the equatorial zone and in the eastern part of the North Atlantic basin.

The AMO is associated with significant anomalies in the net surface heat fluxes. Multidecadal anomalies of the net surface heat fluxes are consistent with the anomalies of horizontal heat advection in the 0–300 m layer. The increase of horizontal heat advection in the negative AMO phase leads to an increase of the ocean's heat release to the atmosphere. This fact confirms the role of advective heat transport in the generation of multidecadal changes of the upper ocean layer temperature.

Multidecadal changes in the velocity of currents and horizontal heat advection in the 0–300 m layer are well coordinated, especially in the western part of the subtropical gyre. The meridional displacement of dynamic system «Gulf Stream – North Atlantic Current» in the south direction is marked in the positive AMO phase. Multidecadal variability of the horizontal heat advection is due to changes in the dynamics of currents, rather than temperature gradients.

Despite the statistically significant acceleration of SSH rise in the North Atlantic, caused by the melting of the Greenland ice sheet [49], after the removal of the linear trend multidecadal SSH changes are well coordinated with the AMO and involve a large-scale increase in SSH in the positive AMO phase. Since SSH is an integral characteristic of the ocean's hydrothermodynamics, its changes in the transition from the cold AMO phase to the warm phase are well coordinated with the thermodynamic and dynamic changes in the 0–300 m layer of the North Atlantic. Changes in SSH also indicate an intensification of the subtropical gyre and, conversely, a weakening of the subpolar gyre. All other composite differences in their own way contribute to this general conclusion. Multidecadal temperature increase corresponds to the weakening (intensification) of circulation in the cyclonic (anticyclonic) oceanic gyre. It should be noted that this conclusion completely confirms the hypothesis of H. Stommel, that when the climate warms, the circulation in subtropical latitudes increases [50]. Thus, multidecadal changes in the upper mixed layer temperature regime are determined to a greater extent not by surface heat fluxes, but by changes in the overall ocean dynamics.

The investigations carried out in this paper also showed that the intrinsic changes in the circulation regime in the North Atlantic are consistent with the hypothesis of Iselin (1940) and Stommel (1958): in the warm period there is an intensification of the subtropical gyre and, conversely, a weakening of the subpolar gyre [50, 51]. But at the same time the Gulf Stream transport itself decreases. Thus, negative feedback in the ocean-atmosphere system is realized, when changes in the ocean circulation play a stabilizing role.

The observed intensification of the subtropical gyre and, conversely, the weakening of the subpolar gyre lead to a weakening of convective mixing in the Labrador Sea and a southern shift of the Gulf Stream and the North Atlantic Current in the warm AMO phase. In accordance with the Stommel hypothesis the southern displacement of the Gulf Stream during the positive AMO phase is caused by the intensification of the subtropical gyre. This conclusion is fully confirmed by our results.

The conclusions obtained in the article are in full agreement with earlier carried out researches. The high positive correlations between multidecadal SSH anomalies and SST anomalies in the North Atlantic are found in [23]. The SSH rise in the positive AMO phase on the eastern coast of the North Atlantic agrees well with the data of tide gauges given in the paper cited above. The model study [52] has shown that multidecadal SSH changes in the Gulf Stream and subpolar gyre are realized due to the variability of meridional heat transport, which reflects the intensity of meridional overturning circulation in the North Atlantic. The revealed variability of horizontal heat advection does not contradict the widely known point of view that ocean circulation determines the phase changes of AMO due to a change in the ocean's heat content [2]. Oceanic advection of temperature and salinity anomalies from low latitudes to high is important in maintaining AMO [33, 34].

Acknowledgments. This study was partially supported by the Russian Scientific Foundation, grant 17-17-01295

References

1. Voskresenskaya, E.N., Polonskii, A.B.: Low-frequency variability of hydrometeorological fields and heat fluxes over the North Atlantic. Phys. Oceanogr. **14**(4), 203–220 (2004)
2. Delworth, T., Mann, M.E.: Observed and simulated multidecadal variability in the Northern Hemisphere. Clim. Dyn. **16**(9), 661–676 (2000)
3. Enfield, D.B., Mestas-Nunez, A.M., Trimble, P.J.: The Atlantic multidecadal oscillation and its relation to rainfall river flows in the continental U.S. Geophys. Res. Lett. **28**(10), 2077–2080 (2001)

4. Guan, B., Nigam, S.: Analysis of Atlantic SST variability factoring interbasin links and the secular trend: clarified structure of the Atlantic multidecadal oscillation. J. Clim. **22**(15), 4228–4240 (2009)
5. Kushnir, Y.: Interdecadal variations in North Atlantic sea surface temperature and associated atmospheric conditions. J. Clim. **7**(1), 141–157 (1994)
6. Schlesinger, M.E., Ramankutty, N.: An oscillation in the global climate system of period 65–70 years. Nature **367**(6465), 723–726 (1994)
7. Ting, M., Kushnir, Y., Seager, R., Li, C.: Forced and internal twentieth-century SST trends in the North Atlantic. J. Clim. **22**(6), 1469–1481 (2009)
8. Kerr, R.A.: A North Atlantic climate pacemaker for the centuries. Science **288**(5473), 1984–1985 (2000)
9. Enfield, D.B., Mestas-Nunez, A.M.: Multiscale variabilities in global sea surface temperatures and their relationships with tropospheric climate patterns. J. Clim. **12**(9), 2719–2733 (1999)
10. Sutton, R.T., Hodson, D.L.R.: Influence of the ocean on North Atlantic climate variability 1871–1999. J. Clim. **16**(20), 3296–3313 (2003)
11. Danabasoglu, G., Yeager, S.G., Kim, W.M., et al.: North Atlantic simulations in Coordinated Ocean-ice Reference Experiments phase II (CORE-II). Part II: inter-annual to decadal variability. Ocean Model. **97**, 65–90 (2016)
12. Knight, J.R., Folland, C.K., Scaife, A.A.: Climate impacts of the Atlantic multidecadal oscillation. Geophys. Res. Lett. **33**(17), L17706 (2006). https://doi.org/10.1029/2006GL026242
13. Sutton, R.T., Hodson, D.L.R.: Atlantic Ocean forcing of North American and European summer climate. Science **309**(5731), 115–118 (2005)
14. Panin, G.N., Diansky, N.A.: On the correlation between oscillations of the Caspian sea level and the North Atlantic climate. Izv. Atmos. Oceanic Phys. **50**(3), 266–277 (2014)
15. Panin, G.N., Diansky, N.A.: Climatic variations in the Arctic, North Atlantic, and the Northern Sea route. Dokl. Earth Sci. **462**(1), 505–509 (2015)
16. Rowell, D.P., Folland, C.K., Maskell, K., Ward, M.N.: Variability of summer rainfall over tropical North Africa (1906–92): observations and modeling. Q. J. R. Meteorol. Soc. **121** (523), 669–704 (1995)
17. Trenberth, K.E., Shea, D.J.: Atlantic hurricanes and natural variability in 2005. Geophys. Res. Lett. **33**(12), L12704 (2006). https://doi.org/10.1029/2006GL026894
18. Häkkinen, S., Rhines, P.B., Worthen, D.L.: Atmospheric blocking and Atlantic multidecadal ocean variability. Science **334**(6056), 655–659 (2011)
19. Polonskii, A.B.: Atlantic multidecadal oscillation and its manifestations in the Atlantic-European region. Phys. Oceanogr. **18**(4), 227–236 (2008)
20. Venegas, S.A., Mysak, L.A.: Is there a dominant timescale of natural climate variability in the Arctic? J. Clim. **13**(19), 3412–3434 (2000)
21. Knight, J., Allan, R., Folland, C., Vellinga, M., Mann, M.E.: A signature of persistent natural thermohaline circulation cycles in observed climate. Geophys. Res. Lett. **32**(20), L20708 (2005). https://doi.org/10.1029/2005GL024233
22. Chylek, P., Folland, C.K., Dijkstra, H.A., Lesins, G., Dubey, M.K.: Ice-core data evidence for a prominent near 20 year time-scale of the Atlantic Multidecadal Oscillation. Geophys. Res. Lett. **38**(13), L13704 (2011). https://doi.org/10.1029/2011GL047501
23. Frankcombe, L.M., Dijkstra, H.A.: Coherent multidecadal variability in North Atlantic sea level. Geophys. Res. Lett. **36**(15), L15604 (2009). https://doi.org/10.1029/2009GL039455
24. Eden, C., Jung, T.: North Atlantic interdecadal variability: oceanic response to the North Atlantic Oscillation (1865–1997). J. Clim. **14**(5), 676–691 (2001)
25. Mantua, N.J., Hare, S.R.: The Pacific decadal oscillation. J. Oceanogr. **58**(1), 35–44 (2002)

26. Liu, Z.: Dynamics of interdecadal climate variability: a historical perspective. J. Clim. **25**(6), 1963–1995 (2012)
27. Timmermann, A., Latif, M., Voss, R., Grotzner, A.: Northern hemispheric interdecadal variability: a coupled air-sea mode. J. Clim. **11**(8), 1906–1931 (1998)
28. Griffies, S.M., Tziperman, E.: A linear thermohaline oscillator driven by stochastic atmospheric forcing. J. Clim. **8**(10), 2440–2453 (1995)
29. Eden, C., Willebrand, J.: Mechanism of interannual to decadal variability of the North Atlantic circulation. J. Clim. **14**(10), 2266–2280 (2001)
30. de Verdière, Colin: A., Huck, T.: Baroclinic instability: an oceanic wavemaker for interdecadal variability. J. Phys. Oceanogr. **29**(5), 893–910 (1999)
31. Frankcombe, L.M., Dijkstra, H.A., Von der Heydt, A.: Noise-induced multidecadal variability in the North Atlantic: excitation of normal modes. J. Phys. Oceanogr. **39**(1), 220–233 (2009)
32. Te Raa, L.A., Dijkstra, H.A.: Instability of the thermohaline ocean circulation on interdecadal timescales. J. Phys. Oceanogr. **32**(1), 138–160 (2002)
33. Polonsky, A.B.: Interdecadal variability in the ocean-atmosphere system. Russ. Meteorol. Hydrol. **5**, 37–44 (1998)
34. Wang, C., Dong, S., Munoz, E.: Seawater density variations in the North Atlantic and the Atlantic meridional overturning circulation. Clim. Dyn. **34**(7–8), 953–968 (2010)
35. Gusev, A.V., Diansky, N.A.: Numerical simulation of the world ocean circulation and its climatic variability for 1948–2007 using the INMOM. Izv. Atmos. Oceanic Phys. **50**(1), 1–12 (2014)
36. Volodin, E.M.: The mechanism of multidecadal variability in the Arctic and North Atlantic in climate model INMCM4. Environ. Res. Lett. **8**(3), 035038 (2013)
37. Frankcombe, L.M., Dijkstra, H.A.: The role of Atlantic – Arctic exchange in North Atlantic multidecadal climate variability. Geophys. Res. Lett. **38**(16), L16603 (2011). https://doi.org/10.1029/2011GL048158
38. Dima, M., Lohmann, G.: A hemispheric mechanism for the Atlantic Multidecadal Oscillation. J. Clim. **20**(11), 2706–2719 (2007)
39. Jungclaus, J.H., Haak, H., Latif, M., Mikolajewicz, U.: Arctic-North Atlantic interactions and multidecadal variability of the meridional overturning circulation. J. Clim. **18**(19), 4013–4031 (2005). https://doi.org/10.1175/JCLI3462.1
40. Wang, C., Dong, S., Evan, A.T., Foltz, G.R., Lee, S.-K.: Multidecadal covariability of North Atlantic sea surface temperature, African dust, Sahel rainfall, and Atlantic hurricanes. J. Clim. **25**(15), 5404–5415 (2012)
41. Balmaseda, M.A., Vidard, A., Anderson, D.L.T.: The ECMWF ocean analysis system: ORA-S3. Mon. Wea. Rev. **136**(8), 3018–3034 (2008)
42. Balmaseda, M.A., Mogensen, K., Weaver, A.T.: Evaluation of the ECMWF ocean reanalysis system ORA-S4. Q. J. R. Meteorol. Soc. **139**(674), 1132–1161 (2013)
43. Carton, J.A., Giese, B.S.: A reanalysis of ocean climate using Simple Ocean Data Assimilation (SODA). Mon. Wea. Rev. **136**(8), 2999–3017 (2008)
44. Chang, Y.-S., Zhang, S., Rosati, A., Delworth, T.L., Stern, W.F.: An assessment of oceanic variability for 1960–2010 from the GFDL ensemble coupled data assimilation Clim. Dyn. **40** (3–4), 775–803 (2013)
45. Köhl, A.: Evaluation of the GECCO2 ocean synthesis: transports of volume, heat and freshwater in the Atlantic. Q. J. R. Meteorol. Soc. **141**(686), 166–181 (2015)
46. Pacanowski, R.C., Philander, S.G.H.: Parameterization of vertical mixing in numerical models of tropical oceans. J. Phys. Oceanogr. **11**(11), 1443–1451 (1981)
47. Large, W.G., Yeager, S.G.: The global climatology of an interannually varying air–sea flux data set. Clim. Dyn. **33**(2–3), 341–364 (2009)

48. Dickson, R.R., Meincke, J., Malmberg, S.-A., Lee, A.J.: The «great salinity anomaly» in the northern North Atlantic 1968–1982. Prog. Oceanogr. **20**(2), 103–151 (1988)
49. Boon, J.D.: Evidence of sea level acceleration at US and Canadian tide stations, Atlantic Coast. North Am. J. Coast. Res. **28**(6), 1437–1445 (2012)
50. Stommel, H.M.: The Gulf Stream: a physical and dynamical description, p. 248. University of California Press, Berkeley and Los Angeles (1958)
51. Iselin, C.O.D.: Preliminary report on long-period variations in the transport of the Gulf Stream System. Pap. Phys. Oceanogr. Meteorol. **8**(1), 1–40 (1940)
52. Häkkinen, S.: Variability in sea surface height: a qualitative measure for the meridional overturning in the North Atlantic. J. Geophys. Res. Oceans (1978–2012) **106**(C7), 13837–13848 (2001)

Reconstruction of Hydrophysical Fields in the Coastal Region of the Black Sea on the Basis of Hydrodynamic Model with Assimilation of Observational Data

Demyshev Sergei and Evstigneeva Natalia(✉)

Marine Hydrophysical Institute RAS, Sevastopol, Russian Federation
nataev2302@gmail.com

Abstract. Hydrophysical fields, continuous in time and space, were reconstructed in the coastal region of the Black Sea (a western coast of the Crimea and a north-western shelf of the Black sea) on the basis of the three-dimensional nonlinear hydrodynamic model and the observational data of temperature and salinity on the research vessel "Experiment" from 16 to 22 of September, 2007. A sequential optimal interpolation of observational data of temperature and salinity was used as an assimilation procedure. A real atmospheric forcing and a high resolution (horizontal grid $\sim 1.6 \times 1.6$ km and 30 vertical layers from 1 m to 1300 m) were used in the calculation. Mesoscale features of currents and thermohaline fields in the coastal zone of the Black Sea were investigated, the coastal upwelling in the Kalamitsky Bay was reconstructed, registered in satellite observations. The influence of the assimilation of observational data on the accuracy of the calculated fields of temperature and salinity was estimated.

Keywords: Numeral modeling · Spatial high resolution · Coastal area Black Sea · Assimilation of data observation · Hydrophysical fields Mezoscale features of circulation

1 Introduction

Reconstruction of coastal circulation is important for shipping, construction and operation of port facilities, mining, ecological monitoring. Synthesis of observational data and the hydrodynamic model gives an opportunity to obtain hydrophysical fields close to observed.

A technology, based on a sequential optimal interpolation of temperature and salinity observations in the hydrodynamic model [1], was proposed in [2] for the investigation of the coastal circulation of the Black Sea, which allows reconstructing the continuous spatio-temporal variability of the level, current, temperature and salinity fields.

The Marine Hydrophysical Institute (MHI) conducts expeditionary research in the coastal zone of the Black Sea. The main task is to obtain experimental data on the vertical and spatial distribution of hydrological characteristics. Hydrologists handle these data, but a more detailed processing of the data is not carried out using

V. Karev et al. (Eds.): PMMEEP 2017, SPRINGERGEOL, pp. 138–147, 2018.
https://doi.org/10.1007/978-3-319-77788-7_15

hydrothermodynamic models. The importance of this research is the ability to reconstruct the circulation and thermohaline fields in the field of survey, using observational data. The results of processing of the data observations give a detailed information on mesoscale structures on the Black Sea shelf.

A complex expedition was conducted by the MHI on the research vessel "Experiment" from 16 to 22 of September 2007 with the aim of studying oceanological processes at the end of the summer warm-up period – the beginning of the autumn period of cooling. In this study, its experimental data on temperature and salinity in the coastal area of the Black Sea, including a western coast of the Crimea and a north-western shelf (Fig. 1), are used.

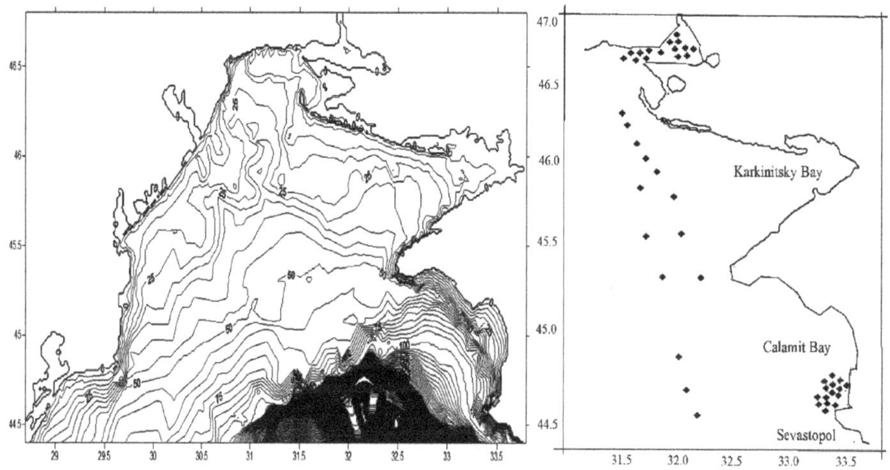

Fig. 1. Bathymetry of the north-western shelf of the Black Sea and the western coast of the Crimea (m) and a scheme of stations, performed on the research vessel "Experiment" on 16 22 September, 2007.

The main objective of the study is to reconstruct three-dimensional fields of currents, temperature and salinity, continuous in time and space, on the basis of assimilation of the observational data in September of 2007 in the hydrodynamic model with a high resolution ~ 1.6 km and to carry out an analysis of mesoscale features of coastal circulation.

2 Description of the Hydrological Survey, Conducted in September of 2007

We give a brief description of the data of temperature and salinity measurements available in the data bank of MHI [3]. The work was carried out using the STD probe (shelf measuring complex), equipped with a cassette of plastic bathometers with a vertical resolution of 0.5 m. The maximum depth, to which the sounding was carried out, varied from 5 to 300 m. There were 44 stations during the expedition from 16 to 22 of September 2007.

The following features of the vertical profiles were noted: a thermocline at the depth of 20–25 m, a temperature varied from 20 to 22 °C in the upper layer of water in the deep water part, surface salinity – from 16.5 to 17‰, in shallow areas, including estuaries, a temperature varied from 17.8 to 18.8 °C, salinity from 0 to 7‰. Zones of water with a temperature from 9 to 13 °C and a salinity of 17.9–18‰ were observed in the upper layer on the ten stations on 22nd of September in the Kalamitsky Bay (Fig. 2).

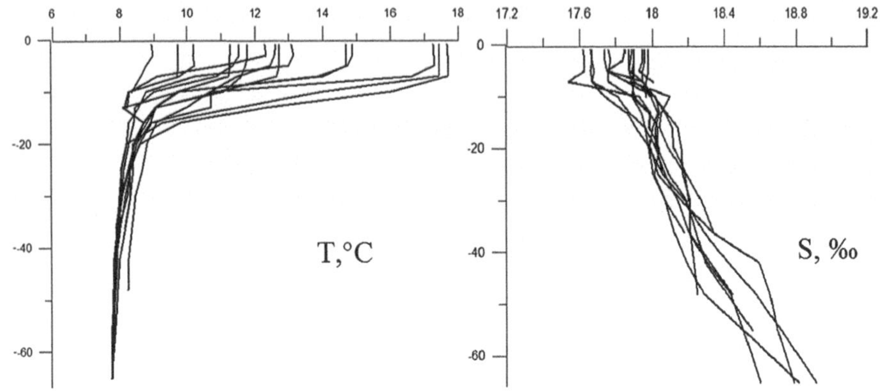

Fig. 2. Temperature and salinity profiles, measured at the research vessel "Experiment" on 22nd of September 22, 2007.

The coordinates of all the stations, where the temperature and salinity were measured, were correlated with the calculated model grid and the data were prepared for the assimilation in the hydrodynamic model.

3 Parameters of the Model and Description of Numerical Experiments

The system of model equations using the Boussinesq approximation, hydrostatic approximation and incompressibility of seawater in the Gromeko–Lamb form, the boundary conditions on the surface, at the bottom, on the solid lateral walls were written as follows [1]. Note that a reduced sea level ζ was calculated from a discrete analog of the continuity equation taking into account the specification of the velocities at the open south boundary of the domain.

In order to adapt the numerical model of the dynamics for the calculation of the coastal circulation of the Black Sea we made the following steps. Data array of the region bathymetry was analyzed and processed, model parameters were chosen on the basis of preliminary experiments, river inflow locations and depths of estuaries were assigned, boundary conditions on the open boundaries of the region were selected and implemented, initial fields as well as fields of wind stress, heat flows, short-wave radiation, precipitation and evaporation were processed in order to be used in the model.

We considered a region of the Black Sea (a north-western shelf and a western coast of the Crimea, Fig. 1), limited by latitude 44.4 °N, located between meridians 28.5° and 33.5 °E (we used a detailed presentation of bottom topography with a resolution ~1.6 km). The numerical experiments were carried out with a resolution ~1.6 km. The time step was equal to 30 s. The choice of the horizontal coefficients of turbulent viscosity and diffusion ($\nu_H = 5 \cdot 10^5$ cm^2/s, $\kappa_H = 5 \cdot 10^5$ cm^2/s) was based on a series of specialized numerical experiments. The total period of integration of model equations 10 days (from 14 to 24 of September 2007). Along the vertical, horizontal components of the current velocity, temperature and salinity were computed at 30 depths: 1; 3; 5; 7; 10; 13; 16; 20; 25; 30; 36; 42; 48; 55; 65; 80; 95; 120; 150; 200; 300… 1300 m. The vertical component of velocity was calculated for intermediate horizons. The vertical coefficients of turbulent exchange of momentum and diffusion were calculated according to the Philander–Pacanowski approximation [4].

Fields of currents, temperature and salinity, obtained from model for the entire sea on a 5 × 5 km horizontal grid within the Operative Oceanography project [5], were used to specify initial and the boundary conditions at the open boundary of the domain. A geostrophic adjustment procedure was carried out in order to adopt the density field (temperature and salinity) to the field of currents on a new spatial grid. The equations of the model were integrated for 5 days with boundary conditions that did not change during the calculating period. It was obtained from the analysis of the graphs of the average in volume and on the horizons of the kinetic energy that the quasi-geostrophic balance was achieved after three days. The obtained fields u, v, ζ, T и S were taken as the initial fields for the basic numerical experiment.

In order to specify conditions on the open southern boundary we used the results of [2], where an efficiency of combined approach was shown on the basis of the numerical experiments. The components of the current velocity, temperature, and salinity (the Dirichlet conditions) were specified in the boundary regions where water flowed into the domain ($v > 0$); conditions $\partial u/\partial n = 0$, $\partial v/\partial n = 0$ for u, v and radiation conditions for T and S were specified in the boundary regions where water flowed out of the domain ($v < 0$).

The fields of tangential wind stress, heat fluxes, short-wave irradiance fluxes, as well as precipitation and evaporation, obtained from data of the regional atmospheric model ALADIN and provided by the department of Marine Forecasts of MHI [5] and linearly interpolated to the selected grid, were specified for each day. We took into account the discharges of rivers: Danube, Dnieper, Dniester, and South Bug.

It was determined from the analysis of wind fields that a south-western wind prevailed with a maximum speed of up to 11 m/s from 14 to 18 of September, a north and north-eastern winds with a maximum speed of 16 m/s were acting from 19 to 24 of September.

A simplified procedure of four-dimensional analysis [6–10], based on the method of sequential optimal interpolation [11], when the covariance functions of the temperature and salinity fields were calculated under the assumption of homogeneity and isotropy, was used to realize the procedure of assimilation of observational data. A correlation radius ~20 km was determined from the analysis of the statistical structure of the fields (calculation of the spatial correlation functions of the temperature and salinity fields).

Based on the results of previous papers [6–10], we approximated the covariance functions of the fields T, S by a function of exponential type $\exp\left(-\lambda\left[(x-x')^2 + (y-y')^2\right]\right)$, where λ – a dimensional parameter (equal to $0{,}016 \cdot (\Delta x)^{-2}$), corresponding to the value of the correlation function 0.1.

All data of hydrological survey were grouped for eight days and assimilation was carried out once a day: September 16 – 3, September 17 – 7, September 18 – 8, September 19 – 7, September 20 – 4 and September 22 – 15 stations. Thus, thermohaline fields were calculated using the equations of the model until the moments of receiving of the data observations. A correction of the temperature and salinity fields was conducted at the moments of assimilation, taking into account the correlation radius.

4 Evaluation of the Influence of the Assimilation of Observational Data on the Accuracy of the Reconstruction of Fields of Currents, Temperature and Salinity

The influence of the use of the assimilation procedure in the numerical model on the formation of current fields, temperature and salinity was analyzed on the basis of comparison of the results of two numerical experiments on the calculation of hydrophysical fields without assimilation and with assimilation of temperature and salinity measurements. Figure 3 presents the fragments of current fields, obtained in the upper layer for two experiments on 20th of September.

Taking into account of observational data could lead to some qualitative and quantitative differences in the structure of fields (change of direction of currents, intensification of currents).

We calculated the modules of mean and mean square errors of the estimation of the temperature and salinity fields at different horizons during integrating the model equations at the moments of assimilation. Tables 1 and 2 show these modules for temperature and salinity fields respectively at the horizons of 1, 10, 20, 30 m for four days.

Analyzing the data presented in the Tables 1 and 2, we note that the biggest values of the errors of the estimation of the temperature fields were observed in the picnoclyne, the biggest values of the errors of the estimation of the salinity fields were in the upper water layer.

Thus, the observational data will allow reproducing the structure of the picnoclyne and propagation of river waters close to the observations. During the calculating period, the values of error modulus decreased at all horizons, which indicates the efficiency of the assimilation procedure.

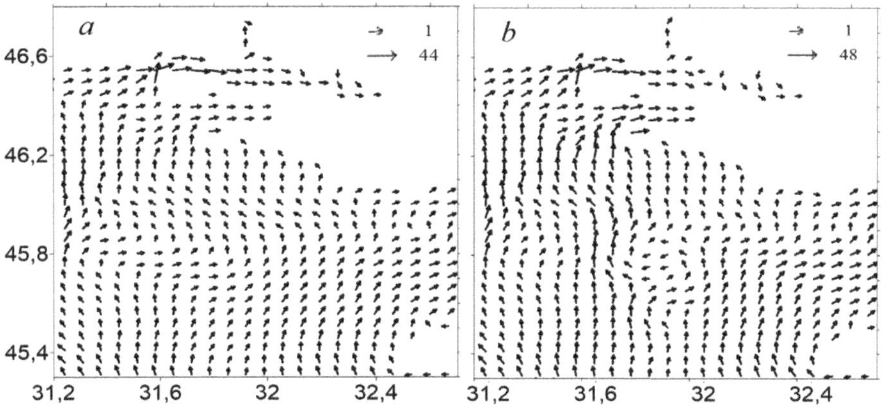

Fig. 3. Fragments of current fields (cm/s) in the upper water layer on 20th of September: a – calculated without taking into account observational data, b – calculated with observational data (every fourth arrow was shown).

Table 1. Values δ_T^n and σ_T at different horizons on 16th, 18th, 20th and 22nd of September 2007.

Depth (m)		16.09.2007	18.09.2007	20.09.2007	22.09.2007
1	δ_T^n	1.06	0.51	0.35	0.24
	σ_T	2.25	2.11	1.41	1.03
10	δ_T^n	1.51	0.44	0.22	0.21
	σ_T	2.38	0.99	0.91	0.82
20	δ_T^n	2.97	2.25	2.15	2.08
	σ_T	4.26	2.25	2.44	2.41
30	δ_T^n	0.64	0.51	0.44	0.11
	σ_T	0.68	0.59	0.42	0.11

Table 2. Values δ_S^n and σ_S at different horizons on 16th, 18th, 20th and 22nd of September 2007.

Depth (m)		16.09.2007	18.09.2007	20.09.2007	22.09.2007
1	δ_S^n	0.70	0.57	0.18	0.11
	σ_S	0.95	0.61	0.49	0.24
10	δ_S^n	0.70	0.32	0.18	0.14
	σ_S	0.93	0.75	0.57	0.35
20	δ_S^n	0.36	0.14	0.09	0.09
	σ_S	0.30	0.39	0.28	0.26
30	δ_S^n	0.09	0.07	0.05	0.04
	σ_S	0.07	0.02	0.01	0.01

5 Fields of Currents

We analyze the fields calculated taking into account the data of observations, which were assimilated once a day. During the calculating period, mesoscale eddies and jets were observed in its structure. We note a domination of the wind component in the formation of water circulation due to the shallow water in the area. Under the influence of the south-western wind, acting from 14 to 18 of September, the main direction of surface currents was the eastern. The following features of circulation were obtained in the upper 30-m layer of water: a cyclonic eddy with a radius of ~ 20 km in the central part of the region, an anticyclonic eddy with a radius of ~ 15 km in the Kalamitsky Bay and near the open boundary, intense jets, directed to the north, along the western coast and in the central part of the region. Note that cyclonic eddy in the center of the region was repeatedly registered in satellite observations. Formation of this eddy was a result of the influence of inhomogeneity of the bottom topography on the jet current. Its lifetime was from 3 to 5 days. Figure 4a presents the field of surface currents for September 16. Jets with a maximum velocity 30 cm/s along the western coast, a cyclonic eddy with a radius of ~ 20 km in the central part of the region, anticyclonic eddies with a radius of ~ 15 km in the Kalamitsky Bay and near the open boundary were observed.

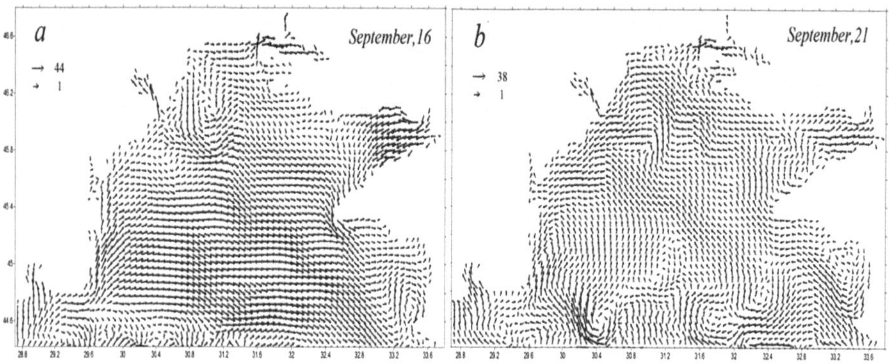

Fig. 4. Fields of currents (cm/s) in the upper layer on 16th and 21st of September (every fourth arrow was shown).

Under the influence of the north and north-eastern winds, acting from 19 to 24 of September, the main direction of surface currents was the western. The features of circulation, associated with the meandering of the Rim Current passing along the depth dump on the southern boundary of the shelf, were formed at the southern boundary. Figure 4b presents the field of surface currents on 21st of September. Along the boundary, the elements of the meanders of the Rim Current were observed, which was located to the south of the selected region.

6 Fields of Temperature and Salinity

Thermohaline fields changed not very intensively during the calculating period. Figure 5 shows the surface temperature fields at the initial moment and after five days of calculation. There was some cooling of the surface waters, which was noted in the estuaries, in the bays and near the cape Tarkhankut (a temperature decreased from 18 to 15 °C). Salinity fields were characterized by the minimum values (0–15‰) in the estuaries and in the north of the region, with maximum values (17–18‰) – near the open boundary. During the calculating period we note some desalination of surface waters near the estuaries and between Odessa and the Dnieper estuary.

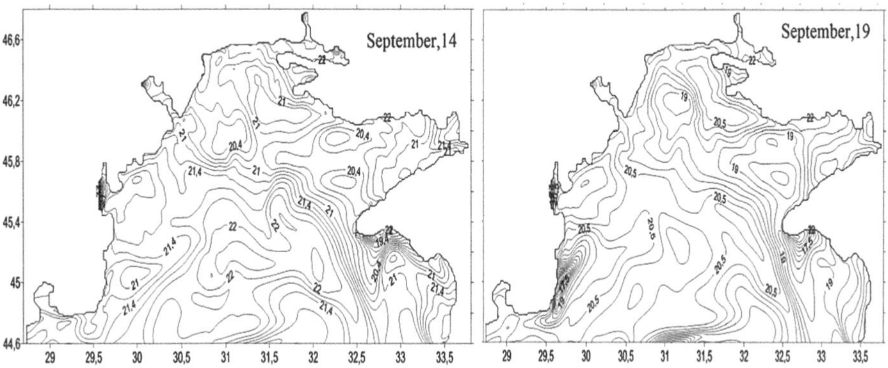

Fig. 5. Fields of temperature (°C) in the upper layer on 14th and 21st of September.

According to the measurements and satellite images, upwelling, caused by the action of the northern and north-eastern winds, was observed on the 22nd and 23rd of September in the southern part of Kalamitsky Bay (a surface temperature was below 14 °C), which was confirmed by the results of our numerical calculations.

We note a distribution of cold water on the model fields of temperature from 21st of September. Figure 6a and b show the reconstructed fields of temperature and vertical velocity for 22nd of September. The hatch marked the areas corresponding to the temperature values from 10 to 13 °C. The maximum values of the vertical velocity (0.01 cm/s) were obtained along the coastline (Fig. 6b, bold).

It is urgent to note that sequential analysis of observational data enabled to reproduce the structure of upwelling, which was also observed in the calculation without data assimilation, more accurately.

Fields of currents, temperatures, salinity were reconstructed on the basis of the hydrodynamic model with the assimilation of observational data of the hydrological survey from 14 to 24 of September 2007, taking into account the real atmospheric forcing and a high resolution ($\sim$1.6 km horizontally and 30 vertical layers) in the coastal area. Calculated fields of currents were characterized by mesoscale eddies and jets. A cyclonic eddy in the central part of the region, an anticyclonic eddy in the Kalamitsky Bay and near the open boundary, intense jets, directed to the north, along

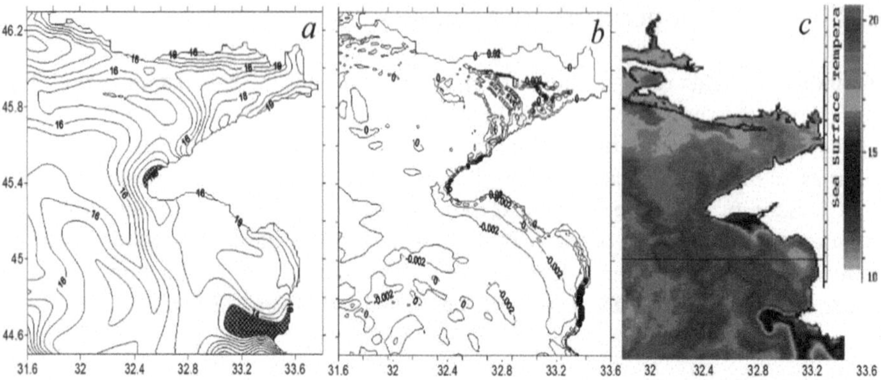

Fig. 6. Fields of temperature (°C), vertical velocity (cm/s) in the upper horizon and a NOAA satellite image of the sea surface on 22nd of September

the western coast and in the central part of the region were obtained in the upper layer of water. In comparison with previous calculations, those features were absent or were much smoother.

There was some cooling and desalination of the surface waters, which was noted in the estuaries, in the bays and near the cape Tarkhankut. As a result of the rise of the underlying cold waters, coastal upwelling in Kalamitsky Bay was reconstructed in September 2007, according to observational data.

Thus, a sequential analysis of observational data on the basis of the assimilation in the numerical model of dynamics enabled to reconstruct the mesoscale features of coastal circulation more accurately in the region of the western coast of the Crimea and the north-western shelf of the Black sea.

The work was carried out within the framework of the state task of the FASO of Russia (the theme "Operative oceanography", No. 0827-2014-0011).

References

1. Demyshev, S.G., Korotaev, G.K.: Numerical energy-balanced model of baroclinic currents of the ocean on the C-grid. In: Numerical Models and Results of Calibration Calculations of Currents in the Atlantic Ocean, pp. 163–231 (1992). (in Russian)
2. Evstigneeva, N.A.: Numerical analysis of the circulation in the coastal zone of the Black Sea on the basis of the hydrodynamic model with the assimilation of observational data. Abstract of the thesis for the degree of candidate of physical and mathematical sciences, Sevastopol (2014). (in Russian)
3. Suvorov, A.M., Andryushchenko, E.G., Godin, E.A., Ingerov, A.V., Kasyanenko, T.E., Plastun, T.V., Khaliulin, A.K.: The bank of oceanological data of the MHI NASU: content and structure of databases, database management system. Environ. Monit. Syst. 130–137 (2003). (in Russian)
4. Pacanowski, R.C., Philander, S.G.H.: Parameterization of vertical mixing in numerical models of tropical oceans. J. Phys. Oceanogr. **11**(11), 1443–1451 (1981)

5. Ratner, Y.B., Martynov, M.V., Bayankina, T.M., Borodin, S.V.: Information flows in the real-time system of rapid monitoring of hydrophysical fields of the Black Sea and automation of their processing. Environ. Control Syst. 140–149 (2005). (in Russian)
6. Knysh, V.V., Moiseenko, V.A., Sarkisyan, A.S., Timchenko, I.E.: Complex use of measurements at the hydrophysical ocean polygons in four-dimensional analysis. Rep. AS USSR **252**(4), 832–836 (1970). (in Russian)
7. Sarkisyan, A.S., Knysh, V.V., Demyshev, S.G.: Multielement four-dimensional analysis of hydrophysical fields on the basis of dynamic-stochastic models. The results of science and technology. Atmos. Ocean Space **9**, 5–64 (1987). (in Russian)
8. Demyshev, S.G., Knysh, V.V.: Model numerical experiments of evaluation the validity of a four-dimensional analysis of the main physical ocean fields. Theory Oceanic Process. 61–69 (1981). (in Russian)
9. Knysh, V.V.: Hydrothermodynamic ocean models in the algorithm of the multielement four-dimensional analysis of hydrophysical fields: a thesis for a scientific degree of doctor of science (1981). (in Russian)
10. Sarkisyan, A.S.: Numerical analysis and forecast of sea currents. L.: Gidrometeoizdat (1997). (in Russian)
11. Gandin, L.S., Kagan, R.L.: Statistical methods for the interpretation of meteorological data. L.: Gidrometeoizdat (1976). (in Russian)

The Vertical Turbulent Exchange Features in the Black Sea Active Layer

A. S. Samodurov(✉) and A. M. Chukharev

Marine Hydrophysical Institute of RAS, Sevastopol, Russia
anatol_samodurov@mail.ru, alexchukh@mail.ru

Abstract. Physical processes of heat and mass vertical transfer in natural stratified basins depend mostly on the intensity of quasi-inertial internal waves breaking, which, in turn, is determined by the intensity of the wave source and local stratification in the basin. There are many research works including estimations of the dependence of the vertical turbulent diffusion coefficient K on the buoyancy frequency N (stratification). In this paper, on the base of a semiempirical model using the analysis of the pulsation data obtained at probing in the lower stratified part of the Black Sea active layer, the dependence of the coefficient of vertical turbulent diffusion K on the buoyancy frequency N is estimated. The measurements were carried out in the deep area of the basin. The measurement data were collected using a high-resolution probe-turbulimeter "Sigma-1". A similar dependence for this layer was built based on a 1.5D model of the vertical exchange in the Black Sea deep region. Previously, the same results were obtained in the upper "strongly" stratified and the lower "weakly" stratified layers in the main pycnocline of the investigated basin. The joint analysis of three pairs of power dependencies $K \cong AN^{\alpha}$ $\mathrm{m^2s^{-1}}$ from different stratified layers showed that the exponents α in each pair were very close to each other, while the estimated (semi-empirical) coefficients A in each pair turned out to be much higher than the pattern coefficients.

Keywords: Stratified layers · Internal waves breaking
Vertical turbulent exchange · Turbulent patches · Energy dissipation
Buoyancy frequency · Measuring complex · Turbulent exchange modeling

1 Introduction

A vertical turbulent exchange in the stratified layers of natural basins plays a significant role in the formation of fluxes of heat, salt, nutrients and other dissolved substances. The stable stratification while inhibiting vertical movements, creates favorable conditions for the development of internal waves. Many factors have an effect on the formation and evolution of internal waves: stratification conditions, bottom topography and presence of background currents. To solve the problem of a vertical exchange it is necessary to establish a physical mechanism (or mechanisms) of the specified process. The nature has a number of such mechanisms. At the same time, based on the analysis of the exchange processes in various areas of the World ocean Wunsch and Ferrari [1] found that in a stable stratified ocean, away from all the sharp frontal zones, areas manifesting double

V. Karev et al. (Eds.): PMMEEP 2017, SPRINGERGEOL, pp. 148–156, 2018.
https://doi.org/10.1007/978-3-319-77788-7_16

diffusion mechanisms etc., the main source of turbulence is the shear instability mechanism in a field of quasi-inertial internal waves. This refers to the internal transverse waves whose phase and group velocities are at the right angles to each other.

As for the quasi-inertial waves that have a frequency close to the inertial frequency, they represent periodic in vertical direction and multidirectional quasi-horizontal flows. Having reached a certain amplitude, these waves evolve in a mode of shear instability according to the criterion of Richardson and break inducing turbulent patches. The turbulent mixing inside such patches generates average vertical fluxes in the stratified layers. The characteristic feature of the stably stratified layers is intermittency – mixed layers with near-zero density gradient are commonly observed in the thicker stratified layers. The spectra based on probing measurements of fluctuations in these layers possess some common features that were described in [2].

2 In-situ Measurements

For a quantitative description of the sea vertical exchange processes the measurements of characteristics that define flows of different substances, namely, fluctuations of velocity, temperature, conductivity etc. are necessary.

The experimental studies in the Black Sea upper layers, including the direct measurements of fluctuations of the basic hydrophysical characteristics have been conducted in the expeditions of Marine Hydrophysical Institute for several years. For these purposes, a specialized measuring complex "Sigma-1", created in 2004 and corresponding to modern speed requirements and measured parameters, described in [3], is used. Its configuration is shown in Fig. 1, and Table 1 lists the main technical characteristics of the device. In addition to the measurement of the marine environment characteristics the device is equipped with a position control system, which allows to take into account the carrier's own motions and to implement a proper correction in the estimation of the turbulent velocity fluctuations (eddy-correlation method).

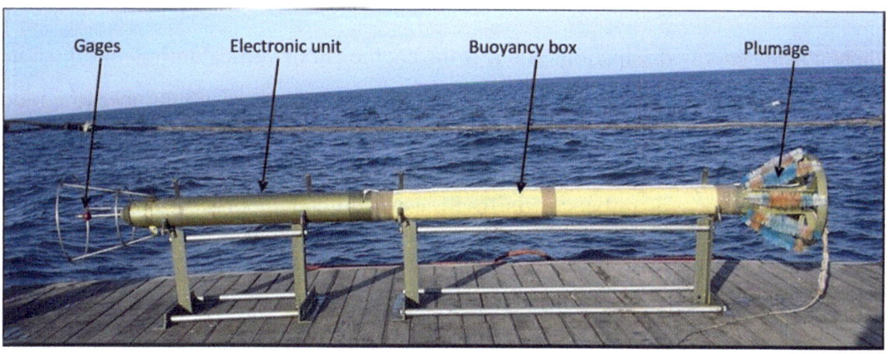

Fig. 1. The probing measuring complex "Sigma-1".

Measurements are performed in the probing mode during which the device falls freely at a speed of about 0.7 m/s. The probe is powered and the information is

Table 1. Technical characteristics of the measuring complex "Sigma-1".

Parameters	Range	Resolution	Accuracy	Sampling frequency
3-D velocity fluctuations, m/s	±2	10^{-3}	±10%	100 Hz for all channels
Temperature, °C	0–30	0.001	±5%	
Relative conductivity	0–0,9	$2.5 \cdot 10^{-5}$	±5%	
3-D acceleration of device, m/s²	±2 g	0.002	0.002	
Roll and pitch, grad	±20	0.01	±1	
Azimuth of device (yaw), grad	0–360	1.0	±5	
Pressure, MPa	0–1	$5 \cdot 10^{-4}$	±1%	

received with the help of a cable veered during probing. The device is given an additional weight at the bottom that together with the tail section increases directional stability during movement. The analysis of the carrier's own motions showed no appreciable contribution to the measured values; and after some appropriate filtering of the original data, the sufficiently objective information can be received.

3 The Turbulence Characteristics Calculation Methods

To estimate the coefficients of the vertical turbulent diffusion and the rate of the turbulent energy dissipation, two different approaches were used.

The first approach is based on the energy analysis of the turbulent patches evolution and convenient in the case when the density gradient is mainly determined by the temperature gradient contribution. In [4, 5] an effective method of calculations using the concept of the "characteristic scale of turbulent patches" L was developed and the following dependences were obtained:

$$\varepsilon \cong 8,2 \cdot 10^{-2} L^2 N^3, \; K \cong \frac{R_f}{1 - R_f} \frac{\varepsilon}{N^2}, \tag{1}$$

where ε is the energy dissipation rate, K is the coefficient of vertical turbulent diffusion, $N = \sqrt{\frac{g}{\rho}\frac{\partial \rho}{\partial z}}$ is the buoyancy frequency, R_f is the dynamic Richardson number (the ratio of the rate of the potential energy increase within the system to the rate of the incoming energy required for mixing) in the acts of the stratified flow shear instability and the breaking of the wave disturbances. In [6–8], using different approaches, the constancy of R_f was established for the considered phenomena. Several approximate values of the quantity R_f (1/3 – in [6] and 1/4 – in [7]) were suggested to be used in calculations, as well as value 0.2 to estimate a common multiplier in the right side of the second equation (1) – in [9].

The approach to determining L is based on the spectra structure analysis of the first differences of the temperature fluctuations measured in the ocean [10]. It was established previously that the effective vertical scale of natural turbulent patches corresponds to the vertical scale of the stable minimum in the small-scale region of the first differences of the vertical spectrum [2].

Using the results of (1) it is shown in the review paper [10] based on our proposed models of the vertical turbulent exchange that in the main pycnocline at the Black Sea deep water region the dependences of K on N have the following form:

$$K \cong 5.6 \cdot 10^{-5} N^{-1} \mathrm{m}^2 \mathrm{s}^{-1} \tag{2}$$

$$K \cong 1.6 \cdot 10^{-2} N \mathrm{m}^2 \mathrm{s}^{-1} \tag{3}$$

Here Eqs. (2) and (3) describe respectively the upper "strongly" stratified and the lower "weakly" stratified layers in the main pycnocline. The coefficients in the presented equations are calculated using the empirical dependence $L(N)$. For case (2) $L \cong 1.4\ N_c^{-1}$m and for case (3) $L \cong 1$ m. Here N_c is the cyclic frequency with the dimension cycles/m. These results and the results of the 1.5D model from [10] will be used to analyze the measurement data in the studied stratified layer above the main pycnocline, Fig. 2.

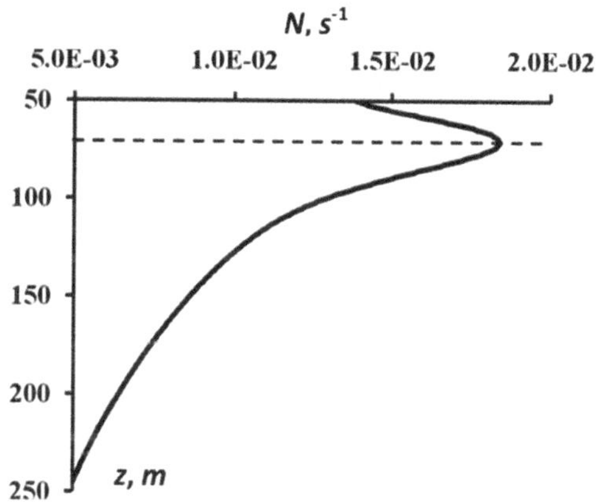

Fig. 2. Average distribution of buoyancy frequency N in the Black Sea. The layer from the 50 m depth to the dashed line is the lower stratified sea area studied in this work. The area below the dashed line corresponds to the main pycnocline of the basin.

In the second method of ε and K estimation, the measured values of velocity fluctuations are used. With the help of these fluctuations, the vertical gradient of horizontal fluctuations is calculated. Vertical fluctuations in this case are not very informative because of the high vertical speed of the device during probing. This method is described in detail in [11] with respect to the device MST Profiler that measures velocity shears with the help of the sensor PNS 93.

The main difference of the complex "Sigma-1" from the said device is that it measures the velocity fluctuations, then extracts the horizontal component and calculates in some layer the smoothed values of $\Delta u/\Delta z$, according to which the rate of turbulent energy dissipation is determined. Therefore, firstly, the spikes and noises caused by the device's own vibrations are removed and then a bandpass filter that limits the frequency of the measured fluctuations of the inertial subrange of the turbulence spectrum from low frequencies (k_l) and Kolmogorov wavenumber k_c from high frequencies is used. The latter can be defined as

$$k_c = (1/2\pi)(\varepsilon/\nu^3)^{1/4}, \tag{4}$$

where ν is the kinematic viscosity.

For calculation of the dissipation rate, an iterative method is used. First, we define the boundary wave numbers k_l and k_{max} and then calculate the spectrum of *du*/*dz* value by the Welch's method. In this technique, the chronograms are broken into overlapping segments, which are multiplied by the Hann time window and then the Fourier transformation with subsequent spectral function averaging for all segments is realized. The dispersion $\left[(du/dz)^2\right]$ is evaluated by integrating the spectrum values in the selected range of wave numbers. The rate of turbulent energy dissipation is determined by the ratio

$$\varepsilon = \frac{15}{2}\nu\left[(du/dz)^2\right]$$

According to this value of ε, the Kolmogorov wave number is calculated using the Eq. (4). If the stop conditions are not realized, the cycle is repeated starting with the spectrum calculation. The stop criteria are a small change in the k_c value (less than a step in the Δk spectrum) and exceeding the k_{max} value. An example of a calculated dimensionless spectrum in the depth range of 72–75 m is shown in Fig. 3. In the same picture, the model spectrum of Nasmyth [12] is demonstrated for comparison.

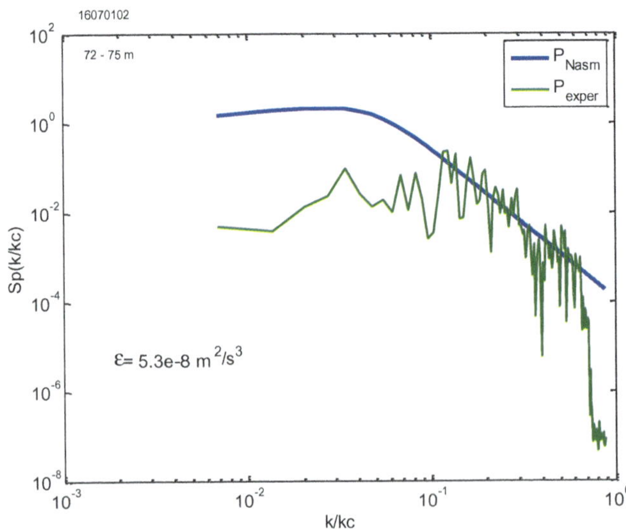

Fig. 3. Comparison of the Nasmyth model spectrum and the *du*/*dz* experimental spectrum. Pulsations are pre-smoothed by a median filter and subjected to bandpass filtering. The dissipation rate estimation is $5.3 \cdot 10^{-8}$ m^2/s^3.

A comparison of the described methods showed good agreement on the calculated energy dissipation rate values and vertical exchange coefficients, which allows to judge the possibility of their use independently, as well as jointly to enhance the estimation objectivity.

4 The Results of Vertical Turbulent Exchange Intensity Calculations in the Studied Layer

The values ε were calculated at three-meter-long intervals in the N-growth layer to the maximum, which in most cases was located at the depths of 60–90 m. At the sampling frequency of 100 Hz and the device speed of 0.75 m/s an acceptable statistical reliability of the results was provided; and with a small assumption of errors constancy N in these segments could be supposed. The depth level everywhere was more than 300 m, i.e. the influence of the bottom on the mixing intensity in the studied layer can be neglected.

The calculation of the vertical turbulent exchange coefficient was carried out according to the formula by Osborne [9]:

$$K = 0,2\,\varepsilon N^{-2}.$$

It is usually used in the analysis of the vertical exchange processes occurring due to the breaking of internal waves.

The calculated coefficient values in the form of a dependence from the buoyancy frequency are shown in Fig. 4. The measurements were performed in the 87-th voyage of R/V "Professor Vodyanitsky", which took place in the summer of 2016 in the Russian economic zone adjacent to the Crimean Peninsula.

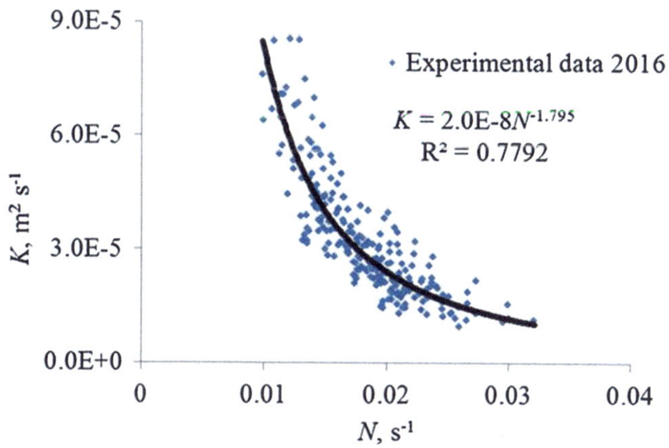

Fig. 4. Experimental dependence of the vertical turbulent diffusion coefficient versus buoyancy frequency for the depth range from 50 m down to the maximum N. The curve is the power approximation, R^2 is the determination coefficient.

5 Comparison of the Field Measurements Analysis Results with the 1.5D Model of Vertical Exchange in the Black Sea

The measurements data analysis allows to judge the nature of the vertical turbulent exchange in the lower part of the deep region of the active layer in the Black Sea during the time spent on measurements. As an element for comparison with other results, we use primarily the results of the 1.5D model of the vertical exchange in the Black Sea deep region [10]. In the above mentioned paper it was shown that power dependences $K \propto N^{\alpha}$ of this model agree well with the results of theoretical and semiempirical models for the main pycnocline. Figure 5 shows the model curve of the vertical turbulent diffusion coefficient K versus buoyancy frequency N in the stratified layer below 50 m (dark line) obtained in 1.5D model.

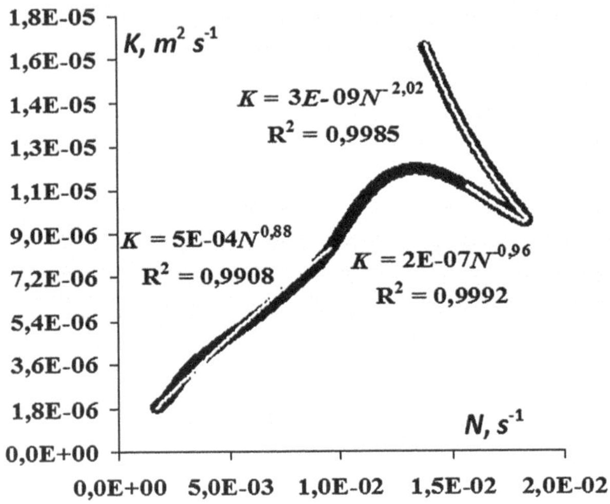

Fig. 5. The coefficient of vertical turbulent diffusion K versus buoyancy frequency N by 1.5D model of the Black Sea [10]. The white lines correspond to power law approximations for the three stratified layers.

If we move along this dark line from the bottom to the top, the three white lines sequentially are detected. These lines represent power approximations of the diffusion coefficient K in the respective layers. Two first lines refer to the lower and upper layers of the main pycnocline (MP). The first line corresponds to the dependence $K(N)$ in the Eq. (3), and the next line is a similar dependence in the Eq. (2). The third upper line corresponds to the similar model power-law approximation $K(N)$ in the lower part of the Black Sea active layer (AL) studied here, Fig. 2.

The model and calculated dependences for the three studied stratified layers in the form $K \cong AN^{\alpha}$ m^{2}s^{-1} are presented below in Table 2.

From the comparison of the presented dependencies two results can be detected. Firstly, the power-law dependences $K(N)$ are very close to the same layers in the

Table 2. Model and calculated parameters in the power dependence $K(N)$.

№	Layer location	The model (Fig. 5)	Calculation
1	Lower layer of MP	$5 \cdot 10^{-4} N^{0.88}$	$1.6 \cdot 10^{-2} N$ (3)
2	Upper layer of MP	$2 \cdot 10^{-7} N^{-0.96}$	$5.6 \cdot 10^{-5} N^{-1}$ (2)
3	Lower layer of AL	$3 \cdot 10^{-9} N^{-2.02}$	$2.0 \cdot 10^{-8} N^{-1.8}$

semi-empirical (estimated) and theoretical local models as well as in 1.5D model presented in [10]. This may indicate that all mentioned models possess the corresponding physics mechanisms. Secondly, there are significant (one-two ranges of the value) differences in the coefficient values A for the same layers. In all the layers the excess is observed in the calculated results. This fact requires additional research.

References

1. Wunsch, C., Ferrari, R.: Vertical mixing, energy, and the general circulation of the oceans. Ann. Rev. Fluid Mech. **36**, 281–314 (2004). https://doi.org/10.1146/annurev.fluid.36.050802.122121
2. Gregg, M.: Scaling turbulent dissipation in the thermocline. J. Geophys. Res. **94**, 9686 (1989). https://doi.org/10.1029/JC094iC07p09686
3. Samodurov, A., Dykman, V., Barabash, V., Efremov, O., Zubov, A., Pavlenko, O., Chukharev, A.: "Sigma-1" measuring complex for the investigation of small-scale characteristics of hydrophysical fields in the upper layer of the sea. Phys. Oceanogr. **15**, 311–322 (2005). https://link.springer.com/article/10.1007/s11110-006-0005-1
4. Samodurov, A.S., Lubitsky, A.A., Panteleev, N.A.: Contribution of breaking internal waves to structure formation, energy dissipation, and vertical diffusion in the ocean. Phys. Oceanogr. **6**, 177–190 (1995). https://doi.org/10.1007/BF02197516
5. Samodurov, A.S., Ivanov, L.I.: Mixing and energy dissipation rate in Mediterranean seas: an intercomparison of existing models. In: Oceanography of the Eastern Mediterranean and Black Sea, Similarities and Differences in Two Interconnected Basins, pp. 369–375. Tübitak Publishers, Ankara (2003)
6. Thorpe, S.A.: Experiments of instability and turbulence in a stratified shear flow. J. Fluid Mech. **6**, 731–751 (1973). https://doi.org/10.1017/S0022112073000911
7. McEwen, A.D.: The kinematics of stratified mixing through internal wave breaking. J. Fluid Mech. **128**, 47–57 (1983). https://doi.org/10.1017/S0022112083000373
8. McEwen, A.D.: Internal mixing in stratified fluids. J. Fluid Mech. **128**, 59–80 (1983). https://doi.org/10.1017/S0022112083000385
9. Osborn, T.R.: Estimations of local rate of vertical diffusion from dissipation measurements. J. Phys. Oceanogr. **10**, 83–89 (1980). https://doi.org/10.1175/1520-0485(1980)010<0083:EOTLRO>2.0.CO;2
10. Samodurov, A.S.: Complimentarity of different approaches for assessing vertical turbulent exchange intensity in natural stratified basins. Phys. Oceanogr. **6**, 32–42 (2016). https://doi.org/10.22449/1573-160X-2016-6-32-42

11. Prandke, H., Stips, A.: Test measurements with an operational microstructure- turbulence profiler: detection limit of dissipation rates. Aquat. Sci. **60**, 191–209 (1998). https://doi.org/10.1007/s000270050036
12. Oakey, N.S.: Determination of the rate of dissipation of turbulent energy from simultaneous temperature and velocity shear microstructure measurements. J. Phys. Oceanogr. **12**, 256–271 (1982). https://doi.org/10.1175/1520-0485(1982)012<0256:dotrod>2.0.co;2

Model of Oscillations of Earth's Poles Based on Gravitational Tides

S. A. Kumakshev(✉)

Institute for Problems in Mechanics of the Russian Academy of Sciences, Moscow, Russia
kumak@ipmnet.ru

Abstract. A model of oscillations of Earth's poles is constructed on the basis of the analysis of the gravitational torques from Sun and Moon. The model reflects physical processes and does not imply using curve fitting techniques, based, for example, on the polynomial approximation. Within the framework of this model, the Chandler frequency is interpreted as the fundamental frequency of oscillations of the mechanical system and the annual frequency as the frequency of the excitation force. A fine mechanism of excitation of the oscillations based on the combination of natural and forced frequencies is revealed. The model has only six parameters that can be identified by applying the least squares technique to the experimental data of the International Earth Rotation and Reference Systems Service. The prediction provided by the proposed model has high degree of accuracy for an interval of several years.

Keywords: Earth's pole oscillations · Gravitational torques

1 Physical Foundations of the Model

If Earth had been an ideal rigid ball and its motion had been unperturbed, then the points of intersection of the rotation axis of Earth with its surface (the poles) would have been fixed. In the late 1700's, Euler, when developing the theory of rotation of a rigid body with a fixed point in the absence of external forces, showed that if the ellipsoidal shape of Earth is taken into account, then the axis of rotation of Earth performs a cyclic motion (precession) with a period of 305 days. This implies periodic changes in the latitudes of localities on Earth, since they are measured relative to "fixed" stars. American astronomer Chandler, when observing the motion of stars, discovered in 1891 that in fact the periodicity has two basic components, the annual period and a floating period of 410 to 435 days (Chandler's period). This phenomenon is now called the polar wander. The swing of the annual oscillations of the pole can reach several tens of meters. Due to this nonuniformity in the rotation of Earth, apart from introducing operational corrections into modern navigation systems GLONASS/GPS, one should also correct the duration of the day. Constructing a simple model of the pole wander phenomenon that would have clear physical sense is a topical problem.

V. Karev et al. (Eds.): PMMEEP 2017, SPRINGERGEOL, pp. 157–163, 2018.
https://doi.org/10.1007/978-3-319-77788-7_17

In terms of the classical theory of oscillations, the pole wander process is composed of natural oscillations and forced oscillations.

In the proposed model, the exciting forces are gravitational forces applied to Earth by other bodies in Solar System. Earth undergoes not only oceanic tides but also daily ascents and descents of continental surfaces with a swing of about 1 m. Such a hump that has been created by a celestial body and is running along the earth surface creates an arm for gravitational torques caused by other celestial bodies. It is just these torques (most important of which are solar and lunar ones) that play the role of excitation forces. Sun exerts stronger influence on Earth (about 200 times as large as that produced by Moon), and for this reason, the most significant excitation force has a period of 1 year that coincides with the period of rotation of Earth about Sun.

Natural oscillations must decay as time elapses, but this is not the case for Earth. The point is that the fundamental frequency of Earth (Chandler's frequency) coincides with a combination of excitation frequencies. This resonance provides energy inflow and sustains the oscillations at the natural frequency. This combination resonance occurs when the Earth's natural frequency coincides with a combination of frequencies of external excitations produced by Sun and Moon. The motion of Moon is rather complex, as compared with the motion of Earth around Sun. One can single out four major features of the motion of Moon.

The orbit of Moon around Earth is an ellipse, with Earth in one of its foci. The eccentricity of this orbit changes by a factor of almost 2 (from 0.04 to 0.07) with a period of 8.85 years. The distance between Moon and Earth is minimal at the perigee point and maximal at the apogee point. The line that connects these two points is called the apse line; it passes through Earth's center and coincides with the ellipse major axis. The apse line rotates and performs its full revolution for 8.85 years.

The plane of Moon's orbit is inclined to the ecliptic plane (the plane of rotation of Earth around Sun). The angle between these two planes oscillates within a range of $4°59'$ to $5°19'$ with a period of 18.6 years. The points of intersection of the plane of Moon's orbit with the ecliptic plane are called the ascending and descending nodes, respectively. The imaginary line that joins these two points is called the line of nodes. The line of nodes performs a full revolution for 18.6 years.

Thus, the combination resonance occurs when the Earth's natural frequency (Chandler's frequency 0.84—here and below all frequencies taken with respect to the time of the year) coincides with the difference of the external excitation frequencies. The external excitations are accounted for by the rotation of Earth around Sun (with a period of 1 year and a frequency of 1) and the periodic perturbations of Moon's orbit (with periods of 8.85 and 18.6 years and frequencies of 0.11 and 0.05, respectively). Thus we have $1 - 0.11 - 0.05 = 0.84$.

The Fourier analysis of the pole oscillations identifies two basic harmonics, with periods of 1 year and 14 months; the one-year harmonic has a sharp peak, while the peak at Chandler's frequency is blurred.

Some researchers think that the one-year frequency has seasonal nature and is accounted for by the motions of masses in the atmosphere and ocean. However, such motions are not regular; they change from year to year and demonstrate rather chaotic behavior. Already for this reason, they cannot provide sharpness for the one-year peak, in contrast to the regular rotation of Earth around Sun.

In accordance with the physical sense of the proposed model, the Chandler's peak should be expected to be blurred, since the natural frequency of Earth depends on the shape of Earth, which changes due to gravitational tides in its mantle. It is apparent that irregular (in particular, chaotic) motions in the atmosphere additionally contribute just to the blur of Chandler's harmonic.

2 Mathematical Model in the First Approximation

Introduce the Earth-attached Cartesian coordinate frame, the axes of which coincide with the principal central axes of inertia of Earth. Let A, B, and C denote the moments of inertia of Earth about these axes. We assume that small deformations of Earth occur mostly in radial directions. To construct the model of the rotation of Earth about its center of mass, we represent the equations of this rotation as classical Euler-Liouville equations with variable tensor of inertia J [1,3,8,9]

$$\begin{gathered} J\dot{\boldsymbol{\omega}} + \boldsymbol{\omega} \times J\boldsymbol{\omega} = \mathbf{M}, \quad \boldsymbol{\omega} = (p, q, r)^T, \quad J = J^* + \delta J, \quad J^* = \text{const}, \\ J^* = \operatorname{diag}(A^*, B^*, C^*), \qquad \delta J = \delta J(t), \qquad \|\delta J\| \ll \|J^*\|. \end{gathered} \tag{1}$$

Here, $\boldsymbol{\omega}$ is the vector of the angular velocity represented in an Earth-attached coordinate frame (reference frame [4]), the axes of which approximately coincide with the principal central axes of inertia J^* of "frozen" Earth [1–4,7]. The additional perturbation terms that appear when differentiating the angular momentum vector for deformable Earth [6] are involved in the vector $\mathbf{M}$. We assume that small variations δJ of the tensor of inertia may involve various harmonic components due to the influence of diurnal tides caused by Sun and Moon and, possibly, the components with other periods (one-year, half-year, one-month, half-month, etc.) We assume, in addition, that major contribution to the external perturbation torque $\mathbf{M}$ that causes the nutation oscillations is due to gravitational forces. Taking into account the term $\dot{J}\boldsymbol{\omega}$ does not increase the accuracy of the first-approximation model.

Euler's kinematic equations that relate the components of the angular velocity in the Earth-attached reference frame to the generalized velocities are given by [8]

$$\begin{aligned} &\dot{\theta} = p\cos\varphi - q\sin\varphi - \omega_0(\nu)\sin\psi, \quad \dot{\nu} = \omega_0(\nu) = \omega_*(1 + e\cos\nu)^2, \\ &\dot{\psi} = \frac{p\sin\varphi + q\cos\varphi}{\sin\theta} - \omega_0(\nu)\cot\theta\cos\psi, \qquad\qquad e = 0.0167, \\ &\dot{\varphi} = r - (p\sin\varphi + q\cos\varphi)\cot\theta + \omega_0(\nu)\frac{\cos\psi}{\sin\theta}. \end{aligned} \tag{2}$$

Here, $\nu(t)$ is the true anomaly, e is the eccentricity of the orbit, and ω_* is a constant defined by the gravitational and focal parameters. For the case of pole wander, the terms in Eq. (2) that are proportional to ω_0 turn out to be substantially larger (approximately by a factor of 300) than p and q; hence, these terms define the derivatives $\dot{\theta}$ and $\dot{\varphi}$. This important property was not noticed in the scientific literature, and the terms indicated above were dropped without justification [1–4].

The expressions for the components of the solar gravitation torque are given by [9]

$$
\begin{aligned}
M_q =& 3\omega^2 \left[(A^* + \delta A - (C^* + \delta C))\gamma_r\gamma_p + \delta J_{pq}\gamma_r\gamma_q \right. \\
&\left. + \delta J_{pr}(\gamma_r^2 - \gamma_p^2) - \delta J_{rq}\gamma_p\gamma_q\right], \qquad \omega = \omega_*(1 + e\cos\nu)^{3/2}, \\
\gamma_p =& \sin\theta\sin\varphi, \qquad \gamma_q = \sin\theta\cos\varphi, \qquad \gamma_r = \cos\theta.
\end{aligned} \tag{3}
$$

Similar expressions for $M_{p,r}$ are obtained by cyclic permutation of indices p, q, and r in (3). Equation (3) implies that the one-year component of the pole oscillations can be accounted for by the term that involves products $\gamma_p\gamma_r$ and $\gamma_q\gamma_r$ of the direction cosines. To calculate these products, one should integrate Eq. (3) in the first approximation to obtain

$$
\begin{aligned}
r = r^0,\ \varphi \approx rt + \varphi^0,\ \nu \approx \omega_* t + \nu^0, \cos\theta(\nu) = a(\theta^0, \psi^0)\cos\nu, \\
\theta(0) = \theta^0 = 66°33', \qquad 0.4 \le a \le 1, \qquad 0 \le \psi^0 \le 2\pi, \\
\cos\theta\sin\theta = b(\theta^0, \psi^0)\cos\nu + d\cos 3\nu + \ldots,\ 0.4 \le b \le \frac{4}{3\pi},\ |d| \ll 1.
\end{aligned} \tag{4}
$$

The second and higher-order harmonics in terms of ν lead to the quantities that are less than the basic quantities by a factor of $10^2 - 10^3$ and, for this reason, are not taken into account. The difference $B^* - A^*$ is substantially (approximately by a factor of 160) less than the difference $C^* - A^*$. Having estimated the terms of Eq. (1) for p and q, taking into account the expressions of (4), after averaging with respect to the fast phase φ, we arrive at a simplified mathematical model given by

$$
\begin{aligned}
\dot{p} + N_p q &= \kappa_q r^2 + 3b\omega_*^2 \chi_p \cos\nu, \qquad N_{p,q} \approx N = \frac{2\pi}{T_1} \approx 0.84\omega_*, \\
\dot{q} - N_q p &= -\kappa_p r^2 - 3b\omega_*^2 \chi_q \cos\nu, \qquad p(0) = p^0, \qquad q(0) = q^0.
\end{aligned} \tag{5}
$$

Here, κ_p and κ_q are the average values of $\delta J_{pr}/B^*$ and $\delta J_{qr}/A^*$; they can be slow functions of time. The quantities χ_p and χ_q are obtained by averaging the coefficients of $\cos\nu$ in the expressions for the solar gravitational torque components with respect to φ. As mentioned above, these components are due to the diurnal tides. The lunar gravitational torques are not taken into account, since their influence on the nutation oscillations is small due to significant frequency difference. The right-hand sides of Eq. (5) involve an explicit harmonic excitation with a period of one year; this excitation accounts for the nutation oscillations that are recorded by International Earth Rotation and Reference

Systems Service. Although the sensitivity of the coefficients $\kappa_{p,q}$ is 5 orders of magnitude higher than that of $\chi_{p,q}$, the explanation of the regular mechanism of one-year excitation with the amplitude estimated as $M_h \sim 10^{20}$ kg m^{2}s^{-2} by means of geophysical influences (atmospheric, oceanic, seasonal, etc.) is inconsistent in terms of mechanics. The frequency analysis of the one-year oscillation component also indicates inconsistency of the geophysical interpretation [1].

3 Numerical Results

The values of the coefficients $\kappa_{p,q}$ and $\chi_{p,q}$, as well as the initial values of p^0 and q^0 in (5) are unknown. These quantities are to be determined on the basis of the observations by International Earth Rotation and Reference Systems Service [2]. Introduce the variables $x(\tau) = p(t)$, $y(\tau) = q(t)$, where $\tau = t/T_h$ is time measured in years, to represent the solution of Eq. (5) as follows [9]:

$$\begin{aligned} x(\tau) &= c_x^0 + c_x^1\tau - a_x^c \cos 2\pi\Omega\tau + a_x^s \sin 2\pi\Omega\tau \\ &\quad - \frac{\Omega}{1-\Omega^2} d_x^c \cos 2\pi\tau - \frac{1}{1-\Omega^2} d_x^s \sin 2\pi\tau, \\ y(\tau) &= c_y^0 + c_y^1\tau + a_y^c \cos 2\pi\Omega\tau + a_y^s \sin 2\pi\Omega\tau \\ &\quad - \frac{\Omega}{1-\Omega^2} d_y^c \cos 2\pi\tau + \frac{1}{1-\Omega^2} d_y^s \sin 2\pi\tau, \\ \Omega &= 0.845. \end{aligned} \tag{6}$$

Here, the coefficients $a_{x,y}^{c,s}$, $c_{x,y}^{0,1}$, and $d_{x,y}^{c,s}$ should be calculated by the least-squares method [10] on the basis of the data provided by International Earth Rotation and Reference Systems Service [2]. These coefficients are uniquely related to the unknowns in Eq. (5). When calculating, one should take into account the relations

$$-a_x^c = a_y^s, \qquad a_x^s = a_y^c; \qquad -\Omega d_x^c = d_y^s, \qquad d_x^s = \Omega d_y^c \tag{7}$$

that characterize a structural property of the model.

In what follows, we present the results of the calculations on the basis of the least-squares method [10]. This method was applied independently to the variables $x(\tau)$ and $y(\tau)$ that were approximated by six-term expressions in accordance with the model of (6):

$$\begin{aligned} x(\tau) &= 0.0810 + 0.0059\tau + 0.0244\cos(2\pi\Omega\tau) - 0.0289\sin(2\pi\Omega\tau) \\ &\quad - 0.0314\cos(2\pi\tau) - 0.0890\sin(2\pi\tau), \\ y(\tau) &= 0.3265 + 0.0079\tau - 0.0309\cos(2\pi\Omega\tau) - 0.0242\sin(2\pi\Omega\tau) \\ &\quad - 0.0818\cos(2\pi\tau) + 0.0276\sin(2\pi\tau), \\ \Omega &= 0.845. \end{aligned} \tag{8}$$

The comparison of the coefficients (in accordance with structural property (7)) that define Chandler's components of the oscillations, as well as the coefficients (taking into account the factor $\Omega = 0.845$) that correspond to the one-year component, confirm the structural property of the model indicated above.

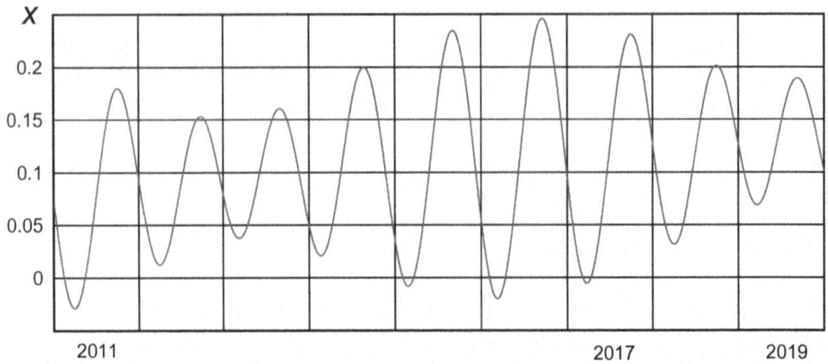

Fig. 1. Component x: experimental data and the theoretical curve that consists of the interpolation in the 7-year time interval, from the beginning of 2011 to the end 2017, and a 2-year forecast

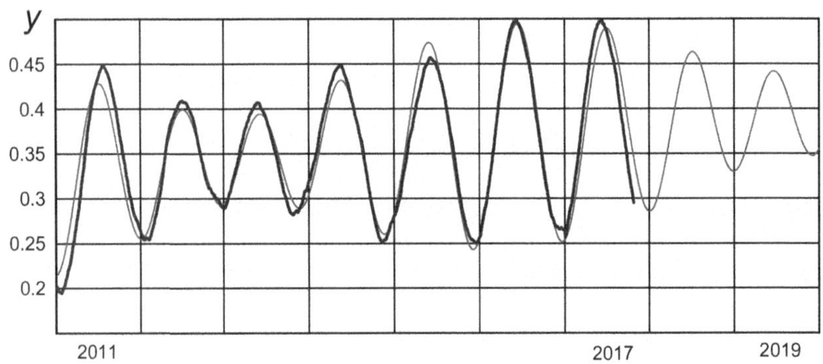

Fig. 2. Component y: experimental data and the theoretical curve that consists of the interpolation in the 7-year time interval, from the beginning of 2011 to the end 2017, and a 2-year forecast

Figures 1 and 2 show the experimental data and the theoretical curves $x(\tau)$ and $y(\tau)$ that consist of the interpolation of daily measurements during 7 years (from 2011 to December 2017) and the forecast up to the end of 2018. The standard deviations of the interpolation curve from the experimental data are given by $\sigma_x = 0.0171$ and $\sigma_y = 0.0182$, which indicates an acceptable accuracy of the constructed model.

4 Conclusion

The reliable forecast of the motion of Earth's pole is highly important for long-term inertial navigation [5] and for solving a number of astrometric and geophysical problems [1–4].

The model described above took part in an international competition [11], where a number of other models, constructed on the basis of various fitting techniques and approximations and containing a large number of fitting parameters, were presented. Despite its simplicity and a small number of parameters (only 6), our model turned out to be among the leaders in terms of the forecast of the pole position.

Acknowledgement. This study was partly supported by the Russian Foundation for Basic Research (project 17-01-00538).

References

1. Munk, W.H., Macdonald, G.T.F.: The Rotation of the Earth. Cambridge University Press, New York (1960)
2. IERS Annual Reports, 1990 July 1991 bis 2000 July 2001. Central Bureau of IERS. Observatoire de Paris
3. Moritz, H., Mueller, I.I.: Earth Rotation: Theory and Observation. Ungar, New York (1987)
4. Avsyuk, Y.N.: Tidal forces and natural processes. Inst. Physics of the Earth RAS, Moscow (1996). (in Russian)
5. Ishlinskiy, A.Y.: Orientation, Gyroscopes and Inertial Navigation. Nauka, Moscow (1976)
6. Ilyushin, A.A.: Continuum Mechanics. Moscow University Press, Moscow (1990)
7. Akulenko, L.D., Kumakshev, S.A., Markov, Y.G., Rykhlova, L.V.: A gravitational-tidal mechanism for the Earth's polar oscillations. Astron. Rep. **49**(10), 847–857 (2005)
8. Beletskii, V.V.: Satellite motion about the center of mass in gravitational field. Izdat. MGU, Moscow (1975). (in Russian)
9. Klimov, D.M., Akulenko, L.D., Kumakshev, S.A.: Mechanical model of the perturbed motion of the Earth with respect to the barycenter. Dokl. Phys. **58**(11), 505–509 (2013)
10. Linnik, J.W.: Method of Least Squares and Principles of the Theory of Observations. Pergamon Press, New York, Oxford, London, Paris (1961)
11. Kalarus, M., Schuh, H., Kosek, W., Akyilmaz, O., Bizouard, C., Gambis, D., Gross, R., Jovanovirc, B., Kumakshev, S., Kutterer, H., Mendes Cerveira, P.J., Pasynok, S., Zotov, L.: Achievements of the Earth orientation parameters prediction comparison campaign. J. Geod. **84**, 587–596 (2010)

Laboratory Modeling of Ring Geophysical Structures

B. Shvilkin(✉)

Physical Faculty, Lomonosov Moscow State University, Moscow, Russia
bshvilkin@polly.phys.msu.ru

Abstract. It is shown that upon a metal explosion of the wire between the electrodes caused by a pulsed arc discharge inside a dielectric ring on the cathode melts with hole formation. Solidification of the cathode metal is accompanied by the appearance of rings on the cathode surface, which may indicate the presence of wave motion. Analogous ring structures of a much larger size can also be found on the Earth surface. Their occurrence may be due to a meteorite explosion in the Earth atmosphere followed by generation of huge electric fields and a breakdown between the meteorite substance and the Earth.

Keywords: Metallic explosion of a wire · Arc discharge · Cathode
Ring structure

1 Introduction

Under an electric explosion of a wire the conductor is heated and disrupt by high-density electric current (see, e.g., [1]). In a pulsed arc discharge, at a low voltage on the electrodes cumulative melted-metal jets appear on the cathode that come from the point of contact between the wire and the cathode [2].

We show that when the contact between the wire and a part of the cathode is surrounded by a dielectric ring, no cumulative jet appears on the cathode. Inside the ring, energy is concentrated sufficient for both the cathode-material surface layer melting and the appearance of round holes in thin metal plates, even in high-melting tungsten plates. When the melted cathode metal solidifies, rings resembling the diffraction-interference patterns can be seen inside the dielectric ring. This may be indicative of the presence of wave motions in the melted metal on the fused cathode of the discharge gap.

2 Experimental Setup

Figure 1 presents the scheme of the experimental setup. Wire 1 joined by holder 2 through ballast resistor 3 with the positive pole of voltage source 4 gets in contact with cathode 5 inside setup the dielectric ring 6. When voltage is applied between the electrodes, wire 1 melts and evaporates, and between the electrodes, a nonself-maintained pulsed arc discharge appears in the atmosphere at low interelectrode voltages [2].

V. Karev et al. (Eds.): PMMEEP 2017, SPRINGERGEOL, pp. 164–170, 2018.
https://doi.org/10.1007/978-3-319-77788-7_18

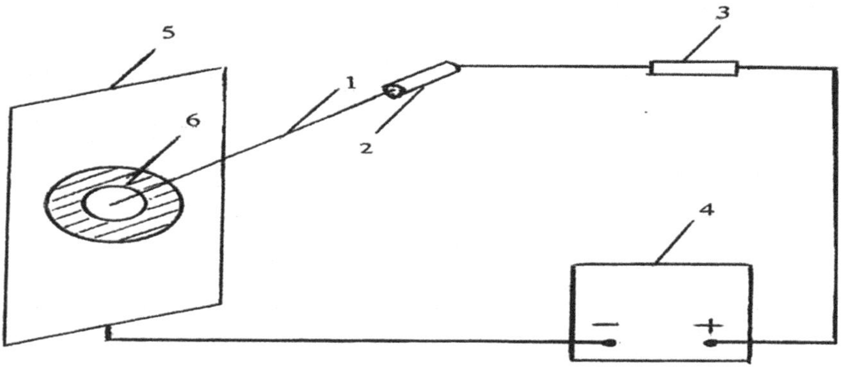

Fig. 1. Experimental setup.

For discharge ignition a rectifier unit "Delfin" with rectified voltage of 240 V was used. The maximum discharge current was changed within the range of 20 to 100 A using resistance 3. Ballast resistance in the discharge was changed within several Ohms. The oscillogram of the process was taken by the device TDS 2024 with a pass band of 200 MHz. The lower oscillogram in Fig. 2 characterizes voltage variation on the ballast resistor of 1 Ω in the discharging circuit.

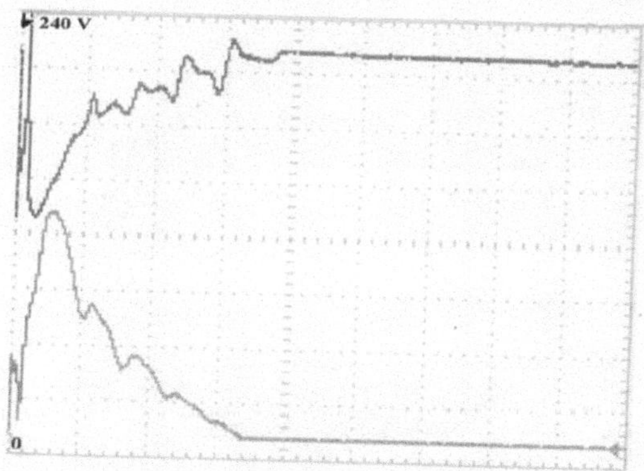

Fig. 2. Oscillograms of resistance (scale on the time axis 5 ms).

The maximum current in this case is nearly 80 A. The upper oscillogram in Fig. 2 shows voltage variation on the same resistor with a series connection with the discharge gap. The pulse duration in the experiments was changed from 0.015 to 0.05 s. With discharge parameters corresponding to Fig. 2 it is equal to about 0.015 s. The cathodes were made of different metals: Fe, Ni, W, Ti, Mo, stainless steel 1X18H9T and others. The wires were made of different metals: Cu, Ni, Fe, W. Their diameter ranged from 0.03 to 0.1 mm and the length from 10 to 30 mm.

3 Results of the Experiment

The applied voltage between the electrodes causes an electric explosion of the wire [1] followed by the occurrence of a nonself-maintained arc discharge in atmosphere and the appearance of a glowing plasmoid in the air [3]. The cathode undergoes a pulsed heating at the point of its contact with the wire.

In the absence of dielectric ring 6 on the cathode (see Fig. 1) a cumulative melted metal jet comes from the point of wire-cathode contact onto the cathode metal surface [2]. The jet is formed by the spreading electrons of the excess space charge and positive ions coming from plasma upon wire explosion. The emergence of a cumulative jet emanating from the point of contact between the wire and the cathode surface under the voltage applied to the electrodes leads to energy scattering about the cathode surface.

In the presence of a dielectric surrounding the wire-cathode contact the discharge energy is concentrated on the cathode inside the slit in the dielectric. This leads to a local metal heating and melting inside the dielectric ring on the cathode. The limitation of the part of the cathode contacting with the wire by the dielectric causes the electron blocking inside the dielectric because of the dielectric charging. The ions from plasma also rush inside the ring like in the case of ambipolar diffusion (see, e.g., [4]) and the cathode region is strongly heated [5]. With the help of a dielectric plate with a round hole on the cathode plate a slot in the metal can be obtained which repeats in shape the slot in the dielectric plate. Figure 3 shows the hole in a tungsten plate thick obtained using a dielectric ring 4 mm in diameter.

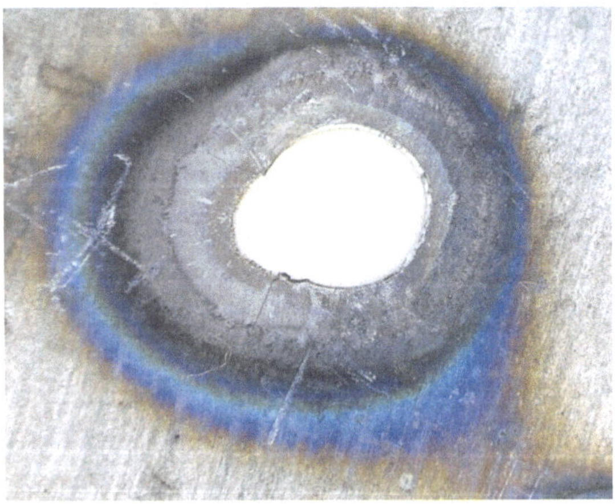

Fig. 3. Hole in a tungsten plate.

When the cathode is represented by two metal or bimetal plates pressed closely to each other, these plates are welded as in a cumulative end welding of thin metal plates [5].

At discharge currents of 30 to 40 A, wire explosions produce no holes in metal plates inside the dielectric ring. In this case clearly pronounced rings are formed during metal solidification. This can be seen in Fig. 4 obtained with the use of

Fig. 4. Rings on a transformer-iron plate.

transformer-iron plate upon an explosion of a copper wire 0.03 mm in diameter at discharge current of 40 A.

Figure 5 presents the spectrogram obtained using the X-ray spectrum analyzer Oxford. The figure shows that only a small amount of copper atoms are found on the cathode inside the ring.

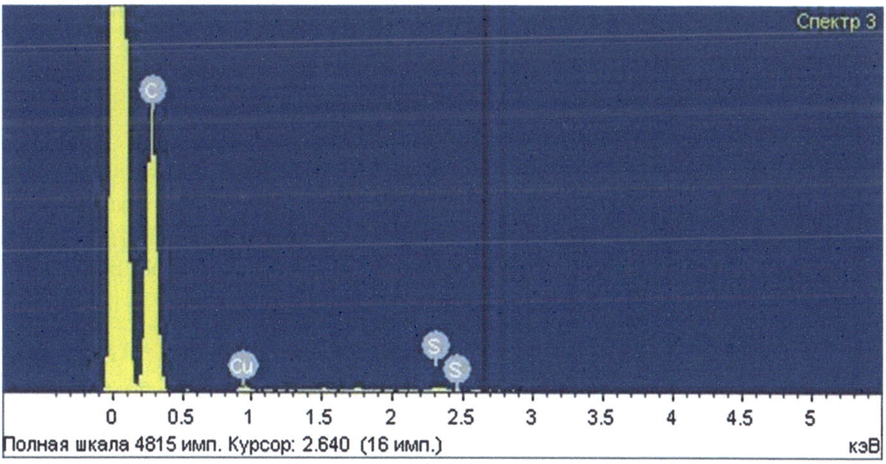

Fig. 5. X-ray photograph: the spectrum of a point on the cathode inside a dielectric ring.

Figure 6 clearly shows the rings obtained inside a dielectric ring 5 mm in diameter on a nickel cathode plate 0.15 mm thick upon a tantalum wire explosion and formed in the process of metal solidification.

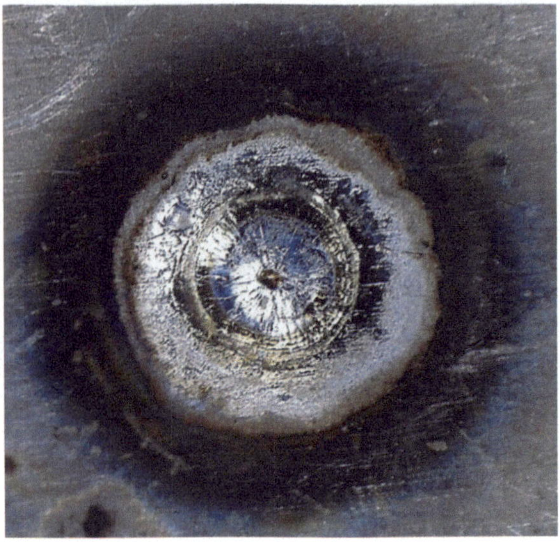

Fig. 6. Rings on a nickel plate.

The rings demonstrated in Figs. 4 and 6 can be identified as traces of heat waves [7]. Note that the formation of analogous rings in a solidifying melted metal flow was reported in [8].

In our case, rings are formed upon a pulse temperature disturbance on the melted metal surface in the discharge due to wire explosion. The heat wave length is

Rings on a transformer-iron plate

$$\lambda = \sqrt{2\pi aT} \quad (1)$$

where a is temperature conductivity, and T is disturbance time. If we assume that $a \sim (1-2)10^{-5}$ $\text{м}^2\text{c}^{-1}$, which is typical of many metals, and the temperature disturbance duration is thought to be 0.015 s, then the heat wave length $\lambda \sim 1.2$ mm, which is consistent with the observed experimental values. The thermal-wave propagation velocity

$$v = 2\sqrt{\pi a/T} \approx 1 \text{ m/s} \quad (2)$$

is also close to the cumulative jet velocity on the metal surface upon wire explosions.

4 Ring Structures

Rings analogous to the observed ones, namely, "enigmatic rings" are seen on the surface of the Earth and other planets. They are called "ring structures". The most well known is the "Richat Structure" or the "Eye of the Sahara" [9]. The diameter of its external contour is nearly 50 km (Fig. 7).

Fig. 7. View of the Richat ring structure from space. Source: NASA.

It was originally considered to be an impact crater (an astrobleme) from a huge meteorite. However, this hypothesis turned out to be unsupported because of the flat bottom and the absence of rock with traces of impact.

At the present time [10] scientists believe that the Richat ring structure is the result of the centuries-old erosion. The most acknowledged explanation at the present time is that the volcanic dome underwent erosion and gradually expanded creating today's shape of the ring structure.

Another assumption concerns the formation of enigmatic ring structures associated with the electric discharge explosion of large meteorites moving in the Earth atmosphere at high velocity. The author of [11] tried to explain the mystery of Tunguska meteorite by such an explosion. The meteorite explosion can be accompanied by an onset of a great potential difference between the meteorite and the Earth and the generation of a giant lightning. A discharge explosion of a meteorite near the Earth surface leads to such a great temperature rise that causes a local melting of the material. Meteorite evaporation provides conditions for the emergence of low-voltage discharges [2]. A local change in the density and temperature induces the appearance of a heat wave of the melted material. A fast cooling of the material causes the onset of concentric surfaces on the Earth which represent Richat-type ring structures. In this context, the emergence of ring structures may be indicative of the wave character of their origin.

5 Conclusion

It is shown that upon a metal explosion of a wire between the electrodes caused by a pulsed arc discharge, inside the dielectric ring on the cathode the energy is concentrated leading to cathode material melting and hole formation in the cathode plates. The rings of solidified metal inside the dielectric holes indicate the existence of wave motions generated by the heat waves in it. We may believe that ring structures analogous to the Richat structure are also formed by the heat waves on the melted Earth surface heated through the electric explosion of a large meteorite in atmosphere.

References

1. Lebedev, S.V., Savvatimskii, A.I.: Metals during rapid heating by dense cur. Phys. Uspekhi **27**(2), 749–771 (1984)
2. Miskinova, N.A., Shvilkin, B.N.: The way of channel formation on the cathode in a nonself-maintained arc discharge. Patent RF № 2537383, 1–3 (2015)
3. Miskinova, N.A., Shvilkin, B.N.: On the penetration of a plasmoid through glass. Phys. Uspekhi **58**, 1215–1217–1335 (2015). https://doi.org/10.3367/UFNe.0185.201512d.1333
4. Miskinova, N.A., Shvilkin, B.N.: Physical electronics in problems. In: Knizhnyi dom "LIBROKOM", pp. 1–256 (2013). ISBN 978-5-397-03908-6
5. Miskinova, N.A., Shvilkin, B.N.: The method of local heating of a cathode surface region. Patent RF № 2483500, 1–3 (2010)
6. Kuz'min, R.N., Miskinova, N.A., Shvilkin, B.N.: Electric welding of thin metal plates by wire "explosions". Elektrichestvo № 7, 61–63 (2011)
7. Grober, H., Erk, S.: Die Grundgesetze der Waermeuebertragung, Berlin (1933)
8. Poznyak, I.M., Sakhronov, V.M., Tsybenko, V.Y.: The motion of a melted metal layer under conditions typical of fast plasma processes. Termoyadernyi sintez **39**(1), 15–21 (2016)
9. Source: Richat Structure in Mauritania. Google Earth
10. Schumaker, D.: Richat Structure in Mauritania. The Geology News Blog, 25 September 2008
11. Nevskii, A.P.: Phenomenon of a positive stabilizing electric discharge and the effect of electric discharge explosion of large meteorite bodies in planet atmosphere. Astron. Vestnik **12**(4), 206–215 (1978)

Principles of Controlling the Apparatus Function for Achieving Super-Resolution in Imagers

E. N. Terentiev[1(✉)], N. E. Terentiev[2], and I. I. Farshakova[1]

[1] Physical Faculty, M.V. Lomonosov Moscow State University, Moscow, Russia
en.teren@mail.ru, en.teren@physics.msu.ru
[2] HiQo Solutions, Moscow, Russia

Abstract. When controlling the Apparatus Function (AF), the size of definition domain of the AF O and the sampling step and conditionality of the AF must be chosen so that its inverse function $pR = pO^{-1}$ obtains a minimum norm. The compensation of the AF O distortions in the measured images is realized point-by-point (without using the Fourier Transform in convolution). The computer of the device uses the resolving function pR, selected by the controlling procedure, for achieving super-resolution in images. Such controlled super-resolution is demonstrated on the Martian images.

Keywords: Regularization method · Super-resolution · Conditionality Invertibility · Modulation · Transfer Function · Convolution · Fourier Transform

1 Introduction

We will say that the New Device (ND) implements the Principles of Controlling (PC) AF O with optimal $pR = pO^{-1}$.

Without the PC, ND = D + computer complexes, as a rule, are not workable.

Regularization methods (in such not workable or non-functional NDs) yield results with a large residual error [1] due to a priori information on the smoothness of the solutions.

With the PC AF O ND complexes are able to realize the super resolution without a residual error.

If it turns out that the step of digitization of AF O, cannot be physically reduced, then in the computer, we perform the corresponding recalculation of the measured data (interpolation) under a smaller step of digitization.

The PC AF O in ND implements the maximum resolution beyond the size of the pixel of the original image.

Under the Characteristics of the Adequacy of the Model (CAM) AF O are understood the mutual dependence of the three функций (in XYZ axes): (X) the noise reactions - Nor (pR), (Y) the errors in the assignment of the AF pO - Err(pO) and (Z) the values of the Indicator of Invertibility - II(pR $*$ O) [2, 3]. If the Modulation Transfer Function

V. Karev et al. (Eds.): PMMEEP 2017, SPRINGERGEOL, pp. 171–182, 2018.
https://doi.org/10.1007/978-3-319-77788-7_19

(MTF) M (O) is bounded: diap $\leq |M(O)| \leq 1$, then $1 \leq |M(R)| = 1/|M(O)| \leq$ DIAP (shorts for the word Diapason).

The limitation in the frequency domain DIAP = 1/diap is the conditionality parameter of the AF O. If the DIAP value decrease, the model of AF becomes coarser, its conditionality increases.

The most conditioned object is the delta Kronecker symbol with DIAP = diap = 1 [2, 3]. In papers [2, 3], one can see a detailed explanation of the notation and meaning of new mathematical constructions, such as the indicator of computational errors: $II(R * O) == M(R)M(O) \leq 1$.

Comment. In the constructions that claim to solve the inverse problem, there must be concepts such as "Conditionality" of the AF, matrix and, most importantly, "Measure or Indicator of Invertibility". AF pO and pR * O, which accompany the obtained results with super resolution imagers, should be presented.

Of course, there must be a construction about the quality of the solution, the inverse problem, which connects the result of the invertibility of the AF with the obtained errors of the type CAM AF O.

To compare the super resolution results, there must be a corresponding numerical value SR that estimates the obtained super resolution in imagers.

If these concepts are not present, then it is unclear what is at stake.

2 Principles of Controlling the AF O, pO in the New Device with AF pR * O

The main task of choosing the AF pO = pR – 1 is set as a minimum problem:

$$\min_{LO}\{\|pR\| \mid Err(pO) \leq err\}, \quad LO = \{Loc,\ dx,\ DIAP\} \tag{1}$$

In (1), the set LO - The AF O Lodge for O contains three sets: Loc is the set of lengths of the domain of definition of the O, dx is the set of steps for digitizing the AO in Loc and DIAP is the set of settings of the conditionality in the controlling AF pO.

Note that the invertibility $pO = pR^{-1}$ in frequency domain corresponds to MTF equality: $M(pR) = 1/M(pO)$.

With the solution of problem (1) we bind the CAM AF O, pO of the Discrete ND:

$$\{x = \|pR\|, \quad y = Err(pO), \quad z = II(pR * O)\} \tag{2}$$

It is well known, that the Fourier coefficients in the data square with side ~40–50 are computed with the error 10^{-13}–10^{-14}. We will call this small value an instrumental error or instrumental zero: $Iz = 10^{-13}$ [2, 3].

In connection with the limitations in the accuracy of the calculations, we will consider the variant CAM AF O, taking into account the Iz inversion of M(O):

$$M(zR) = \begin{cases} 1/M(O), & if\ |M(O)| > Iz \\ M(O) \end{cases} \tag{3}$$

$$\{x = \|zR\|, \quad y = Err(zO), \quad z = II(zR * O)\} \tag{4}$$

PC AF O with CAM (2, 4) Discrete ND allow us to control the settings of AF O.

3 Tools of PC AF O

The band path of the Discrete ND will be estimated by the quantities $BP = \Sigma M(pR) * M(O)$ and $BPz = \Sigma M(zR) * M(O)$.

The efficiency of (using the Discrete ND) bandwidth will be estimated by the quantities $M(pR)M(O) = BP/\Sigma 1$ and $M(zR)M(O) = BPz/\Sigma 1$.

Theorems *on invertibility in the bandwidth* of Discrete ND. Given AF O, if the computation of the inverse R is realized with an instrumental error Iz, then with this instrumental error the equality and inequality are preserved: $II(zR * O) = \underline{M(zR)M(O)} \leq 1$.

Inequalities occur in cases with irreversibility $II(zR * O) < 1$. This means that inversion takes place only in the $\underline{M(zR)M(O)}$ part of the bandwidth of the Discrete ND.

The property of invertibility: For all values of Loc, dx = 1 and DIAP in LO, the set AFO is mapped to the set of invertible $pO = pR^{-1}$, $M(pR) = 1/M(pO)$, and the Invertibility Indicator $II(pR * pO) = 1$.

The invertibility theorem: The Iz inversion (3) is given, then the invertibility $II(zR * O) = 1$ implies the we have ordinary invertibility $II(R * O) = 1$, $zR = R = O^{-1}$ with the use of the full band pass $\underline{M(zR)M(O)} = 1$ and the identity $M(zR)M(O) == 1$.

3.1 Estimation of the Super Resolution Value

If there is a normalization of the AF O: $\Sigma O = 1$, then at the zero of the MTF $M(O)(0) = 1$. In this case, we estimate the super resolution by value $SR = BP/\Sigma M(pO) \leq \Sigma 1/\Sigma M(pO) \geq 1$, i.e. $SR \geq 1$. If the AF O = D_K-delta Kronecker symbol, then SR = 1. It is clear that a super resolution problem is raised for the real Devices with AF O $\neq D_K$.

The value of SR does not depend on the size of the AF O domain of definition and sizes of images.

3.2 Errors of AF pO

The error in the AF pO was estimated as $Err\,(pO) = SD(O - pO)/\max(O)$. SD is the Standard Deviation of the value $O - pO$ or $SD(O - pO) = \text{sqrt}(\Sigma(O - pO)^2)/\Sigma 1$, where $\Sigma 1$ is the number of points in the domain of definition of O, pO.

3.3 The Convolution Theorem

For even A and B, there are equalities in the spatial domain $A * B = B * A = F^{-1}(M(A)F(B)) = F^{-1}(M(B)F(A))$ and in the frequency domain $F(A * B) = F(B * A) = M(A)F(B) = F(A)M(B)$. M(A) and M(B) are the spectra or MTF of AF A and B, M(A) and M(B) are even in the frequency domain, the functions F() is Discrete FT and $F^{-1}()$ is Inverse Discrete FT.

Suppose that we have an even O, dependent on the difference of the arguments (O moves). Let A be given, containing even and odd parts. Then in the spatial domain the convolution $O * A = F^{-1}(M(O)F(A))$, in the frequency domain $F(O * A) = M(O)F(A)$.

In applications, the last result is almost impossible to use because of the boundary effects in F(A), see the examples at the end of the paper. Therefore, the convolution operation must be used point-by-point, without applying the FT.

3.4 MTF in Convolution

MTF M (O) from AF O is obtained by solving the eigenvalue problem: $O * H = M(O)H$, where the orthonormal Fourier harmonics are located in the rows of the matrix H. For an even AF O the MTF M (O) is even, for odd O, M (O) is odd [2, 3].

For the delta Kronecker symbol, D_K MTF M $(D_K) == 1$ is even. For D_K we will use the other notation D_C (c from the word cos). The fact is that there is an odd analogue of the Kronecker D_S symbol (s from the word sin). The following result holds: the convolution of two odd $D_S * D_S = \underline{D}_C = D_C - 1/\Sigma 1$ is a delta Kronecker symbol without constant.

3.5 The Finite-Dimensional Sampling Theorem for Interpolation

Sampling Theorem FDST: Given a two-dimensional array of samples $D = f(x0, y0)$, two-dimensional matrices of discrete Fourier harmonics $H(kx, x0)$, $x0 = 0 : N - 1$, $H(ky, y0)$, $y0 = 0 : M - 1$, and the "continuous" Fourier harmonics $H(kx, x)$, $x = 0 : dx : N - dx$, $H(ky, y)$, $y0 = 0 : dy : M - dy$, $dx < 1$, $dy < 1$,

$$f(x, y) = \sum_{k_x, k_y = 1}^{N,M} C_{k_x k_y} * H(k_x, x) * H(k_y, y) \tag{5}$$

$$\begin{aligned} c_{k_x k_y} &= (f(x0, y0), H(k_x, x0) * H(k_y, y0)) \\ &= \sum_{x0, y0 = 1}^{N,M} f(x0, y0) * H(k_x, x0) * H(k_y, y0), \quad k_x = 1 : N, \quad k_y = 1 : M \end{aligned} \tag{6}$$

then "continuous function" passes through the sample points f (x0, y0). The scalar products (6) realize a Direct Discrete FT, and the Fourier series (5) is an Inverse Discrete FT with interpolation if $dx < 1$, $dy < 1$. Such interpolation gives good result, if there are no differences in values at the boundary of the arrays data. Otherwise, the interpolation should be performed in a special protected mode [2, 3].

4 An Example of the Implementation of an ND with a PC AF

In the low-contrast fragment of the Martian byte image, the relative errors in small brightness are of the order of 3–10%, due to the fact, that we use one byte for brightness storage or we have luminance in the range 0–255.

Therefore, the reaction to noise Nor (pR) should be of the order of ~ 10. It is precisely this that determined the coarsening of the AF pO model by the conditionality parameter DIAP = 10 (Fig. 2).

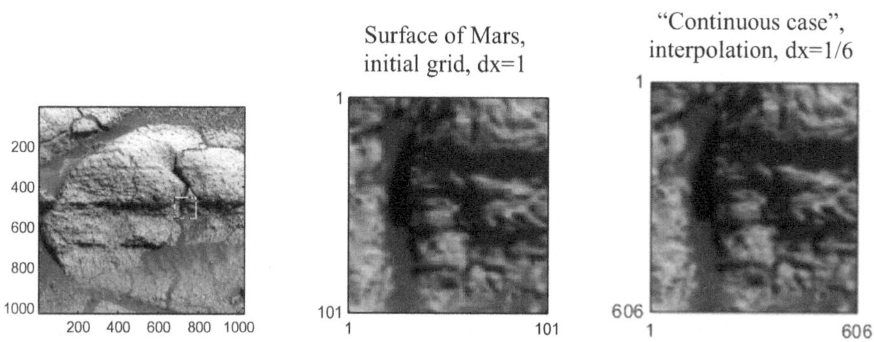

Fig. 1. Fragment of the image of the Martian surface presented in the original grid dx = 1 and recalculated with help the FDST to "continuous case" with a grid dx = 1/6.

The detailed graphic data in Figs. 2, 3 are given in green g color. By using three rgb colors we show values: noise reactions - Nor (pR), SR - super-resolutions and BP - band paths.

Conditionality with DIAP = 10 gives us an acceptable result, see images in Figs. 2, 3.

The super-resolution in the "continuous case" is approximately on 30% higher than in the discrete case, compare SR values in Figs. 2, 3.

4.1 Examples of CAM AF O, pO for ND

In this section, we will present HAM AF O, depending on the values of the parameters from $LO = \{Loc, dx, DIAP\}$ (1).

AF O with Loc = 4 corresponds to the length of the side of the square of the domain of definition of AF O with sides [−2, 2] and on the graphs of Figs. 4, 5 this is the left points of the graphs, which are labeled with the number 1.

Similarly, for the length Loc = 12 for squares with sides [−6, 6], the right-hand points of the graphs are labeled with the number 2, see Fig. 4.

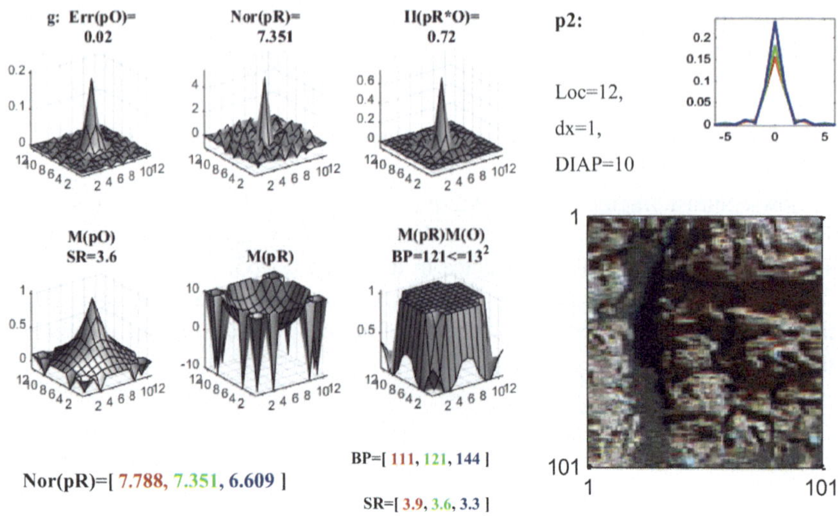

Fig. 2. Image with super-resolution is in the original grid with Loc = 12 with DIAP = 10 for AF O.

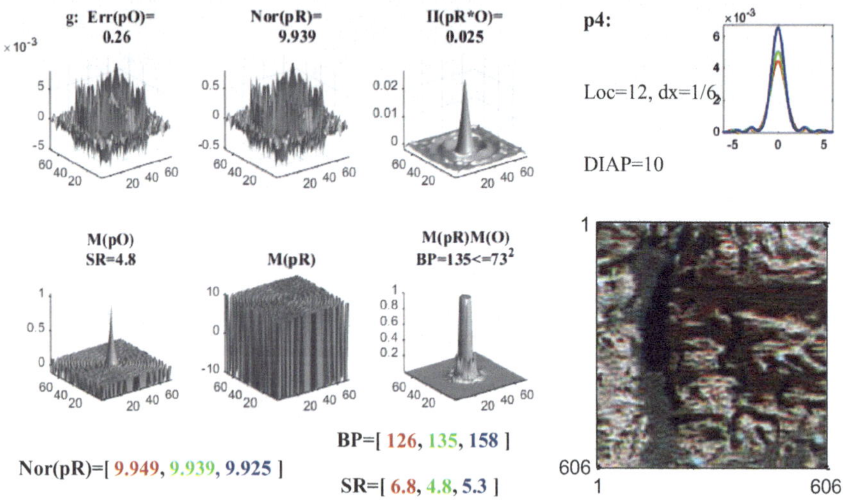

Fig. 3. "Continuous" case with Loc = 12, dx = 1/6, DIAP = 10.

The lengths Loc = [4, 6, 8] in the original grid are associated with a noise reaction from 17 to ~50. This is acceptable for high-precision measurements. With the DIAP = 10 conditionality, the noise reaction are down to ~7.35, but this decreases the invertibility and increases errors, see Fig. 4.

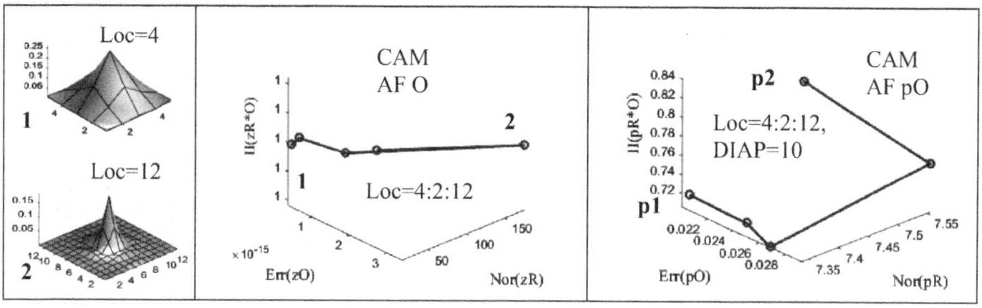

Fig. 4. The effect of DIAP = 10 conditionality on the noise reaction Nor(zR), Nor(pR), invertibility II(zR ∗ O), II(pR ∗ O) and errors Err(zO), Err(pO) depending on different lengths of the Loc = 4:2:12 definition domains in the original grid dx = 1.

In a "continuous case" with the changing in the lengths Loc = 4:dx:12, dx = 1/6 on the graph CAM AF O, the left point with Loc = 4 is labeled with the number 3, and the right-hand point with Loc = 12 is labeled with the digit 4, see Fig. 5.

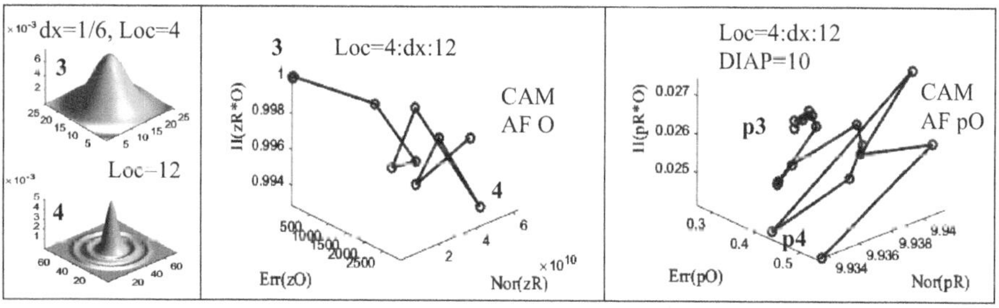

Fig. 5. The effect of DIAP = 10 conditionality on the noise reaction, invertibility and errors, depending on the different lengths of the definition domains Loc = 4:dx:12 in the "continuous case" with dx = 1/6.

The prefix "p" before numbers 2 and 3 in Figs. 4, 5 means that conditionality with DIAP = 10 is activated. In Figs. 2, 3, the results are presented by the super resolved images in the cases of **p2** and **p4**.

The conditionality parameter DIAP = 10 with Loc = 12 gives us the resolving functions pR c Nor(pR) ~7.6 and 9.9, with which the super resolved results are obtained in discrete and "continuous" cases.

5 Where Does the Invertible Function AF O Exist?

In this section, we will compare CAM AF O with Iz invertibility (3) with CAM AF pO for a given conditionality DIAP = 10 (2) in the "continuous case" with dx = 1/6. In Fig. 6, we see on the left, marked by double arrows, a very small region of AF O within the main lobe. In the domain of definition with Loc = 12, the noise reaction is Nor (zR) ~ $6 * 10^{10}$. In order to see this all, we use Iz inversion.

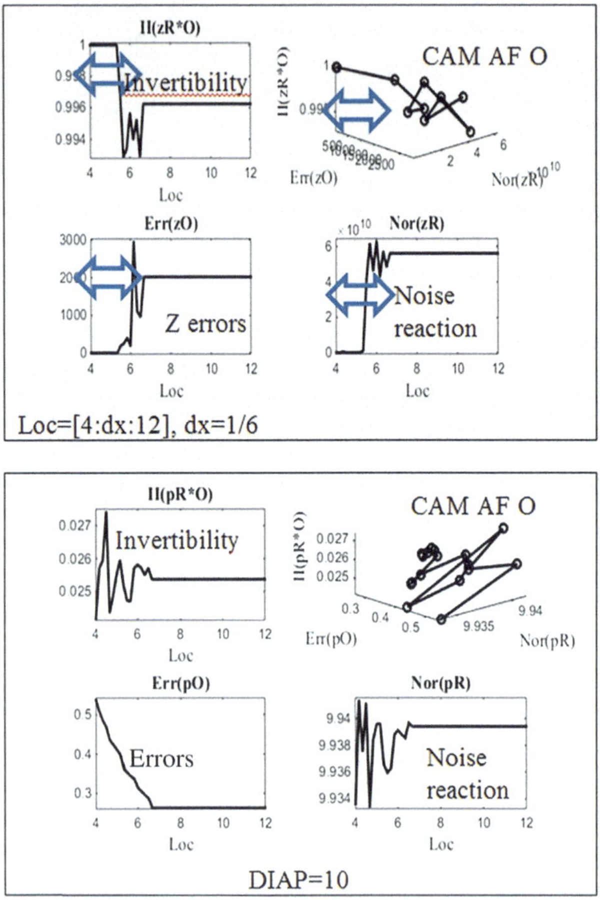

Fig. 6. In the "continuous case" **3:** with Loc = [4:dx:12], dx = 1/6, we are forced to increase the conditionality to DIAP = 10 to obtain a practical result with the definition domains of arbitrary length.

The result is obtained with Nor(pR) $\sim$ 10 if we include conditionality with DIAP = 10. In this case, the reversibility falls and errors are fixed; see Fig. 6 graphics on the right.

It is curious, but in the methods of regularization in this situation, it is proposed to seek smooth solutions. In the regularization method, there is no indicator of invertibility and the conditionality parameter.

6 Resolution of Objects Smaller Than a Pixel

In the highlighted area of the stone, an area with a suspicion of sand in the dunes is visible. Below you can see the possibility of selecting individual grains, smaller than a pixel size.

In the "continuous" case, the super-resolution of SR is $\sim$ on 30% higher than in the discrete case, compare the values of SR in Figs. 2, 3 and the images in Figs. 6, 7.

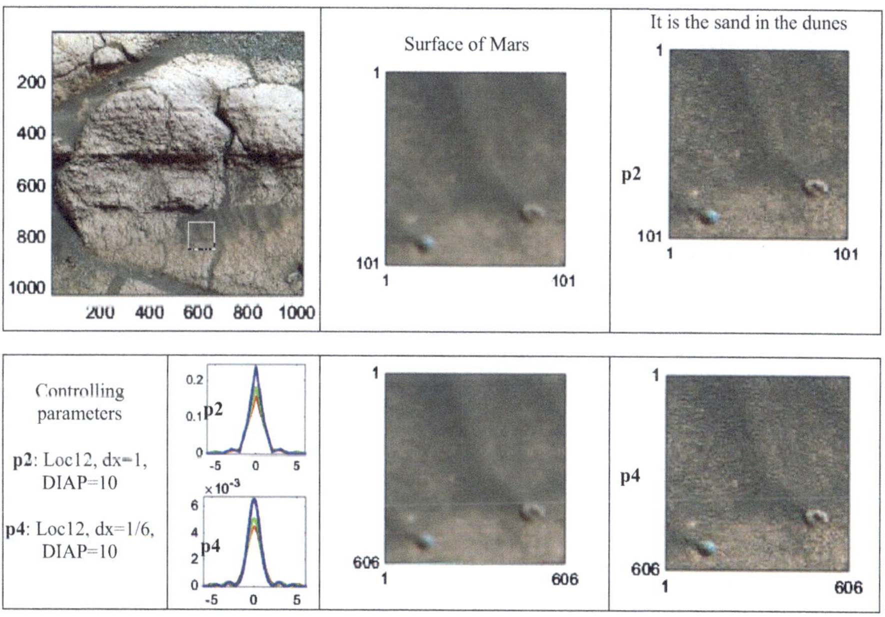

Fig. 7. Resolution of individual grains in the discrete case **p2** and continuous case **p4.**

7 Structure of Stone Similar to Fossilized Coral

In the selected area of the stone, we found structures, in our opinion, similar to the fossilized coral. Of course, specialists must make the final decision (Fig. 8).

8 The Application of the Convolution Theorem

The convolution operation together with the Fourier Transformation is usually used in many cases.

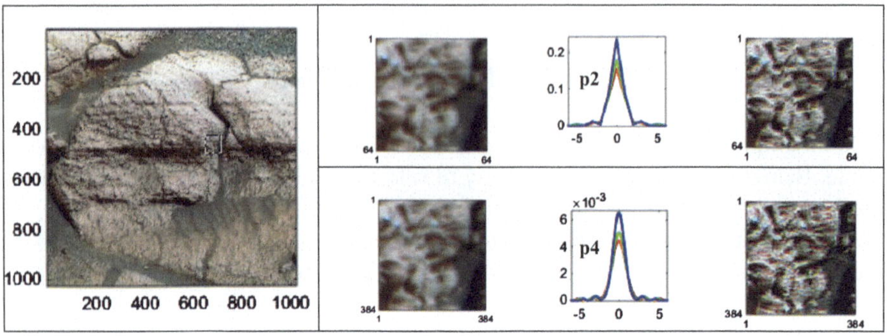

Fig. 8. Resolution of the suspected region to organic origin in the discrete case **p2** and the continuous case - **p4**.

The result is completely incomprehensible if we try to realize this through Discrete FT. The well-known result that convolution is the Inverse of the Fourier Transform of the Direct FT functions is valid with certain conditions for even functions.

The fragment of the image is not an even function and brightness jumps occur on the boundaries of the images.

We do not use FT for solving inverse problems; see boundary effect in imagers in Figs. 9, 10.

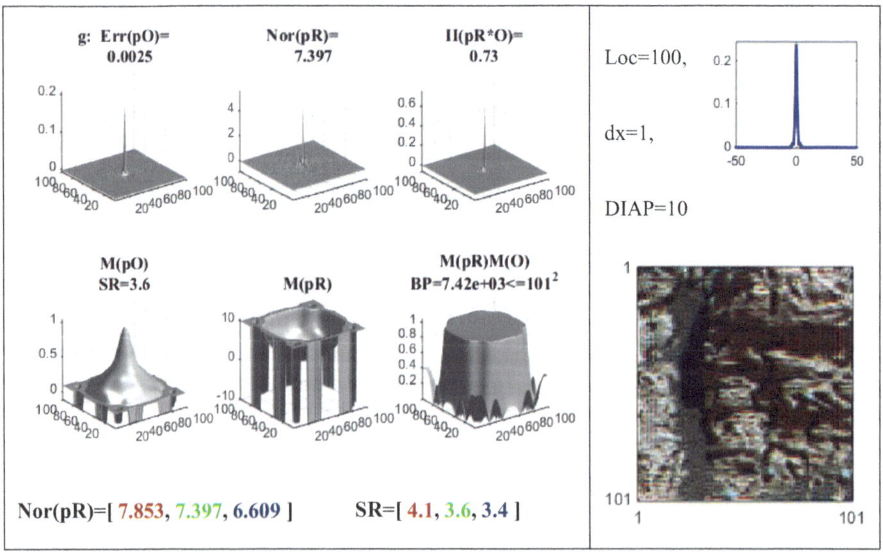

Fig. 9. The AF O definition domain is the entire image, discrete case.

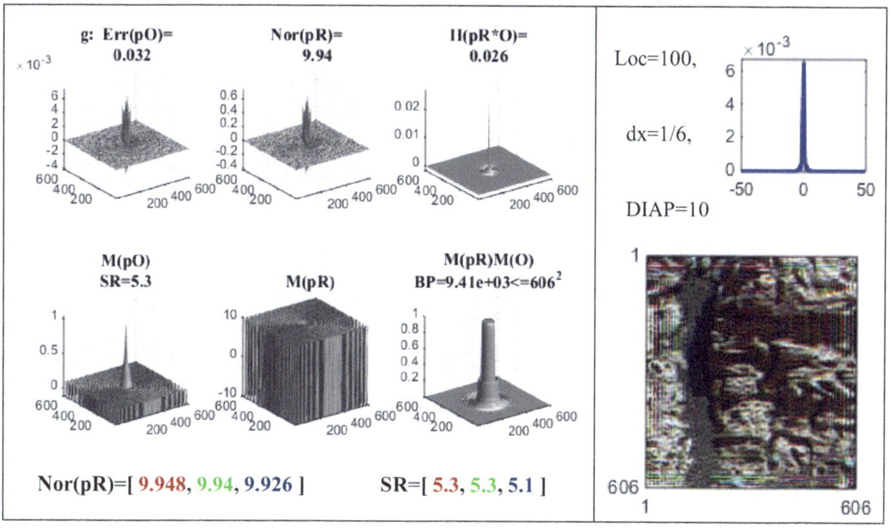

Fig. 10. The AF O definition domain is the entire image, continuous case.

9 Discussion

For single-byte color images, the SR values vary from [3.9, 3.6, 3.3], dx = 1 is the discrete case **to** [6.8, 4.8, 5.3], dx = 1/6 is a continuous case.

Transition to a continuous case (for one-byte image) gives us an increase in the SR value by about 30%.

For double-byte images with a similar AFO, the values SR ~ 10–15 are expected in the continuous case.

In super-resolved images, jumps and brightness differences are possible, that is, the solutions of the inverse problem to compensate for the distortion of AFO are not smooth.

The non-smoothness of the solutions allows us to apply this method reasonably and safely, for example, in the reactor control loops.

It is desirable that the ND "can be automatically tuned" according to AF, that is, it can measure with a smaller step and if step reduction is impossible, then the ND should implement interpolation.

If there is a need, then the solutions – super-resolved images – can be "smoothed" according to a priori information.

Controlling the AF of devices with FDST on CAM AF O is the basis for creating an ND with pR * O AF with embedded processors in various fields.

It is possible to use PC AF with interpolation to achieve the maximum resolution in electron microscopy, atomic force microscopes, radar, telescopes, tomography, etc. To increase resolution in tomography requires a very powerful processor.

Interesting modifications of the PC AF O methods are expected in new radar technologies, synthesized aperture, X-ray tomography, telescopes, etc.

In this paper, we considered problems of increasing the resolution only due to the methods [4, 5] of controlling the AF O, mathematical methods proved much more complicated than the methods of classical regularization [1].

References

1. Tikhonov, A.N., Ufimtsev, M.V.: Statistical Processing of Experimental Results. Publishing House of Moscow University, Moscow (1988). (in Russian)
2. Terentiev, E.N., Terentiev, N.E.: Izvestiya RAN. Physics **79**(12), 1633–1637 (2015). (in Russian)
3. Terentiev, E.N., Terentiev, N.E.: Bulletin of the Russian academy of science. Physics **79**(12), 1427–1431 (2015). https://doi.org/10.3103/S1062873815120229
4. Terentiev, E.N., Terentiev, N.E., Farshakova, I.I.: Scientific conference "Lomonosov Readings", section of physics, pp. 187–190 (2017). (in Russian)
5. Terentiev, E.N., Terentiev, N.E., Farshakova, I.I.: Proceedings of the school-seminar ≪Waves-2017≫. Mathematical modeling in radio physics and optics, pp. 56–58 (2017). (in Russian)

A Regular System of Vortices in a Circular Stratified Flow Behind the Edge of a Rotating Disk

Roman N. Bardakov(✉)

Ishlinsky Institute for Problems in Mechanics RAS, Vernadskogo. 101 k 1, 119526 Moscow, Russia
bard@ipmnet.ru

Abstract. In this paper the spatial pattern of vortex flow around the rotating disk studied experimentally. The disk is mounted in a continuously stratified fluid. The phenomenon of breaking the annular flow near the disk edges to a regular sequence of vortex loops is found. The governing parameters are found and the range of their values is determined. Observations have been made both for strongly stratified and for weakly stratified media. The influence of stratification on the evolution of the current is determined.

Keywords: Rotating disk · Vortex · Stratified fluid

1 Introduction

The investigation of the flow created by a uniformly rotating disc in a homogeneous fluid is one of traditional problems in mechanics of fluids. A great influence on the systematic study of currents on the disk was provided by von Karman's paper where the velocity field was calculated over an infinite uniformly rotating disk [1]. The results were repeated and re-interpreted in various ways [2]. In addition to studying flows on a smooth disk surface the influence of the flat lateral surface roughness [3] and the possibility of partial slip [4] are analyzed. The flow pattern proposed in [1–3] formed the basis of modern experimental researches.

Later it was shown that in the vicinity of the plane of the disk rotating in a stationary liquid the flow loses its heterogeneity and converts into a system of spiral vortex arms that leave a characteristic typical trace on the special paint layer [5, 6]. Spiral vortices were also observed in the pattern of particle distribution near the surface of the disk rotating inside a cylindrical container filled with glycerin solution [7].

The thermoanemometric measurements showed the presence of spiral structures on the plane of a disk rotating in the air [8]. The stability of spiral vortex pattern on a disk rotating in the air was studied in a wide range of parameters including transition to turbulence [9]. However, the flow pattern behind the edge of the disk rotating in free space was not previously observed.

A convenient object for studying a spatial structure of flows is stratified fluids whose density is linearly related to the optical refraction coefficient [10]. Shadow visualization of nonstationary flows near the torsional vibration disk showed in its

V. Karev et al. (Eds.): PMMEEP 2017, SPRINGERGEOL, pp. 183–190, 2018.
https://doi.org/10.1007/978-3-319-77788-7_20

neighborhood a presence of internal waves and a toroidal circular vortex with a spiral structure [11–13]. However, the main attention in this series of papers was devoted to a study of internal waves. The evolution of spatial structure of toroidal vortex behind the edge of a uniformly rotating disk was not also considered.

The aim of this paper is an experimental study of the spatial structure of the flow near the rotating disk edge in a stratified medium. Stratification makes a flow pattern near the rotating disk essentially dependent on the orientation of the disk relative to the gravity. The main attention is paid to the analysis of the flow pattern in the case of the disk vertical orientation.

2 Experimental Setup

A disk of diameter D uniformly rotating with angular velocity Ω in the viscous continuously stratified fluid forms a flow. Homogeneous continuous stratification is characterized by the scale $\Lambda = |\rho/\Delta\rho|$, the buoyancy frequency $N = \sqrt{g/\Lambda}$ and buoyancy period $T_b = 2\pi/N$ that serve as a characteristic frequency and time scales. Stratification is created by changing the concentration of a water solution of common salt with the diffusion coefficient $\kappa_S = 1.4 \cdot 10^{-5}\text{cm}^2/\text{s}$ and the kinematic viscosity $\nu = 1.0$ cSt. A uniformly moving body in a continuously stratified fluid generates attached waves with the wavelength $\lambda = U\,T_b = 2\pi U/N$ that also serve as a characteristic scale.

A set of own time and spatial scales determines the size of current observation area that should include the main large-scale components such as internal waves and vortices. The resolution of the measuring instruments must be taken into account when carrying out the experimental procedure.

Experiments were performed at the basin sizes 0.7 × 0.25 × 0.7 m. High-quality optical windows were inserted into the side walls of the basin (see, Fig. 1). Before beginning the experiments, the basin was filled continuously with stratified liquid by the method of continuous displacement [14].

Observations of the flow pattern were carried out using a shadow instrument with a view field of 23 cm. Registration of the flow pattern was carried out using the camera in single-frame shooting and video modes with a frequency of 25 fps and an exposure of up to 1/1000 s. The optical scheme of the experiment provided a spatial resolution up to 0.05 mm.

The value of buoyancy period was determined from the recording of the liquids oscillations created by the immersed salt crystal or a popping gas bubble.

In the experiments we used flat plastic and metal disks with a diameter of 6 to 14 cm located in the center of the basin. The disks were driven by a geared motor which was placed on a special platform installed above the water surface. The movement from the output shaft of the motor was transmitted by steel shafts and bevel gears to the axis of the disk which was set horizontally at the selected depth. The rotational speed of the disc could be continuously adjusted in the range from 0.03 to 2.17 1/ s.

Fig. 1. Photo of the experimental setup.

The design of the drive mechanism made it possible to place the disk at an arbitrary angle to the optical axis of the shadow device. In most experiments, the disk is located so that the axis of rotation and the sight coincide.

After adjusting the shadow device, the pool was filled with continuously stratified fluid with a selected value of the buoyancy period in the range from 4.0 to 27.1 s.

The installation of the drive mechanism and the disk was accompanied by the mixing of the liquid and the generation of non-stationary internal waves. Observations showed that within 1–1.5 h all the disturbances faded and the density distribution became stable. This state was taken as the initial state in all subsequent experiments.

3 Main Results

Examples of flow patterns around a disk that is uniformly rotated counterclockwise and located along the normal to the optical axis of the shadow instrument are shown in Fig. 2. With the beginning of the movement the disk start to drags the liquid and dumps it outside the neutral buoyancy horizon. The growing annular vortex is contoured with a light border on Fig. 2a. Previously performed experiments showed that a double toroidal current forms at the edge of a rotating disk.

Over time, the developing current loses its uniformity and turns into a system of vortex loops with a pitch of 5–11° mean 7° (see Fig. 2b). In the left part of the figure 12 loops are counted from the upper pole of the disk. Gradually stratification suppresses vortex motion and vortex remnants can be seen in the angular sector from

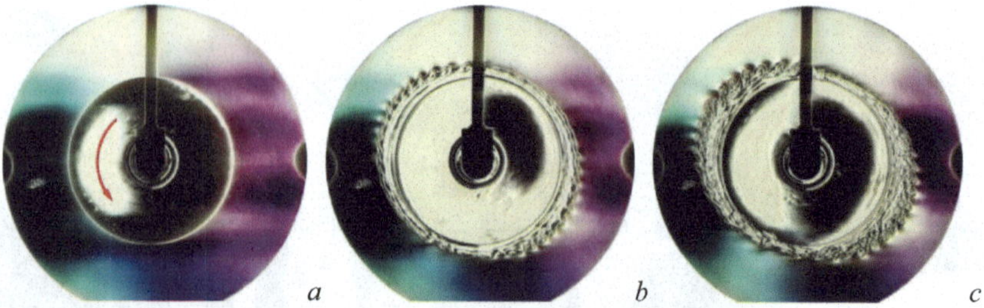

Fig. 2. Dynamics of breaking of a circular edge vortex into a system of vortex loops in a strongly stratified fluid ($D = 12$ cm, $T_b = 8$ s, $\Omega = 18$ rpm): (a, b, c) – $t = 0, 2, 3$ s.

"8" to "6" h. In this case, attention should be paid to the fact that neither the step between the vortices nor even their number is constant in time. In the right part of the figure, the vortex system develops from the lower pole of the disk. 16 vortices are also represented here and transverse structures are also visible in the remaining part of the annular flow. In general, the pattern of flow has an elliptical shape, the minimum size of the region is 14 cm and is oriented at an angle of 139° to the vertical. The maximum size is 15 cm along the line inclined at an angle of 50° (see Fig. 2b). Vortex loops have a pronounced fine structure with the size of individual elements is 0.1 cm. From the outside to the individual compact vortices are adjacent light spots that visualize the field of the forming internal waves.

Over time, the general geometry of the vortex current region changes so the maximum size of which increases and at t = 3 s is equal to 15 cm. The minimum size of the structure is practically unchanged and is 12.5 cm. The angular position of the line of the minimum size of the vortex system tends to shifts to the vertical on Fig. 2c. The vortex structure, expressed practically on the entire left side, is weakened by the action of gravity in the sector with an angular dimension of 92 adjacent to the vertical axis. However, here again, transverse elements of the vortex flow are observed near the edge of the disk. The mean step between these vortices is 12°.

The rate of formation and the size of the region of the vortex structure behind the edge of the rotating disk with constant diameter and stratification value depend substantially on the angular velocity of rotation of the disk (In dimensionless variables from the Reynolds number $\mathrm{Re} = D^2\Omega/\nu$). With slow rotation ($\mathrm{Re} < 2500$), the only expressed structural element is the elliptical high-gradient shell of the outer ring, which is at a distance from the disk edge by a distance of a maximum 1.26 cm. Major axis 10.5 cm inclined at an angle 154° and minor axis inclined at an angle 64°(see Fig. 3a).

The deviation of the shape of the envelope of the circular vortex from the circular vortex is due to the action of the buoyancy forces, which are most clearly manifested in the vicinity of the poles of the disk. In areas where the axial velocity is oriented horizontally, the buoyancy forces do not inhibit the movement of entrained liquid particles, which are able to move as far as possible from the edge of the disk.

Even a small increase in the angular velocity of rotation in about critical conditions is accompanied by a qualitative change in the flow structure. In the regions of maximum deletions of the edge of the annular vortex to the right and left of the disk, eleven

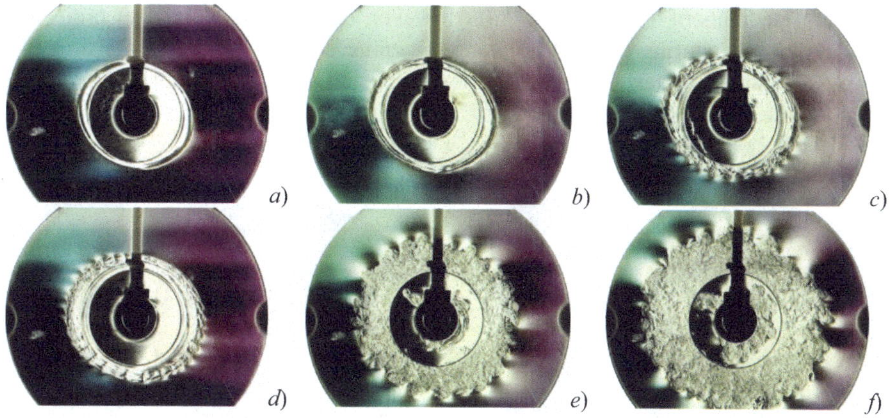

Fig. 3. The structure of the flow at various angular speeds of rotation of the disk ($T_b = 8$ s, $D = 8$ cm, $t = 2$ s, $\tau = 0.25$): (a – f) – $\Omega = 0.35, 0.42, 0.47, 0.57, 1.28, 2.18 s^{-1}$.

thin shells of vortex loops are located at a distance of 0.7–1.0 cm from each other Fig. 3b. Major axis 11.5 cm inclined at an angle 160° and minor axis is 70°. The area of the vortex current region here is noticeably larger than in Fig. 3a. Near the edge of the disc, a shell of a new circular vortex is visible, which slowly advances to the outer edge of the current. Vortex loops are presented here only individual high-gradient filaments.

At an angular velocity of $\Omega = 0.47$, practically the entire annular vortex is covered by vortex loops, each of which is delineated by a pair of vortex fibers elongated at the disk edge in the direction of rotation Fig. 3c. At the same time, the filament structure of both the main annular flow and the transverse vortex loops are retained. Here the size of flow region are 12.5 cm by 10.2 cm. The total number of vortex loops is 29, the distance between them is 0.7–1.6 cm.

The spatial symmetry of the flow begins to be more pronounced with an increase in the angular velocity of $\Omega = 0.57$, when the regular sequence of vortices deforms and partially merges in the first quadrant in Fig. 3d. The dimensions of the flow region in this case reach 12.7 cm by 10.5 cm. The total number of vortex loops 33, the distance between them is 0.5–1.5 cm.

As the angular velocity increases, the flow pattern becomes more and more complicated and homogenized. However, the vortex spots on its periphery remain pronounced sources of short internal waves (see Fig. 3e). The number of distinguishable vortex loops here is 25, the size of the region in this case increase to 14.8 by 13.3 cm.

At large angular velocities of rotation, the individual structural elements do not differ in the annular flow region (see Fig. 3f). However, regular inhomogeneities of the sequence at its outer edge indicate the existence of separate elements in a small-scale vortex flow. The number of distinguishable vortex elements here 23 the dimensions of the area, 17.0 by 15.5 cm.

The stages of breaking up of the annular vortex are shown in Fig. 4 in different projections. Here it can be noted that initially loops are visible in the upper right and lover left quadrants of flow (see Fig. 4a). However, observations in another projection show that the structuring takes place along the entire edge of the disk. Over time, the size of the annular vortex, individual vortex loops and the entire region of the vortex flow increase (see Figs. 4c, d).

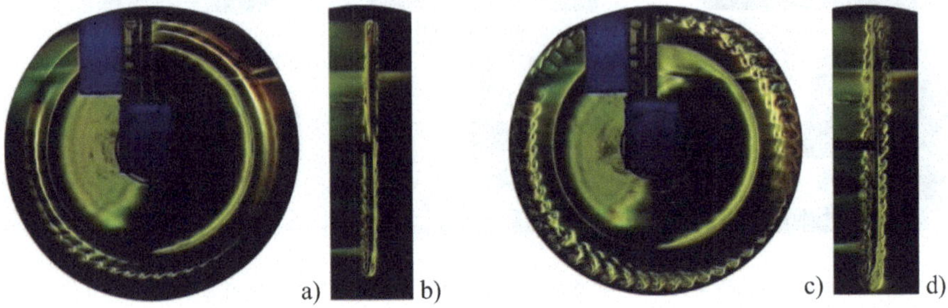

Fig. 4. Annular vortex breaking in different projections ($T_b = 24$ s, $D = 12$ m, $\Omega = 0.3\ s^{-1}$): frontal view $\Theta = 0°$ (*a*, c), side view $\Theta = 90°$(*b*, d); (a, b) – $\tau = 0.067$; (c, *d*) – $\tau = 0.1$.

The evolution of the fine structure of individual vortex loops in the band located to the left of the center of the disk is shown in Fig. 5 (the arrow indicates the direction of rotation of the vortex system). At the initial stage of the formation of vortex loops on an almost annular vortex forming asymmetric wave-like disturbances with a wide depression of 0.7 cm in length and a narrow crest of 0.4 cm appear on the leeward side of which the layer thickens (Fig. 5a). On the vertices on the disc side visible there are weak diffuse inhomogeneities that are the precursors of the vortex loops. The lower edges of the inhomogeneities are connected by segments of highly graded filaments which are bent and even torn in the vicinity of the loops.

Over time, the thickening on the wave-like boundary of the annular vortex with the leeward side of the wave crest widens and transforms into a nucleus of a double helical structure, the formation of which is accompanied by the rupture of the high-gradient envelope of the current (Fig. 3b). The thickness of the boundary between the vortex and the undisturbed medium, as well as the minimum thickness of individual fibers in the vortex loops observed with the further development of the vortex system, is 0.04–0.05 cm. The thinnest structures with a thickness of 0.02–0.03 cm are observed in the cores of double helices formed by the convolution of the windward and leeward parts of the vortex flow boundary.

Over time, the dimensions of the vortex nuclei, the thickness of the vortex loops, whose boundaries, like the binding filaments become more distinct (see Fig. 3c). The outer part of the loops is formed by the shell of the windward part of the disturbance, the inner part of the double helix is the liquid from the leeward side.

A regular system of vortex loops in a circular vortex around a disc rotating in a stratified fluid is visualized in a wide range of angular velocities of the disk rotation.

Vortex loops were observed in fluids with strong and weak stratification.

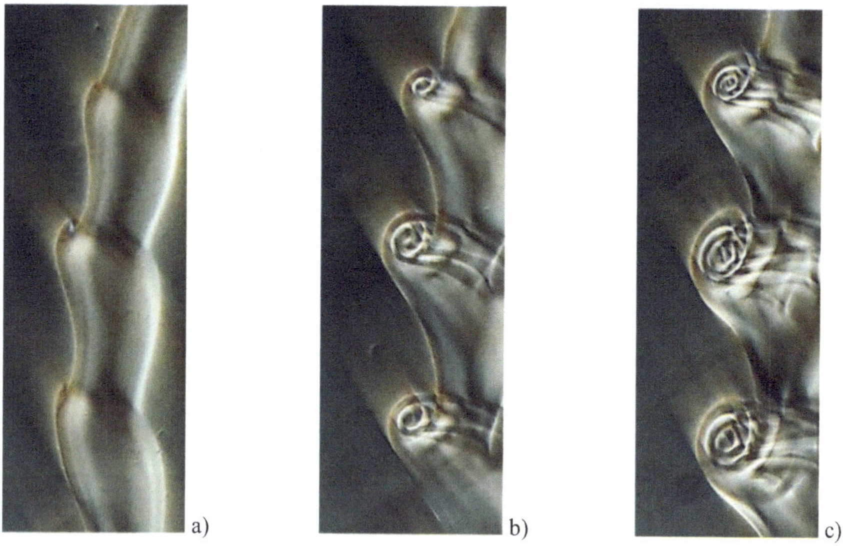

Fig. 5. Vortex loop forming ($R = 6$ cm, $\Omega = 2.2\ s^{-1}$, $T_b = 21$ s, Re = 8000, Fr = 7.3): (a – $в$) – $t = 5.5$, 6.9, 8.5 s.

References

1. von Karman, T.: Uberlaminare und turbulence reibung. ZAMM, No. 1, pp. 52–233 (1921)
2. Landau, L.E.M.: Fluid Mechanics. Course of Theoretical Physics, vol. 6, 2nd edn. Pergamon Press, Nauka (1986)
3. Shahmohamadi, H., MohammadpourA, M.: Series solution for three-dimensional navier-stokes equations of flow near an infinite rotating disk. World J. Mech. **4**, 117–127 (2014)
4. Miklavcic, M.M., Wang, C.Y.: The flow due to a rough rotating disk. JouZ. Angew. Math. Phys. **55**(2), 235–246 (2004)
5. Gregory, N., Stuart, J.T., Walker, W.S.: On the stability of three-dimensional boundary layers with application to the flow due to a rotating disk. Phil. Trans. Roy. Soc. Lond. Ser. A, **248**, 155–199 (1955)
6. Schlichting, H.: Boundary Layer Theory, McGraw-Hill, New York. (1968). Nauka, Moscow (1969)
7. Furukawa, H., Tsutsui, H., Aoi, K., Watanabe, T., Nakamura, I.: Visual study of variation of flow patterns around a disk in a casing; effects of effects of the disk. J. Phys. Conf. Ser. **14**, 220–227 (2005)
8. Imayama, S., Alfredsson, H.P., Lingwood, R.J.: A new way to describe the transition characteristics of a rotating-disk boundary-layer flow. Phys. Fluids **24**, 031701 (2012)
9. Siddiqui, M.E., Mukund, V., Scott, J., Pier, B.: Experimental characterization of transition region in rotating-disk boundary layer. Phys. Fluids **25**, 034102 (2013)
10. Mowbray, D.E.: The use of schlieren and shadowgraph techniques in the study of flow patterns in density stratified liquids. J. Fluid Mech. **27**(Pt. 2), 595–608 (1967)

11. Il'inykh, Y.S., Chashechkin, Y.D.: Generation of periodic motions by a disk performing torsional oscillations in a viscous, continuously stratified fluid. Fluid Dyn. **39**(1), 148–161 (2004)
12. Il'inykh, Y.S., Kistovich, Y.V., Chashechkin, Y.D.: Comparison of the exact solution of a problem of excitation of periodic internal waves with experiment. Izv. Ross. Akad. Nauk. Fizika Atmosfery i Okeana **35**(5), 649–655 (1999)
13. Chashechkin, Y.D., Kistovich, Y.V.,. Il'inykh, Y.S.: Experimental investigation of generation of internal waves by a boundary flow on a rotating disk. Dokl. Ross. Akad. Nauk **375**, 338–342 (2000)
14. Bardakov, R.N., Mitkin, V.V., Chashechkin, Y.D.: Fine structure of a stratified flow over a plate. Zh. Prikl. Mekh. Tekh. Fiz. **48**(6), 77–91 (2007)

Comparison of Empirical Sea-Surface Slopes Probability Densities for the Purposes of Satellite Sounding

Nick Evgenievich Lebedev(✉) and Alexandr Sergeevich Zapevalov

Marine Hydrophysical Institute RAS, Sevastopol 299029, Russian Federation
nick_leb@mail.ru

Abstract. Based on a comparison of Cox-Munk and Bréon-Henriot empirical models of the sea-surface slopes probability density that are more close to the results of satellite observations as compared to other empirical models, an estimate of their practical error for winds equal to 1, 3, 7, 14 m/s is presented. Also, values of the discrepancy between the Cox-Munk model and that frequently used simplified forms are calculated, along with the discrepancy between the Cox-Munk model and its modification that, unlike the original one, is non-negative for all possible slopes and wind speeds. The physical causes of the discrepancy between the Cox-Munk and Bréon-Henriot models typical for small and large wind speeds are considered. It is shown that to reduce the systematic errors of empirical models of the sea-surface slopes probability density, it is necessary to study and take into account the effect of thermal stability of the marine boundary layer on their parameters.

Keywords: Sea surface slopes · Probability density · Error · Sun glint

1 Introduction

Data of the sea surface satellite images are used to determine sea color, amount of phytoplankton, content of water vapor and atmospheric gases, optical properties of atmosphere, sea wind velocity, etc. [1]. Commonly, substantial part of these images is occupied by the zone of sun glint produced by the sea surface mirror-like reflection of solar radiation towards a satellite. For some tasks the sun glint presence is an interfering factor that must be rejected; whereas for others it is informative. In both cases it is more preferable to have that with maximal adequacy. For this it is necessary to know the probability density function of the sea surface slopes P and its possible errors.

At present, about a dozen empirical P models are known based on observations of the reflected solar/lidar or the reflected/refracted laser radiation [2]. These models include the Cox-Munk P_{CM} [3] and the Bréon-Henriot P_{BH} [4] ones.

Based on the results of comparing the model calculations related to the upward radiation of the ocean-atmosphere system with the relevant satellite measurements of this value, it is shown that the P_{CM} and P_{BH} models provide the closest correspondence between model calculations and observational data [5]. In more earlier similar work

V. Karev et al. (Eds.): PMMEEP 2017, SPRINGERGEOL, pp. 191–200, 2018.
https://doi.org/10.1007/978-3-319-77788-7_21

before P_{BH} presentation, P_{CM} was found to be most adequate [6, 7]. Therefore, a quantitative comparison of P_{CM} and P_{BH} can be considered as a practical error of these models. Since the glint component of the upward radiation is directly proportional to the P value, the error of its calculation includes the error of the empirical P model; hence, its values could be applied to estimate errors of calculated glint component of the upward radiation.

The purpose of this paper is to analyze the values of relative differences between the P_{CM} and P_{BH} models for typical values of wind speed W_{10} at standard 10 m level being equal to 1 m/s (near total calm), 3 (weak wind), 7 (close to the average for the Black and Mediterranean Sea), and 14 m/s (maximal value for these models).

2 Cox-Munk and Bréon-Henriot Empirical Models of the Sea-Surface Slopes Probability Density

Observations have revealed that the surface slopes distribution deviates from the Gaussian one. These deviations, which increase as wind increases, effect on the reflecting properties of the ocean-atmosphere boundary [8, 9].

Quasi-Gaussian distributions are usually described by the Gram-Charlier series [10]; the Cox-Munk and Bréon-Henriot models of the sea-surface slopes distribution both have the form of a truncated Gram-Charlier series:

$$P\left(\tilde{\xi}_c, \tilde{\xi}_u\right) = P_G\left(\tilde{\xi}_c, \tilde{\xi}_u\right)\begin{bmatrix} 1 - \frac{1}{2}C_{21}H_2\left(\tilde{\xi}_c\right)H_1\left(\tilde{\xi}_u\right) - \frac{1}{6}C_{03}H_3\left(\tilde{\xi}_u\right) + \\ + \frac{1}{24}C_{40}H_4\left(\tilde{\xi}_c\right) + \frac{1}{24}C_{04}H_4\left(\tilde{\xi}_u\right) + \\ \frac{1}{4}C_{22}H_2\left(\tilde{\xi}_c\right)H_2\left(\tilde{\xi}_u\right) \end{bmatrix}, \quad (1)$$

where $P_G\left(\tilde{\xi}_c, \tilde{\xi}_u\right) = \frac{1}{2\pi\sigma_c\sigma_u}\exp\left(-\frac{\tilde{\xi}_c^2 + \tilde{\xi}_u^2}{2}\right)$

is the two-dimensional Gaussian slopes distribution corresponding to a linear anisotropic wave field, σ_u^2 and σ_c^2 are dispersions of the slopes components along and across the direction of the wind respectively, $\tilde{\xi}_u = \xi_u/\sigma_u$ and $\tilde{\xi}_c = \xi_c/\sigma_c$ are the normalized slopes of the given surface element (facet) along and across the wind; H_i are the i-th order orthogonal Hermite polynomials. The empirical coefficients σ_u^2, σ_c^2, C_{21}, C_{03}, C_{40}, C_{22} and C_{04} of each model are determined by the least-squares method on the differences in the measurements of the upward radiation (with simultaneous wind measurements) and theoretical ones based on the given model P distribution [3, 4].

Summands to one in square brackets are due to existence of various nonlinear mechanisms that lead to deviations in the distributions of sea surface slopes from the Gaussian one. The values of both models parameters are given in Table 1. The amount of observational data forming P_{BH} is approximately three orders of magnitude larger (24000 satellite images from all possible areas of the World Ocean) than those forming P_{CM} (29 aerial photographs near the island Maui, Hawaii). This results in less values of coefficients error estimations.

Table 1. Parameters of P_{CM} [3] and P_{BH} [4] models

Parameters	P_{CM} [3]	P_{BH} [4]
σ_c^2	$0.003 + 0.00192W_{12.5} \pm 0.002$	$0.003 + 0.00185W_{10} \pm 0.0005$
σ_u^2	$0.00316W_{12.5} \pm 0.004$	$0.001 + 0.00316W_{10} \pm 0.0005$
C_{21}	$0.01 - 0.0086W_{12.5} \pm 0.03$	$-0.0009W_{10}^2 \pm 0.01$
C_{03}	$0.04 - 0.033W_{12.5} \pm 0.12$	$-0.45[1 + \exp(7 - W_{10})]^{-1} \pm 0.01$
C_{40}	0.4 ± 0.23	0.3 ± 0.05
C_{22}	0.12 ± 0.06	0.12 ± 0.03
C_{04}	0.23 ± 0.41	0.4 ± 0.1

In the Cox-Munk experiment, wind sensor was located at 12.5 m height, whereas in the Bréon-Henriot measurements, wind values at 10 m height were used. Therefore, for a more correct comparison, the wind values at these altitudes were linked. On the base of [11], this link can be expressed as

$$W_{12.5} = 1.026W_{10} - 0.016.$$

3 Differences Between Cox-Munk and Bréon-Henriot Models

In Fig. 1 the calculated relative differences $r_{BH} = P_{BH}/P_{CM} - 1$ for $W_{10} = 1, 3, 7, 14$ m/s, depending on the components β_u, β_c of the zenith angle of the facet normal β along (β_u) and across (β_c) the wind are shown. Here and after, the horizontal axis, which is an axis of symmetry, is directed along the wind, and the vertical axis is across the wind. Because of the axial symmetry, the sketches for negative vertical components are omitted. The approximately rectangle lines denoted by dots in the figures, delineate the areas where the conditions

$$|\xi_u| \leq 2.5, \quad |\xi_c| \leq 2.5 \tag{2}$$

are satisfied, within which models of the form (1), are adequate [3]. This is due to the approximate nature of such models.

As can be seen, the minimal $|r_G|$ values are located in the central regions of the areas (2) with characteristic values up to 0.05 for W_{10} = 3, 7 m/s and up to 0.1 for W_{10} = 1, 14 m/s. Maximal values are near the edges of the areas (2), up to 0.4–0.7 for W_{10} = 1 m/s and up to 0.25–0.7 for W_{10} = 14 m/s. In general, for the entire areas (2), the $|r_{BH}|$ values are minimal for W_{10}=7 m/s and increase consecutively for W_{10} = 3, 14 and 1 m/s.

The r_{BH} value (as well as values of similar parameters further) for any specific measurement is determined by its projections β_u, β_c, which are fully determined by the geometry of the given measurement, i.e. satellite altitude and mutual location of the sub-sun, sub-satellite and the facet points. A set of expressions to calculate the β_u, β_c values is presented in [4, 11].

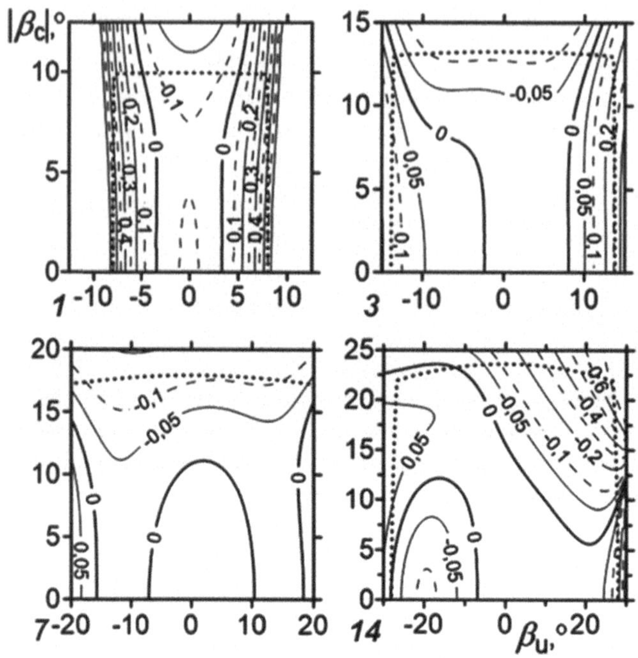

Fig. 1. The r_{BH} values for $W_{10} = 1, 3, 7, 14$ m/s (W_{10} values are marked near the left graph corners), depending on β_u (horizontal axes), $|\beta_c|$ (vertical). Areas within conditions (2) are satisfied, are bordered by dotted lines.

4 Differences Between Cox-Munk Model and Its Simplifications

In the literature, as the probability density function of sea surface slopes, the two-dimensional Gaussian $P_G\left(\tilde{\xi}_c, \tilde{\xi}_u\right)$ and its isotropic variant $P_{G0}\left(\tilde{\xi}\right)$ obtained from the two-dimensional one by substitution σ_c^2 and σ_u^2 on $\sigma^2 = \sigma_c^2 + \sigma_u^2$ and under the assumption $\tilde{\xi}_c = \tilde{\xi}_u = \xi$, are often used. The advantage of the P_G is that it is always positive outside the areas (2), the disadvantage is that it does not take into account the asymmetry of the sea surface slope distribution. Running a little forward, we'll mention that in [12] a combined model joining the merits of P_{CM} and P_G models is presented. It is close to the P_{CM} within the areas (2), and approaches the P_G beyond them.

The values of $r_G = P_G/P_{CM} - 1$ and $r_{G0} = P_{G0}/P_{CM} - 1$ depending on the $\beta_u, |\beta_c|$ are shown in Figs. 2, 3. Here also the minimal $|r_G|$ values are located in the central regions of the areas (2); their characteristic values are up to 0.1 for W_{10} = 1, 3, 7 m/s and up to 0.2 for 14 m/s; maximal values are also about the edges of the areas (2). The $|r_G|$ values are approximately two times higher than $|r_{BH}|$.

The $|r_{G0}|$ values are minimal and about equal to $|r_G|$ for W_{10} = 3 m/s. For W_{10} = 1, 7 14 m/s at whole $|r_{G0}|$ are noticeably higher than $|r_G|$. They are about 0.15–0.3 in the central regions and up to 0.5–1 at the edges of the areas (2). The regions for these

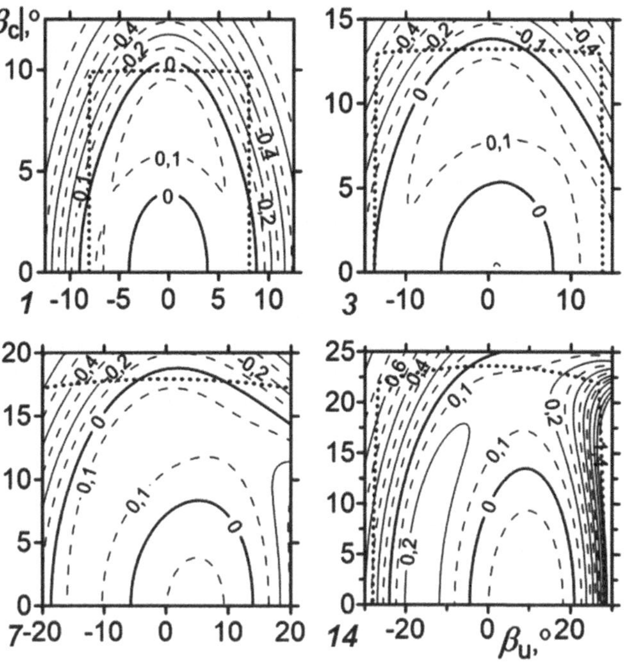

Fig. 2. The r_G values for $W_{10} = 1, 3, 7, 14$ m/s depending on $\beta_u, |\beta_c|$.

winds where $|r_{G0}| > 0.2$ are several times wider than those where $|r_G| > 0.2$. Thus, such simplification may turn too rough. When using it, it is advisable to make sure that the β_u, β_c values do not correspond to large values of the $|r_{G0}|$.

5 Differences Between Cox-Munk Model and Its Modification

Since the P_{CM}, P_{BH} models have the form of a truncated Gram-Charlier series, which is characterized by distortions on the "wings" of the distribution, up to the appearance of negative values [10], then their application to development of satellite data in the full range of angles of sea surface observation can lead to additional errors. In the [12], a $Pm\left(\tilde{\xi}_c, \tilde{\xi}_u\right)$ which is modification of $P\left(\tilde{\xi}_c, \tilde{\xi}_u\right)$ distributions of the form (1) is presented, which removes this problem. Mathematically, this modification is carried out by multiplication of all the terms in square brackets of Eq. (1), depending on $\tilde{\xi}_i$, on $\exp\left(-\left|\tilde{\xi}_i/3.4\right|^3\right)$ factor.

Figure 4 shows the $r_m = Pm_{CM}/P_{CM} - 1$ values. As can be seen, $|r_m|$ values do not exceed 0.02 in most regions of areas (2), which indicate a good proximity of Pm_{CM} to P_{CM} up to the "wings" of the distribution.

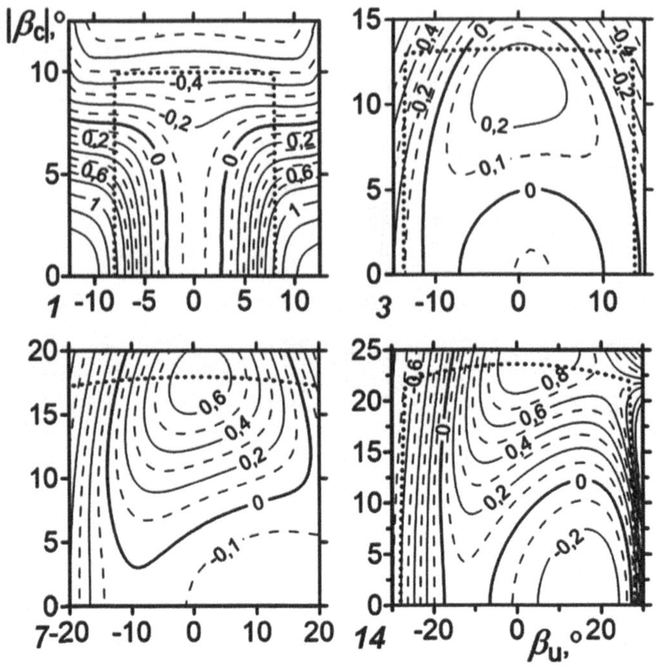

Fig. 3. The r_{G0} values for $W_{10} = 1, 3, 7, 14$ m/s depending on $\beta_u, |\beta_c|$.

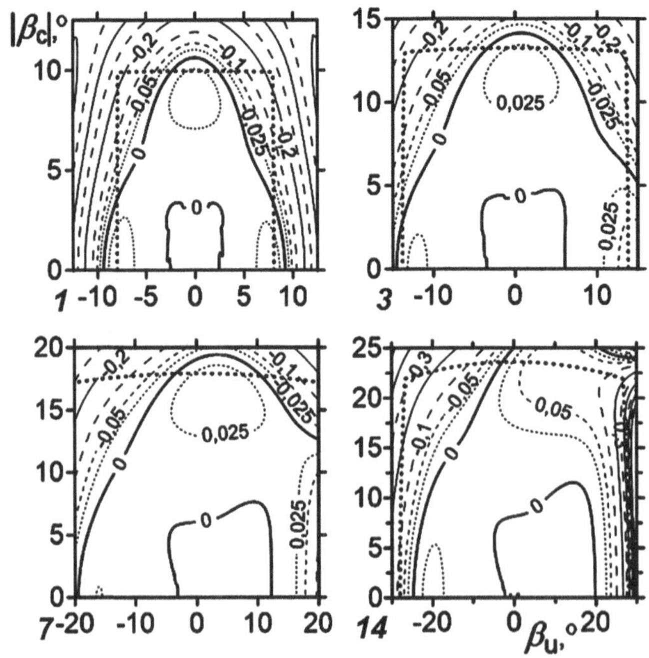

Fig. 4. The r_m values for $W_{10} = 1, 3, 7, 14$ m/s, depending on $\beta_u, |\beta_c|$.

6 Discussion

The authors of $P_{\rm CM}$, $P_{\rm BH}$ models give estimations of errors in their parameters; they are presented in Table 1. These errors characterize the integral proximity of the aggregation of calculated radiation values with the use of these models to reference measurements - a set of experimental radiation values, according to which model parameters were calculated. But these errors cannot fully characterize the errors of the corresponding models. The fact is that for small or large winds there are specific factors that affect the surface roughness or its reflecting properties, and they cannot be sufficiently taken into account upon development of such models. With another set of wind conditions accompanying the reference measurements, both the values of the model parameters and their errors would be different.

In very weak winds, the sea surface roughness is noticeably and uncontrollably affected by the surface currents, internal waves and thin surface films. It is therefore natural that in these cases the relative error of the $P\left(\tilde{\xi}_c, \tilde{\xi}_u\right)$ determination is relatively large; with wind increasing, the relative contribution of these factors to the roughness level decreases.

With wind increasing, a part f of the surface covered with foam increases sharply. The reflecting properties of the foam are about an order higher than of a clean surface. Numerous observations have shown that the factor f is very variable - its estimations made by different authors, are differed by more than an order [13]. Therefore, the magnitude of radiation reflected by the sea surface at high winds can also be very different from measurement to measurement. In addition, with wind increasing, the probability of events when the sub-pixel size cloud fragments will fall into the field of view of satellite device, increases, too. Such situations, if left undetected, lead to an apparent non-controlled increase of radiation reflected by the sea surface. These two mechanisms explain the fact of relatively large values of $|r_{\rm BH}|$ for $W_{10} = 14$ m/s (Fig. 1).

Relatively large $|r_{BH}|$ values at slopes close to their maxima (on the "wings" of the distributions) and for W10 = 14 m/s upon large β_u, $|\beta_c|$ may be due to the fact that the presence of such situations in the development of both models was significantly smaller than of those with moderate slopes of the sea surface.

It is impossible to state that the models cover "from all sides" the true unknown P, since the sources of their methodological errors are mainly the same [2]. In particular, one of them is the neglecting of the effect on the surface roughness of the thermal stability of the sea-air layer. In the work [14] on near-surface laser sounding of sea slopes for winds between 4 and 9 m/s it was shown that with the change in sea-air temperature difference, the change in slopes dispersion can reach 30% and more.

Figure 5 shows the results of $r_{30\%} = P_{\rm CM}\left(1.3\sigma^2_{u,c}\right)/P_{\rm CM}\left(\sigma^2_{u,c}\right) - 1$ calculations for 30% increasing of σ^2_u, σ^2_c. As one can see, for all wind speeds and β_u, β_c these values are noticeably large, reaching the values 0.4 in the central regions of areas (2) and 0.8–2.5 on their periphery. It should be noted [2] that the statistical characteristics of slopes by active methods being made near the surface and by passive ones, at a considerable

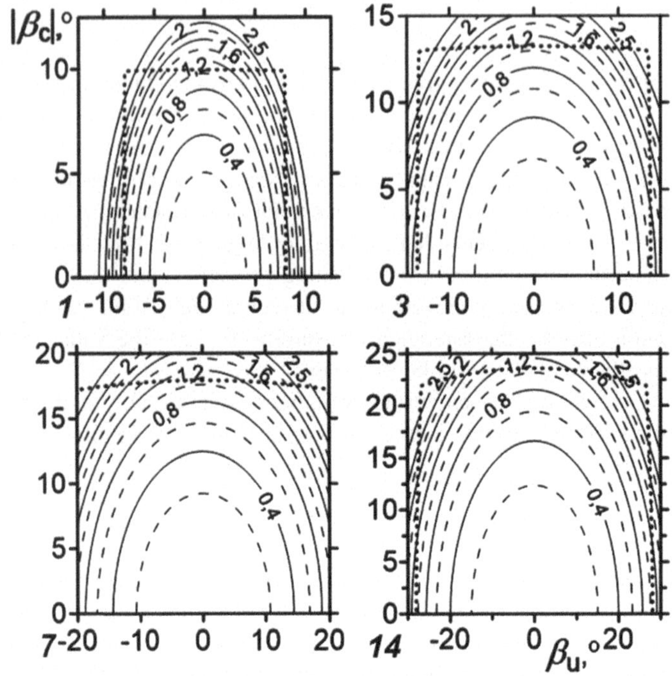

Fig. 5. The $r_{30\%}$ values for $W_{10} = 1, 3, 7, 14$ m/s (left to right), depending on the β_u, $|\beta_c|$.

distance from it, differ significantly and the results obtained in [14] cannot be directly applied for remote sensing purposes. But they show the existence of a factor not taken into account by Cox, Munk and Bréon, Henriot, which can lead to systematic errors of their models.

Therefore, the discrepancies between P_{CM} and P_{BH} represent only a certain part of their practical errors. But at present it is almost impossible to give such estimation in other ways. A more complete understanding of the P_{CM}, P_{BH} errors can be obtained only by comparing a large number of satellite measurements of the upward sea surface radiation with those calculated based on these models over a wide range of winds and with full knowledge of the above factors, which were not taken into account in [3, 4] (sea and air temperatures are among them).

7 Conclusions

- The absolute relative difference $|r_{BH}|$ between the Cox-Munk P_{CM} and Bréon-Henriot P_{BH} models in very weak winds of 1 m/s is relatively large and within the area (2) of adequate P_{CM} and P_{BH} applying it is up to 0.15 when facet slope is across the wind direction ($\beta_u = 0$) and 0.7 - along the wind ($\beta_c = 0$). In central region of the area (2) this difference is up to 0.1.

- With winds ≥ 3 m/s, within the areas (2) $|r_{\mathrm{BH}}|$ is up to 0.1 across the wind ($\beta_u = 0$) and 0.17 along ($\beta_c = 0$), while in the central regions of the areas (2) its characteristic values are about 0.05–0.1. For $W_{10} = 1$ and 14 m/s close to edges of these areas, there are zones of large $|r_{\mathrm{BH}}|$ up to 0.7.
- The absolute relative difference $|r_{\mathrm{G}}|$ between the two-dimensional Gaussian distribution and the P_{CM} is less than 0.15–0.2 in the central parts of the areas (2) and about 0.4–0.6 in their corners. This difference is taking more place in area (2) for $W_{10} = 14$ m/s. In whole $|r_{\mathrm{G}}|$ is about twice higher than $|r_{\mathrm{BH}}|$.
- The absolute relative difference $|r_{\mathrm{G0}}|$ between the isotropic probability density model and the P_{CM} is most significant than $|r_{\mathrm{G}}|$, especially at $W_{10} = 1$ and 14 m/s. This difference is about 0.15–0.3 in the central regions and up to 0.5–1 at the edges of the areas (2). The isotropic model might be used with caution, i.e. in that range of the β_u, β_c angles, where it does not strongly diverge from the P_{CM} model. This caution is also actual in similar situations with big $|r_{\mathrm{G}}|$ and $|r_{\mathrm{BH}}|$.
- The modified Cox-Munk model Pm_{CM} differs from the P_{CM} by less than 0.02 in most regions of areas (2), which indicates a good proximity of Pm_{CM} to P_{CM} up to "wings" of the distribution (where usage of P_{CM} is problematic).
- Substantial systematic errors of P_{CM} and P_{BH} models can be related with initial ignoring of thermal stability effect on sea roughness in [3, 4]. To exclude this kind of errors, it is necessary to study more extensively the relationship between parameters of these models and thermal stability of the marine boundary layer and to introduce it into these models.

The work was done within the framework of the state task on the topic № 0827-2014-0011.

References

1. Barale, V., Gower, J.F.R., Alberotanza, L. (eds.): Oceanography from Space: Revisited, 1st edn. Springer, Netherlands (2010). https://doi.org/10.1007/978-90-481-8681-5
2. Lebedev, N.E., Aleskerova, A.A., Plotnikov, E.V.: The development of optical methods for sea surface slope measurement (In Russian). Sovremennye problemy distantsionnogo zondirovaniya Zemli iz kosmosa **13**(3), 136–149 (2016). https://doi.org/10.21046/2070-7401-2016-13-3-136-149
3. Cox, C., Munk, W.: Measurements of the roughness of the sea surface from photographs of the sun glitter. J. Optical Soc. Am. **44**(11), 838–850 (1954)
4. Bréon, F.M., Henriot, N.: Spaceborne observations of ocean glint reflectance and modeling of wave slope distributions. J. Geoph. Res. Oceans **111**(C6), C06005 (2006). https://doi.org/10.1029/2005JC003343
5. Zhang, H., Wang, M.: Evaluation of sun glint models using MODIS measurements. J. Quant. Spectrosc. Radiat. Transf. **111**, 492–506 (2010). https://doi.org/10.1016/j.jqsrt.2009.10.001
6. Ross, V., Dion, D.: Assessment of sea slope statistical models using a detailed micro-facet BRDF and upwelling radiance measurements. In: Proceedings of the SPIE 5572, Optics in Atmospheric Propagation and Adaptive Systems VII, pp. 112–122 (2004). https://doi.org/10.1117/12.565127

7. Hu, Y., Stamnes, K., Vaughan, M., Pelon, J., Weimer, C., Wu, D., Cisewski, M., Sun, W., Yang, P., Lin, B., Omar, A., Flittner, D., Hostetler, C., Trepte, C., Winker, D., Gibson, G., Santa-Maria, M.: Sea surface wind speed estimation from space-based lidar measurements. Atmos. Chem. Phys. **8**, 3593–3601 (2008). https://doi.org/10.5194/acp-8-3593-2008
8. Liu, Y., Yan, X.-H., Liu, W.T., Hwang, P.A.: The probability density function of ocean surface slopes and its effects on radar backscatter. J. Phys. Oceanogr. **27**, 782–797 (1997). https://doi.org/10.1175/1520-0485(1997)027<0782:tpdfoo>2.0.co;2
9. Zapevalov, A.S.: Probability of mirror reflection glitters during oblique sounding of the sea surface (In Russian). Oceanology **45**(1), 11–15 (2005)
10. Kendall, M.G., Stuart, A.: The Advanced Theory of Statistics: Distribution theory, vol. 1. C. Griffin, London (1977)
11. Smith, S.D.: Coefficients for sea surface wind stress, heat flux, and wind profiles as a function of wind speed and temperature. J. Geophys. Res. **93**(C12), 15467–15472 (1988). https://doi.org/10.1029/JC093iC12p15467
12. Zapevalov, A.S., Lebedev, N.E.: Simulation of statistical characteristics of sea surface during remote optical sensing (In Russian). Atmos. Oceanic Optics **27**(6), 487–492 (2014). https://doi.org/10.1134/S1024856014060220
13. Anguelova, M.D., Webster, F.: Whitecap coverage from satellite measurements: a first step toward modeling the variability of oceanic whitecaps. J. Geophys. Res. **111**(C3), C03017 (2006). https://doi.org/10.1029/2005JC003158
14. Shaw, J.A., Churnside, J.H.: Scanning–laser glint measurements of sea–surface slope statistics. Appl. Optics **36**(18), 4202–4213 (1997). https://doi.org/10.1364/AO.36.004202

Mathematical Modeling of Thermomechanical Behavior of Porous Impermeable Medium with Active Filler

M. V. Alekseev[1], E. B. Savenkov[1](✉), and N. G. Sudobin[2]

[1] Keldysh Institute of Applied Mathematics of Russian Academy of Science, Moscow, Russian Federation
e.savenkov@googlemail.com

[2] Autonomous Non-profit Organization Scientific and Technical Association "ITIN", Moscow, Russian Federation

Abstract. In this paper we consider a self-consistent mathematical model and numerical simulation techniques which are supposed to be suitable to analyze behavior of the impermeable porous media with isolated pores filled with chemically active multiphase multicomponent substance under thermal loads. The porous matrix is described by linear thermomechanical equations. The substance in pores is described by lumped model which includes chemical (pseudo) components mass conservation equations and energy conservation equation for the mixture. Amount of components can change due to chemical reactions induced by heating of the media. Lumped energy balance equations account for matrix/pores heat transfer and heat produced by chemical reactions. Composition of phases is governed by phase equilibrium conditions with an arbitrary number of components and four phases (solid, liquid hydrocarbon, gas and liquid water phases). The two groups of equations (for the matrix and for the pores) are coupled by suitable interface conditions at the "reservoir"/"pore" interfaces.

The purpose of the model is a validation of the basic mechanisms of the formation of connected porosity and permeability in the initially impermeable heavy hydrocarbon reservoirs treated by the modern thermal recovery techniques such as, e.g., *in citu* combustion.

Preliminary numerical results are presented for synthetic but realistic test case.

Keywords: Thermomechanics · Chemical reactions
Phase equilibrium · Pore-scale simulation

1 Thermoelastic Model of Solid Medium

Consider space domain $\tilde{V}$ which consists of impermeable solid matrix V_s and N_v isolated pores V_i, $i = \overline{1, N_v}$, $\tilde{V} = V_s \bigcup \left[\bigcup_{i=1}^{N_v} \right]$. Assume that pore volumes are filled

V. Karev et al. (Eds.): PMMEEP 2017, SPRINGERGEOL, pp. 201–206, 2018.
https://doi.org/10.1007/978-3-319-77788-7_22

with specified amount of chemically active multicomponent substance which can undergo chemical and phase transformations.

The matrix medium V_s is supposed to be described by linear thermomechanical model, see, e.g., [1]:

$$\mathrm{div}\boldsymbol{T} = \rho \boldsymbol{b}, \quad \rho c_\epsilon \frac{\partial T}{\partial t} + \mathrm{div}\boldsymbol{Q} = \rho q, \quad \boldsymbol{x} \in V_s, \tag{1}$$

where ρ is the density, $\boldsymbol{b}$ is external forcing term, $\boldsymbol{T} = \boldsymbol{C} : (\boldsymbol{\epsilon} - \boldsymbol{\epsilon}_T)$ is stress tensor, $\boldsymbol{C}$ — elastic coefficients tensor, $\boldsymbol{\epsilon} = \left[\nabla \otimes \boldsymbol{u} + (\nabla \otimes \boldsymbol{u})^T\right]/2$ — strain tensor, $\boldsymbol{u} = \boldsymbol{u}(\boldsymbol{x})$ — displacement field, $\boldsymbol{\epsilon}_T = \boldsymbol{A}(T - T_0)$ — thermal strain tensor, $\boldsymbol{A}$ — tensor of thermal expansion coefficients, T — temperature, T_0 — reference temperature, $\boldsymbol{Q} = -\boldsymbol{\Lambda} \cdot \nabla T$ — heat flux, $\boldsymbol{\Lambda}$ — thermal conductivity tensor. Boundary conditions for the system of Eq. (1) are imposed on the complete boundary $S_s = \partial V_s$ of the matrix volume V_s, $S_s = S_0 \bigcup \left[\bigcup_{i=1}^{N_v} S_i\right]$, where $S_0 \equiv \tilde{S}$ is a boundary of $\tilde{V}$, $S_i \equiv \partial V_i$.

Boundary conditions at S_0 are defined by thermodynamical and geomechanical conditions in the environment. Boundary conditions on S_i are defined by the PVT state of substance inside pores V_i. For the momentum Eq. (1) they are given by:

$$\boldsymbol{T} \cdot \boldsymbol{n} = P_i \boldsymbol{n}, \quad \boldsymbol{x} \in S_i, i = \overline{1, N_v}, \tag{2}$$

where $\boldsymbol{n}$ is the vector of the external unit normal to S_i, P_i — pressure inside ith pore. The boundary conditions for the energy equation will be considered below.

2 Thermochemical Model for Pores

Consider that the mixture average momentum inside each single pore volume V_i is vanishing and the mixture behavior is described by lumped energy and components mass conservation equations. In the case when thermodynamic equilibrium within the pores is prescribed, the aforementioned lumped equations are accompanied by algebraic phase equilibrium conditions. It is assumed that components are distributed among three mobile fluid phases (liquid hydrocarbon phase, gas and liquid water phases denoted by Greek indices $\alpha = W, L, G$, respectively) and one immobile solid phase. The mobile components are assumed to exist in all fluid phases and not in the solid phase. Components in the solid phase are assumed to do so exclusively. Amount of the components can vary due to chemical reactions between components in pores. Mixture energy changes due to heat transport from matrix to pore volume and due to chemical reactions. The inflow of mass across the boundary of the pores is not considered, i.e., the matrix is supposed to be chemically inert.

Since pore volumes V_i are isolated to each other it is enough to consider the case of a single pore with volume V. By assumption V is occupied by the solid and three fluid spaces, hence $V = V_S + V_W + V_L + V_G = V_S + V_f$. Define fluid porosity (i.e., volume concentrations of the mobile phases) as $\phi_f = V_f/V$. This allows to define phase saturations $S_\alpha = V_\alpha/V_f$, $\alpha = W, L, G$, and $\tilde{S}_S = V_S/V = 1 - \phi_f$, $\tilde{S}_f = 1 - \tilde{S}_s$. The saturations of the mobile components can then be re-defined relatively to the total volume V as $\tilde{S}_\alpha = V_\alpha/V = S_\alpha\phi_f$.

Let n_i be the mole number of ith component, $i = \overline{1, N}$; n_α — phase mole number, $n_{i\alpha}$ — ith component mole number in phase α. Then the molar concentration of component i in phase α is given by $x_{i\alpha} = n_{i\alpha}/n_\alpha$ if $n_\alpha > 0$, molar concentration of mobile phase is given by $C_\alpha = n_\alpha/n_f$. Molar density of ith component of the mobile phase is $m_i = n_i/V_f$, $i = \overline{1, N_f}$, and of the solid phase — $m_i = n_i/V$, $i = \overline{N_f + 1, N}$. Molar density of phase is defined as $b_\alpha = n_\alpha/V_\alpha$.

The saturation of mobile phases are expressed as $S_\alpha = m_{\text{tot}}C_\alpha/b_\alpha$, $\alpha = W, L, G$ which leads to

$$m_i = \sum_{\alpha=W,L,G} b_\alpha S_\alpha C_\alpha x_{i\alpha}, \quad i = \overline{1, N_f}; \quad m_i = b_S x_{iS}\tilde{S}_S, \quad i = \overline{N_f + 1, N}. \tag{3}$$

Thermodynamic properties of phases and its component composition can be defined from equation of state for phases and components. Thermodynamic equilibrium calculation is complicated but rather standard issue and many efficient algorithms are available [2].

Finally, the lumped components and energy conservation equations are given by

$$\frac{d}{dt}(\phi_f m_i) = Q_i^{(m)}, \; i = \overline{1, N_f}; \; \frac{d}{dt}(m_i) = Q_i^{(m)}, \; i = \overline{N_f + 1, N}; \tag{4}$$

$$\frac{d}{dt}E = Q_E, \; E = \sum_{\alpha=W,L,G} b_\alpha S_\alpha U_\alpha + b_S U_S, \tag{5}$$

where m_i are expressed by (3), $Q_i^{(m)}$ is rate of change in ith component due to chemical reactions, U_α — molar energy densities of phases, $Q_E = Q_E^{\text{ext}} + Q_E^{\text{ch}}$, Q_E^{ch} — energy inflow due to chemical reactions, Q_E^{ext} — inflow of energy from the enclosing volume V of the medium. All chemical reactions are assumed to be kinetically driven and are governed by Arrhenius rate relations with parameters dependent on the component concentrations in a specific phase and PVT conditions in pore.

3 Matrix/Pore Interface Conditions

Let the temperature distribution in the matrix is $T = T(\boldsymbol{x}, t)$, the temperature in the pore is constant in space and is given by $T_f = T_f(t)$. Let generic pore occupies volume V with boundary ∂V. At each point of ∂V the flux $Q_s \equiv Q_{s\to f}$ from matrix to pore and *vice versa*, $Q_f \equiv Q_{f\to s}$, is defined. From energy conservation considerations it follows that at each $\boldsymbol{x} \in \partial V_f$ it holds $Q_{s\to f} + Q_{f\to s} = 0$. Consider $Q_s(\boldsymbol{x}) = -\boldsymbol{\Lambda}\partial T/\partial \boldsymbol{n}_s(\boldsymbol{x})$, $Q_f(\boldsymbol{x}) = -\alpha(T(\boldsymbol{x}) - T_f)$, where α is a heat transfer coefficient to obtain $-\boldsymbol{\Lambda}\cdot\partial T/\partial \boldsymbol{n}_s - \alpha(T - T_f) = 0$, from where

$$\bar{Q}_s = \frac{1}{|\partial V|}\int_{\partial V} -\boldsymbol{\Lambda}\frac{\partial T}{\partial \boldsymbol{n}_s}\, dV_{\boldsymbol{x}}, \; \bar{T} = \frac{1}{|\partial V|}\int_{\partial V} T(\boldsymbol{x})\, dV_{\boldsymbol{x}}, \; \bar{Q}_f = \alpha(\bar{T} - T_f). \tag{6}$$

The last equation defines a flux from matrix to pore using surface average values, that is, $Q_E^{\rm ext} = \bar{Q}_f|\partial V|$ in (5). Together with the local equations of energy balance in the matrix and pore, the relation (6) provides *global* energy balance in the system.

4 Algorithm for the Numerical Solution of the Problem

An outline of the numerical solution algorithm for the described model is presented below for the case of a single pore.

Consider displacements $\boldsymbol{u}(\boldsymbol{x},t)$, temperatures $T(\boldsymbol{x},t)$, $T_p(t)$, pore pressure $P(t)$ and volume $V(t)$, mass of components inside pore m_i, $i = \overline{1, N_c}$, volume concentrations $\tilde{S}_s$, S_α, $\alpha = W, L, G$ and molar concentrations $x_{i\alpha}$ of components in phases as primary unknowns.

Denote Δt as the time step, $f = f(t)$, $\hat{f} = f(t+\Delta t)$ for generic time-dependent quantity f. Let all primary unknowns are known at time t. To advance primary fields from time t to time $t+\Delta t$ the following sequence of steps is used [3]:

1. Compute temperature field $\hat{T}$ in the matrix integrating heat Eq. (1) on the interval $[t, t+\Delta t]$ with a given pore temperature T_p.
2. Compute displacements $\hat{\boldsymbol{u}}$ as a solution of the elasticity Eq. (1) for given $\hat{T}$ and P.
3. Update flux $\bar{Q}_s$ from matrix to pore at given temperature distribution $\hat{T}$ in matrix using (6).
4. Compute pore volume $\hat{V}$ using displacements $\hat{\boldsymbol{u}}$ obtained in step 2
5. Compute the amount of components in the pore integrating Eq. (4) from t to $t+\Delta t$ at the given T_p.
6. Update pore energy by numerical integration of the Eq. (4) on the interval $[t, t+\Delta]$ with given PVT parameters in the pore computed at the previous time step t.
7. Compute the temperature in the pore and pressure inside it, saturation of phases and their component composition using pore energy, volume and substance amount at time step $t+\Delta t$ obtained at the previous step.

To simulate thermomechanical behavior of the matrix the standard finite element method utilizing equidistant rectilinear cartesian (i.e., the "voxel") FE meshes was used, which makes it possible to analyze μCT images of the rock samples. Phase equilibrium computations naturally suitable for the described algorithm relies on the UV-flash (i.e., phase equilibrium solution procedure with prescribed internal energy of the mixture and its volume).

5 Numerical Results

The domain of the interest is "synthetic" micro-specimen of core sample which is a cube of the rock with an edge of 1 mm with three spherical pores with radii 0.125 mm containing kerogen (Fig. 1, top left). The initial temperature of

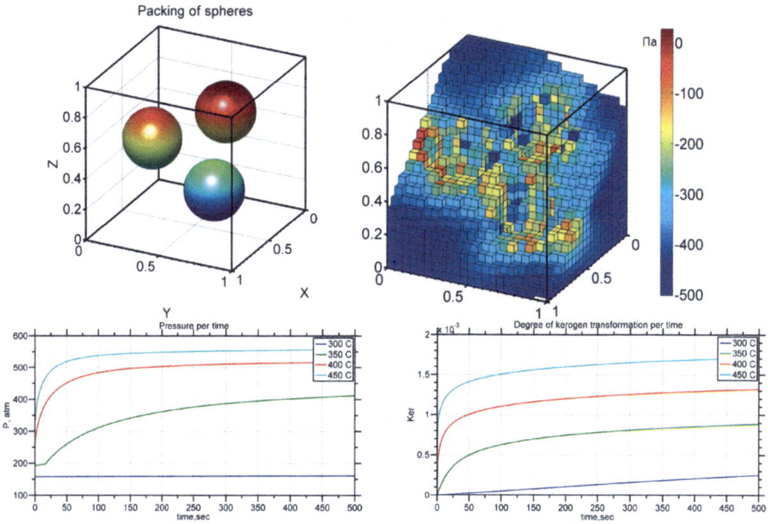

Fig. 1. Pore geometry (top left), maximal principal stresses (top right), pore pressure (bottom left) and kerogen decomposition degree (bottom right) vs. time.

the rock and the component composition of the kerogen-containing pores were set as the initial conditions. The model system of chemical reaction consists of 4 reactions for 9 components [4]: kerogen, asphaltene, resins, saturated paraffins, residual coke, light oil, water, aromatic hydrocarbons and carbon dioxide. Maximal stress field distributions is shown on Fig. 1 (top right). Temporal dynamics of pore pressure and kerogen decomposition degree (for one pore) are shown on Fig. 1 (bottom left and right, respectively).

Simulation results clearly demonstrates that if the initial temperature of the specimen is sufficiently high, the thermal destruction (i.e., kerogen pyrolysis) reactions in the pores are activated. The resulting pressure in the pores is sufficiently high to produce stresses in the solid matrix large enough to induce matrix damage and further micro-scale fracturing. As a result, pores may become hydrodynamically connected, i.e., the formation of permeability occurs.

Acknowledgement. The work was supported by the Russian Foundation for basic research, project No. 16-29-15078 ofi m.

References

1. Ziegler, H.: An Introduction to Thermomechanics. North-Holland Publishing Company, Elsevier, Burlington (1983)
2. Reid, R.C., Prausnitz, J.M., Sherwood, T.K.: The Properties of Liquids and Gases. McGraw-Hill, New York (1977)

3. Alekseev, M.V., Kuleshov, A.A., Savenkov, E.B.: A mathematical model for impermeable porous media under thermomechanical loads. Preprint of KIAM RAS, No. 35, Moscow (2017). https://doi.org/10.20948/prepr-2017-35. In Russian
4. Alekseev, M.V., Kuleshov, A.A., Savenkov, E.B.: Thermomechanical model for impermeable porous medium with chemically active filler. In: Mathematical Models and Computer Simulations, vol. 29, No. 12 (2017). Accepted

Evaluation of the Temporal Dynamics of Oceanic Eddies with Initial Peripheral Rate Shift

Alexander Alexeyevich Solovyev[1] and Dmitry Alexandrovich Solovyev[2(✉)]

[1] Faculty of Geography, M.V.Lomonosov Moscow State University, 1, Lenin Hills, Moscow 119991, Russia
[2] Shirshov Institute of Oceanology, Russian Academy of Sciences, 36 Nakhimovsky ave, Moscow 117858, Russia
solovev@ocean.ru

Abstract. We consider the problem of the development of two-dimensional vortex with initial vorticity in the peripheral ring area of the circularly rotating incompressible viscous fluid. As a mechanism for the generation of vortex motion, we propose the model of plain circular rotation of incompressible viscous fluid with initial perturbation localized at the peripheral ring zone with rate shift. Based on the analytical solution axisymmetric nonstationary equation of plain evolution of vortex in viscous fluid we obtained the resulting integral for the vorticity in the whole area of rotating fluid. This allowed us to study the dynamic characteristics of oceanic vortex formations and the time duration of the vortex motion conservation in a viscous liquid. The presented calculations given here show that the process of vorticity generation is characterized by the complex of interacting movements from the main vortex in the center of the rotating system and from the peripheral secondary perturbations. Vortex formation occurs in such direction that the central vorticity intensifies due to the secondary perturbations, as a result forming two areas – vortex and potential – of the rotating system. The obtained calculation results for the vorticity generation time in rings forming at the meanders of Gulf Stream jets, when compared to the observational data, show that lifespan of rings of different scales distributes rather variously according to their sizes and viscosity, the value of which correspond to turbulent regimes of jet streams. We discuss the prospect of its possible applications for predictive analysis of the development dynamics of large-scale ocean eddies.

Keywords: Eddy growth · Vortex line stretching · Finite-time singularities

1 Introduction

There are still some problems in the theory of ocean jet stream circulation concerning detection of the influence of initial vorticity sources on the development dynamics of large-scale eddies [1]. There are alternative approaches to determine the cause of the ocean eddies (rings). Some studies associate the formation of large-scale eddies (rings) with baroclinic instability of the fluid stratified by density [2–4]. Barotropic instability

V. Karev et al. (Eds.): PMMEEP 2017, SPRINGERGEOL, pp. 207–216, 2018.
https://doi.org/10.1007/978-3-319-77788-7_23

is also considered to be possible cause of the formation of vortices in the jet currents [5]. However, the development of vortices initiated by the barotropic vorticity localized in the center of the circular area of rotating fluid, as shown in [6, 7], may result only in diffusion and their rapid disappearance due to the viscosity of fluid. Further research of two-dimensional vortex structures showed the necessity of considering the possibility of the effect of the initial vorticity localized at the edges of streams with horizontal rate shifts on the generation and long-term evolution of large-scale vortices [8]. Moreover, such study seems also appropriate to describe temporal dynamics of the development of vortex rings in the system of horizontal rate shifts at the edge of oceanic jet streams [7, 9–13]. In light of this, here we consider the problem of two-dimensional vortex with vorticity localized in the peripheral ring area of the circularly rotating incompressible viscous fluid. The goal of our study is to deduce an analytic solution of two-dimensional vortex equation in the circle of rotating incompressible viscous fluid with initial peripheral horizontal rate shift. We discuss the prospect of the possible application of solutions achieved for predictive analysis of the development dynamics of ocean eddies forming in the ocean jet streams.

2 Analytical Vortex Model

As a mechanism for the generation of vortex motion we propose the model of plain circular rotation of incompressible viscous fluid with initial perturbation localized at the peripheral ring zone with rate shift. To determine how perturbation stimulated by vorticity generated in some specific area of rotating fluid affects the flow of surrounding fluid in the course of time, we use the axisymmetric nonstationary equation of plain evolution of vortex ω in viscous fluid.

$$\frac{\partial \omega}{\partial t} = \nu\left(\frac{\partial^2 \omega}{\partial r^2} + \frac{1}{r}\frac{\partial \omega}{\partial r}\right). \tag{1}$$

And boundary conditions in the following form:

$$t = 0\begin{cases} \omega_0 = 0 & 0 \le \mathrm{r} < a \\ \omega_0 = const & a \le r \le b, \\ \omega_0 = 0 & r > b. \end{cases} \tag{2}$$

where ω_0 is constant vortex value at the circle periphery in the ring zone with radius r from $r = a$ to $r = b$.

Separate the variables and integrate to obtain the general equation integral (1) with the integration constant *C(k)* in the following form:

$$\omega = \int_0^\infty \mathrm{C}(k) e^{-k^2 \nu t} J_0(kr) dk. \tag{3}$$

Rewrite (3) by replacing *C(k)* function with $k \cdot A(k)$, then:

$$\omega = \int_0^\infty A(k) \mathrm{e}^{-k^2 \nu t} \mathrm{J}_0(kr) k dk. \tag{4}$$

Assume that the value of initial peripheral vortex perturbation is the function of radial coordinate, i.e. $\omega = f(r)$, then, acc. to (4)

$$f(r) = \int_0^\infty A(k) \mathrm{J}_0(kr) k dk. \tag{5}$$

Using Hankel's inverse dependence formula [14] we can rewrite the last equation as follows:

$$A(k) = \int_0^\infty f(\alpha) \mathrm{J}_0(kr) \alpha d\alpha. \tag{6}$$

As a result, the general integral (4) reduces to the following form:

$$\omega = \int_0^\infty e^{-k^2 \nu t} J_0(kr) k dk \int_0^\infty f(\alpha) J_0(k\alpha) d\alpha. \tag{7}$$

Introduce the initial vorticity conditions instead of $f(a)$ into the integral (7) in the form of rectangle with height A and width $(b-a)$ (Fig. 1.):

$$\omega = \mathrm{A} \int_0^\infty e^{-k^2 \nu t} J_0(kr) k dk \int_0^\infty J_0(k\alpha) d\alpha. \tag{8}$$

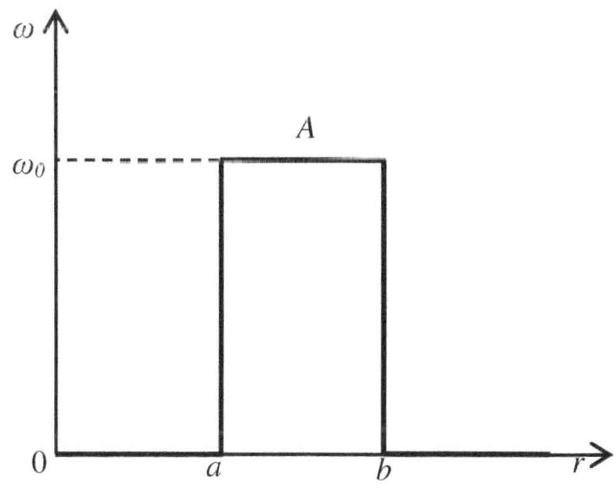

Fig. 1. Schematic drawing of initial vorticity conditions in the form of rectangle with height A and width $(b - a)$.

Based on the known recurrent equations, obtain:

$$z^n \mathrm{J}_n(z) = \int z^n \mathrm{J}_{n-1}(z)dz; \tag{9}$$

$$\int_a^b \mathrm{J}_0(k\alpha)\alpha d\alpha = \frac{b}{k}\mathrm{J}_1(kb) - \frac{a}{k}\mathrm{J}_1(ka). \tag{10}$$

Integral (8) may be equated as follows:

$$\omega = \mathrm{A}\cdot b\int_0^\infty \mathrm{e}^{-k^2 vt}\mathrm{J}_0(kr)\mathrm{J}_1(kb)dk - \mathrm{A}\cdot a\int_0^\infty \mathrm{e}^{-k^2 vt}\mathrm{J}_0(kr)\mathrm{J}_1(ka)dk = S_1 - S_2 \tag{11}$$

For further development of integrals in (11) use the following equation:

$$J_0(kz_1)J_1(kz_2) = \frac{1}{\pi}\int_0^\pi J_1(ky)\frac{z_2 - z_1\cos\theta}{y}d\theta. \tag{12}$$

The first integral (11) is developed as follows:

$$S_1 = \mathrm{A}\cdot b\int_0^\infty e^{-k^2 vt}J_0(kr)J_1(kb)dk = \frac{Ab}{\pi}\int_0^\pi \frac{b - r\cos\theta}{R}d\theta\int_0^\infty e^{-k^2 vt}J_1(kR)dk, \tag{13}$$

where $R = b^2 + r^2 - 2rb\cos\theta$. Hankel integral equals $\int_0^\infty e^{-k^2 vt}J_1(kR)dk = \frac{1-e^{-\frac{R^2}{4vt}}}{R}$, then (13) $S_1 = \frac{Ab}{\pi}\int\limits_0^\pi\left(\frac{1-e^{-\frac{R^2}{4vt}}}{R}\right)(\frac{b-r\cos\theta}{R})d\theta$. Now we can equate the integral (13) as two integral equations:

$$S_1 = S_{11}' - S_{11}'' = \frac{Ab}{\pi}\int_0^\pi (\frac{b - r\cos\theta}{R^2})\,d\,\theta - \frac{Ab}{\pi}\int_0^\pi e^{-\frac{R^2}{4vt}}\frac{b - r\cos\theta}{R^2}d\theta. \tag{14}$$

After integrating two terms included in S_1, by using Hansen formula [15] we obtain as follows:

$$S_1(b) = \omega_0\left[1 - \mathrm{e}^{-\frac{b^2+r^2}{4vt}}\sum_{m=0}^{\infty}(\frac{r}{b})^m J_m\left(\frac{br}{2vt}\right)\right]. \tag{15}$$

The first integral easily integrates while the second one can be calculated using Hansen formula. Finally obtain for $r=b$:

$$S_1(b) = \omega_0 \left[1 - \mathrm{e}^{-\frac{b^2+r^2}{4vt}} \sum_{m=0}^{\infty} \left(\frac{r}{b}\right)^m J_m\left(\frac{br}{2vt}\right) \right]. \tag{16}$$

The second integral of Eq. (7) after the integration for $r=b$ can be put in the following form:

$$S_2(b) = \omega_0 \mathrm{e}^{-\frac{b^2+r^2}{4vt}} \sum_{m=0}^{\infty} \left(\frac{r}{b}\right)^m J_m\left(\frac{br}{2vt}\right). \tag{17}$$

Similar expressions $S_1(a), S_2(a)$ obtain from (16) and (17) by changing the variable.

$$S_1(a) = \omega_0 \left[1 - \mathrm{e}^{-\frac{a^2+r^2}{4vt}} \sum_{m=0}^{\infty} \left(\frac{r}{a}\right)^m J_m\left(\frac{ar}{2vt}\right) \right], S_2(a) = \omega_0 \mathrm{e}^{-\frac{a^2+r^2}{4vt}} \sum_{m=0}^{\infty} \left(\frac{r}{a}\right)^m J_m\left(\frac{ar}{2vt}\right).$$

The resulting integral (11) for the vorticity in the whole area of rotating fluid has the following form:

$$\begin{aligned}
\omega &= \omega_0 \left[1 - e^{-\frac{a^2+r^2}{4vt}} \sum_{m=0}^{\infty} \left(\tfrac{r}{a}\right)^m J_m\left(\tfrac{ar}{2vt}\right) \right] - \omega_0 \left[1 - e^{-\frac{b^2+r^2}{4vt}} \sum_{m=0}^{\infty} \left(\tfrac{r}{b}\right)^m J_m\left(\tfrac{br}{2vt}\right) \right] && r<a, \\
\omega &= \omega_0 \left[1 - e^{-\frac{a^2+r^2}{4vt}} \sum_{m=0}^{\infty} \left(\tfrac{r}{a}\right)^m J_m\left(\tfrac{ar}{2vt}\right) \right] - \omega_0 e^{-\frac{b^2+r^2}{4vt}} \sum_{m=0}^{\infty} \left(\tfrac{r}{b}\right)^m J_m\left(\tfrac{br}{2vt}\right) && a \le r \le b, \\
\omega &= \omega_0 e^{-\frac{a^2+r^2}{4vt}} \sum_{m=0}^{\infty} \left(\tfrac{r}{a}\right)^m J_m\left(\tfrac{ar}{2vt}\right) - \omega_0 e^{-\frac{b^2+r^2}{4vt}} \sum_{m=0}^{\infty} \left(\tfrac{r}{b}\right)^m J_m\left(\tfrac{br}{2vt}\right) && r > b.
\end{aligned} \tag{18}$$

3 Discussion

We use the obtained analytical Eq. (18) to answer the questions about the dynamics of vortex generation and the duration of vortex movement in the viscous fluid. Calculate the distribution of the relative vorticity of vortex ω/ω_0 by radius r/a, normalized to the border of the peripheral ring vorticity area, for different times $t* = \frac{a^2}{nv}$. Calculations are performed with $a^2/v = 1$, n = 12, 8, 4, 2. As follows from Fig. 2, the vorticity of the initial vortex perturbation ω_0 limited by ring area $1 \le r/a \le 1,4$ first moves during a certain time towards the vortex rotation center and then diffuses towards the periphery with simultaneous weakening along the whole circle of rotation.

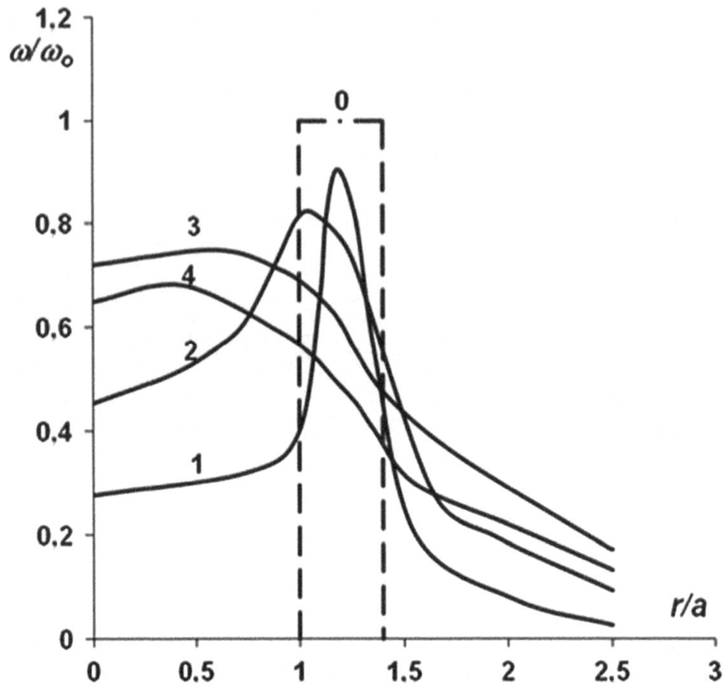

Fig. 2. Space-time dynamics of the vorticity development after the initial shift perturbation of the vortex in the ring area of the rotating fluid with $0 - t* = 0$; 1 − 1/12; 2 − 1/8; 3 − 1/4; 4 − 1/2.

The following expression is the measure of intensification and weakening of the vortex:

$$\frac{\omega}{\omega_0} = 1 + \mathrm{e}^{-\frac{b^2}{2\nu t}} - e^{-\frac{a^2}{2\nu t}}. \tag{19}$$

To illustrate the process of vortex evolution the timespans of the perturbations development can be expressed, for example, using the following equation: $\frac{\tau_b}{\tau_a} = \frac{b^2}{a^2} = 2$. Here we obtain the following value of the width of the ring zone of the initial vortex perturbation from which the vorticity will be distributed at the particular moment in time:

$$\frac{b - a}{a} = \left(\sqrt{2} - 1\right) = 0,4. \tag{20}$$

Calculations given here show that the process of vorticity generation is characterized by the complex of interacting movements from the main vortex in the center of the rotating system and from the peripheral secondary perturbations. Vortex formation occurs in such direction that the central vorticity intensifies due to the secondary perturbations, as a result forming two areas – vortex and potential – of the rotating system.

According to the calculations by Eq. (18), the dynamics of the variation in time of relative vorticity $\omega^* = \omega/\omega_o$ of different areas of the rotating liquid from center to border of the ring area $0 \le r/a < 1$ and then beyond the shift area border at $r^* = r/a > 1,4$ is characterized by the vortex intensification achieving a stable maximum at t* = 1/4 (Fig. 3).

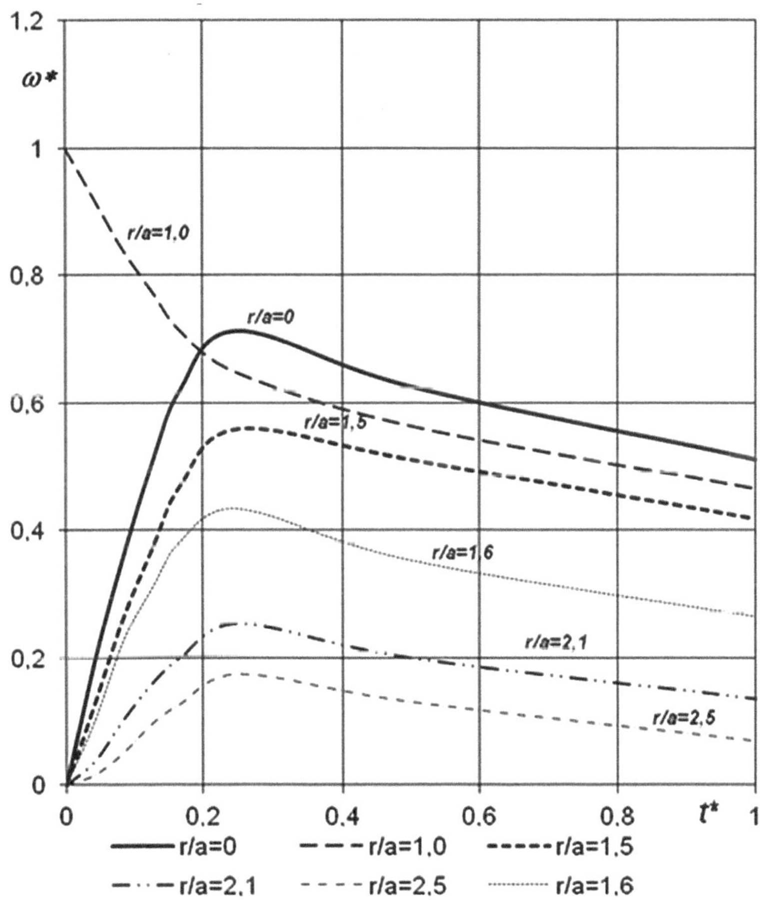

Fig. 3. Variation in time of the vorticity in different areas of the rotating viscous fluid before and after the initial vortex perturbation at the border at $r^* < 1$; $r > 1,4$.

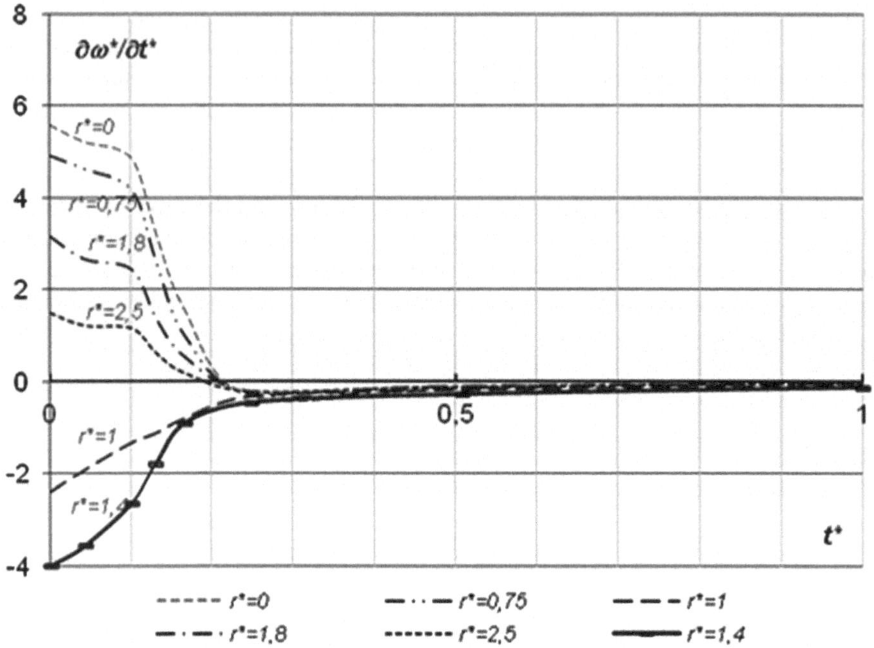

Fig. 4. Distribution rate of the relative vorticity $\omega* = \omega/\omega_0$ in time $t*$ normalized at the moment of shift effect for different distances from the border of initial vortex perturbation.

Due to monotonous in time viscous dissipation, in the center the intensity of the vortex increases to the maximum values and decreases after that. At the same time the vorticity increases with lower intensity beyond the ring area, with the transition to systematic viscous dissipation. This is a very important result, as it shows the role of the initial source of the vorticity forming in the streams with rate shift.

Vortex rate $\frac{\partial\omega*}{\partial t*}$ distributes both sides from the ring area, and the rate decreases unequally in time. In the direction from the ring zone inside the vortex the rate decrease is small, while in the outside direction, according to Eq. (18), the rate changes in time with significant negative acceleration (Fig. 4).

The fact that vorticity after the perturbation distributes due to horizontal advection allows for realistic predictive estimates of the lifespan of oceanic vortex rings (Fig. 5).

Calculation results for the vorticity generation time in rings forming at the meanders of Gulf Stream jets, when compared to the observational data, show that lifespan of rings of different scales distributes rather variously according to their sizes and viscosity, the value of which correspond to turbulent regimes of jet streams.

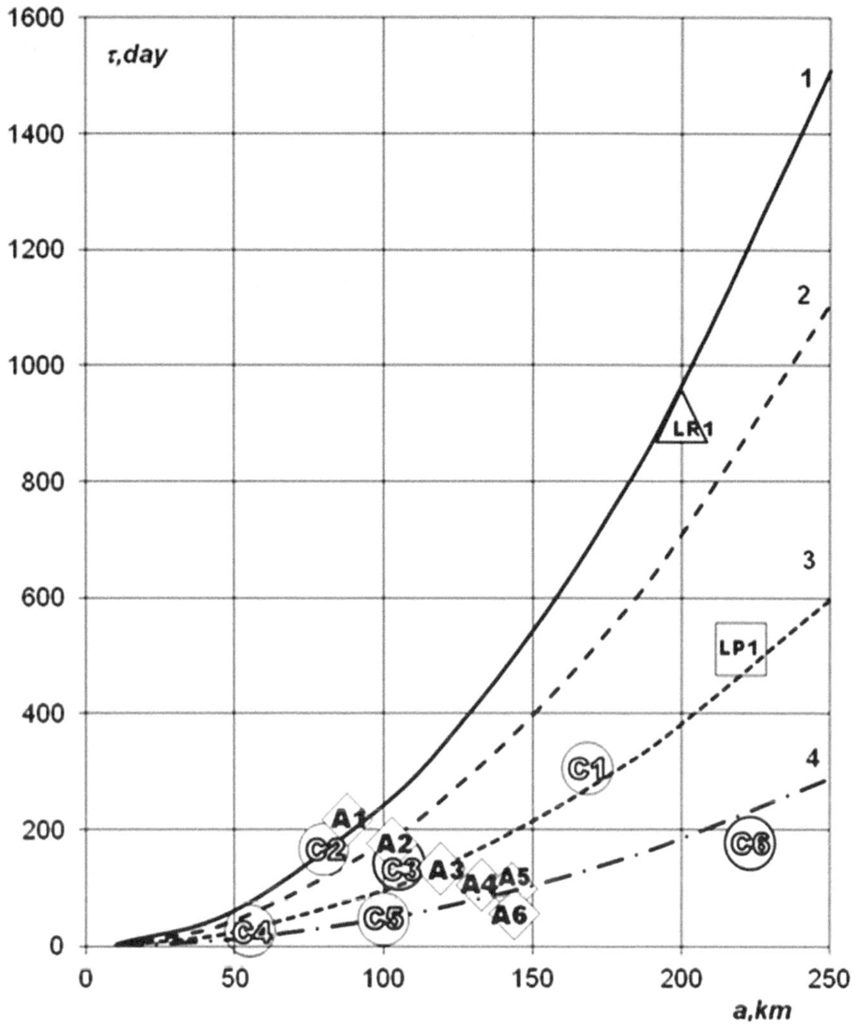

Fig. 5. Lifespans of ocean rings with different diameters a: C1, C2, C3, C4, C5, C6 - [16, 17]; A1, A2, A3, A4, A5, A6 - [18]; LR1 - [19]; LP1 - [20]. Curves No. 1–4 are calculated acc. to Eq. (18) with viscosity ν: 1–20 m^2/s; 2–27 m^2/s; 3–51 m^2/s; 4–105 m^2/s.

4 Conclusion

Observational data and calculation results show the possibility of such processes of vortex formation in which the horizontal advection forming at rate shifts prevents the dissipative distribution of the vorticity over the whole area of rotation due to diffusion.

Acknowledgment. The work was supported by the Ministry of Education and Science of the Russian Federation (Agreement No. 14.616.21.0035, unique identifier of the project RFMEFI 61615X0035).

References

1. North, G.R., Pyle, J.A., Zhang, F.: Encyclopedia of Atmospheric Sciences. Elsevier (2014)
2. Fedorov, K.N.: The Physical Nature and Structure of Oceanic Fronts. Springer, New York (1986)
3. Shevchenko, I., Berloff, P.: On the roles of baroclinic modes in eddy-resolving midlatitude ocean dynamics. Ocean Model. **111**, 55–65 (2017). https://doi.org/10.1016/j.ocemod.2017.02.003
4. Uchida, T., Abernathey, R., Smith, S.: Seasonality of eddy kinetic energy in an eddy permitting global climate model. Ocean Model. **118**, 41–58 (2017). https://doi.org/10.1016/j.ocemod.2017.08.006
5. Monin, A.S., Zhikharev, G.M.: Ocean vortexes. Phys. Usp. **160**, 1–47 (1990)
6. Nekrasov, A.N.: Diffusion of the vortex. Coll. op. T.1, pp. 92–116. The USSR Academy of Sciences Publishing House, Moscow (1961)
7. Terazawa, K.: On the decay of vortical motion in viscous fluid. Report of the Aeronautical Research Institute. Tokyo Imperial University, vol. 1 (1922)
8. Maddison, J.R., Marshall, D.P., Shipton, J.: On the dynamical influence of ocean eddy potential vorticity fluxes. Ocean Model. **92**, 169–182 (2015). https://doi.org/10.1016/j.ocemod.2015.06.003
9. Batchelor, G.K.: An Introduction to Fluid Dynamics. Cambridge University Press, Cambridge (2000)
10. Solovyev, A.A., Solovyev, D.A.: The barotropic instability of the oceanic jet currents. Atmos. Ocean. Sci. **2**, 80–84 (2017). Science Publishing Group, New York. https://doi.org/10.11648/j.aos.20170203.13
11. Gulev, S., Freeman, E.: Tracking progress in marine climatology (2017)
12. Alekseev, V.V., Kisileva, S.V., Lappo, S.S.: Laboratory models of the physical processes in the atmosphere and ocean. Nauka, Moscow (2005)
13. Zhang, Y., Afanasyev, Y.D.: Baroclinic turbulence on the polar $β$-plane in the rotating tank: down to submesoscale. Ocean Model. **107**, 151–160 (2016)
14. Polyanin, A.D., Manzhirov, A.V.: Handbook of Integral Equations. CRC Press, Boca Raton (2002)
15. Smirnov, V.I.: A Course of Higher Mathematics. Elsevier (2014)
16. Barton, K.W., Sano, M.H.: Anticyclonic warm-core Gulf Stream rings off the northeastern United States during 1985. NAFO SCR Doc. 86, 77 (1986)
17. Chelton, D.B., Schlax, M.G., Samelson, R.M., de Szoeke, R.A.: Global observations of large oceanic eddies. Geophys. Res. Lett. **34** (2007). https://doi.org/10.1029/2007gl030812
18. Auer, S.J.: Five-year climatological survey of the gulf stream system and its associated rings. J. Geophys. Res. Ocean **92**, 11709–11726 (1987). https://doi.org/10.1029/JC092iC11p11709
19. Lai, D.Y., Richardson, P.L.: Distribution and movement of gulf stream rings. J. Phys. Oceanogr. **7**, 670–683 (1977). https://doi.org/10.1175/1520-0485(1977)007<0670:DAMOGS>2.0.co;2
20. Leterme, S.C., Pingree, R.D.: The Gulf stream, rings and North Atlantic eddy structures from remote sensing (Altimeter and SeaWiFS). J. Mar. Syst. **69**, 177–190 (2008). https://doi.org/10.1016/j.jmarsys.2005.11.022

Reservoir Proxy Model as a Part of Geo-Technological Model of Gas Fields and Underground Gas Storages

Sergey A. Kirsanov[1], Andrey V. Chugunov[2], Oleg S. Gatsolaev[2], Yan S. Chudin[2], Ivan A. Fedorov[2(✉)], Aleksey A. Kontarev[2], and Alexandra P. Popovich[2]

[1] Gazprom PJSC, 16 Nametkina str., GSP-7, Moscow 117997, Russian Federation

[2] Gazprom VNIIGAZ LLC, Razvilka settlement, Moscow Region 142717, Russian Federation

I_Fedorov@vniigaz.gazprom.ru

Abstract. This article includes an issue of applying a reservoir proxy model as a component part for geo-technological model of gas fields and underground gas storages. Authors have proposed an algorithm of applying the created proxy model into intelligent management system for gas fields and UGS. The task of proxy model creating is considered by the example of an underground gas storage "X" (the name of the object is encrypted commercial confidentiality reasons). All characteristics of the object "X" and the data on its actual operation mode, whereon model adaptation was performed, are given in the article. Authors have presented a detailed mathematical description of reservoir proxy model and considered the process of its creation, adaptation and integration into a geo-technological model complex of an object. Also, the limits of applicability of this method was determined. A Fully Functional Hydrodynamic (FFH) model of the "X" object was created in the commercial simulator«Schlumberger Eclipse»in order to assess the correctness of the reservoir proxy model, and a detailed comparison of the actual data and model calculations was conducted. In addition, the issue of using hydrodynamic models as a core of intelligent management system for reservoir development was considered in detail. The conclusions made by authors tell about the necessary of computing capacities, both for operation of the proxy model and the FFH model based on the commercial simulator. As a result of the research, conclusions about the applicability of the reservoir proxy model as an integral part of geo-technological model of gas fields and UGS were made.

Keywords: Proxy model · Hydrodynamic model · Reservoir Intelligent management system

V. Karev et al. (Eds.): PMMEEP 2017, SPRINGERGEOL, pp. 217–232, 2018.
https://doi.org/10.1007/978-3-319-77788-7_24

1 Introduction

Presently the oil-gas industry has a tendency to decline of the main developed fields production, as well as to development of difficult extracted hydrocarbon reserves in newly developed objects. At the same time expenses for field construction and hydrocarbon production cost can essentially increase, as well as economic risks in making decisions. Due to high volatility of prices for hydrocarbons at global market, requirements for decision-making time in operative management of UGS are increased. Therefore, of special urgency became issues of increasing monitoring quality of operational geo-technological parameters in gas fields and UGS in real time mode, and optimization of process mode parameters in well operations based on adaptive mathematical models, active and forward planning of production and technological indices.

Also, it is worth to note essential growth of data volume on dynamics of changing operation parameters in gas fields and UGS due to introduction of telemetry systems. Development of methods for complex application of obtained information may allow increasing efficiency of object operation and reducing the first cost of natural gas production and storage.

Therefore, creation and improvement of intelligent management systems for gas fields and UGS with regard to active geo-technological monitoring and forward planning become an important task. Intelligent management system for gas fields and UGS means such a system, wherein subsystems for intelligent decision support and risk assessment are employed in object and/or process models, during formulation and implementation of control actions upon extraction from/injection to the reservoir and preparing product transportation [1]. This system must support effective decision-making management on object level and on producing branch level and the whole company level.

Intelligent system complex should allow analysis of real time field information, detect any deviations from designed (defined) parameters, compose choice of control actions, generate optimal decisions, and in some cases, implement these decisions in automated mode [1]. Geo-technological model of a field should be the core for such complex, and it must implement prediction of operation parameter dynamics in system "reservoir – wells – pipelines – processing equipment". Availability of such model may allow implementing proactive management with feedback [2].

At present, hydrodynamic simulators are applied to creation of geo-technological models of gas fields and UGS, most common of them being Schlumberger Eclipse, Roxar Tempest, and RFD tNavigator.

Using the above models as a core of intelligent management system imposes relevant requirements to necessary computing power in order to assure operability of the whole complex in real time mode. Real time mode [3] means a data handling mode, wherein interaction is provided between information processing system and extrinsic processes at a rate comparable to progress of these processes.

At that, main load on computing power represents object's reservoir parameters calculation. In this connection, suggestions on creation of simplified reservoir models – proxy models, become more and more often.

General approach to reservoir proxy model creation is enlarged size and reduced number of cells in computational grid, while solving simplified set of equations for calculating fluid flow and pressure distribution. However, even subject to considerable simplification, the model should respond to calculating accuracy requirements, whereof basic is coincidence between factual and computed values of reservoir pressure with admissible tolerance.

Follows below a detailed review of construction, adaptation, quality assessment, and integration of reservoir proxy model into geo-technological model by example of a real Underground Gas Storage – UGS "X".

2 Object of Study

2.1 Process Scheme of the Object Operation

Natural gas is stored in UGS "X" in salt caverns and depleted gas field within porous reservoir that are connected by common gas-collecting pipeline system. Injection or extraction of gas is performed either directly due to pressure drop, or using compressor units. UGS "X" is connected to 3 different main gas pipelines of different operation pressure.

Main operation parameters of UGS "X" process scheme are shown in Table 1, and schematic diagram for the Object units connection is shown on Fig. 1.

Table 1. Operation parameters of the object UGS "X"

Parameters	Values
Number of wells in reservoir part, pcs.	14
Number of wells in cavern part, pcs.	17
Number of compressor units	4

2.2 Properties of the Object Reservoir Part

The reservoir part of UGS "X" is peculiar for its block structure (blocks being conventionally divided by fractures) and lacking of water-bearing stratum. Main characteristics of the reservoir are presented in Table 2, and its structure map of reservoir roof by depth parameter is shown on Fig. 2.

2.3 Factual Data on the Object Operation Mode

For research purposes and construction of digital models, factual data on the object operation were obtained (for 5 month of gas extraction and 2 month of gas injection), recorded at 1 day intervals, with regard to the following:

- Gas extraction/injection rate;
- Pressure at manifold of pipelines system at the reservoir part;
- Well gas extraction/injection rates;

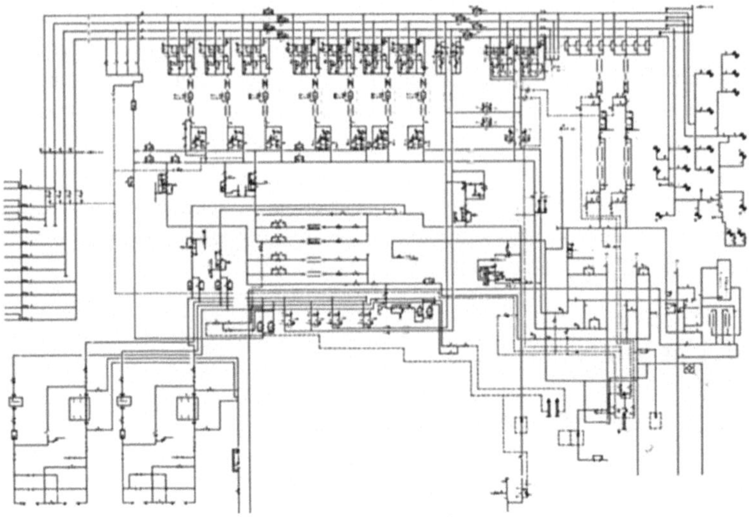

Fig. 1. Schematic diagram for the object UGS “X” units connection

Table 2. Characteristics of the reservoir part for the Object UGS “X”

Parameters	Values
Maximum/minimum depth, meters	1180.27/577.75
Absolute permeability, mD	10.16
Porosity	0.02–0.15
Pore volume, mln m^3	3.05
Maximum operation gas volume, mln m^3	440.31
Presence of impermeable fractures	yes

- Well head pressure rates;
- Reservoir pressure from depth pressure gauge in the monitoring well;
- *a* and *b* gas flow coefficients, obtained by results of well studies.

3 Hydrodynamic Modelling of the Reservoir

At present, the most commonly used software for fully functional hydrodynamic models construction is hydrodynamic simulator Eclipse by Schlumberger Company, so we employed it in this study. In addition, calculations were carried out on the Russian software RFD tNavigator.

Using mathematical tools for proxy modelling is proposed by authors as an alternative method for hydrodynamic modelling of reservoir systems [4].

Detailed description of creation processes for both fully functional hydrodynamic model of the Object UGS “X” and reservoir proxy model presented in sections below.

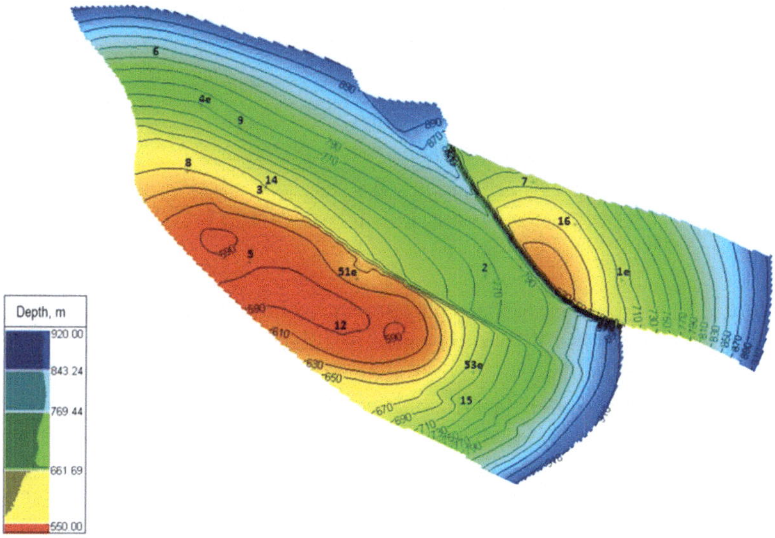

Fig. 2. Structure map by roof of the reservoir part for the Object UGS "X" (well-names is encrypted for reasons of commercial secrecy)

3.1 Creation of Fully Functional Hydrodynamic Model of the Object

Creation of geological model is the first step and the basis for subsequent construction of fully functional hydrodynamic model of the object. Three-dimensional, digital geological model was created in Schlumberger Petrel software package, and it was constructed based on the following geological-geophysical field data:

- Well-specific coordinates and altitudes;
- Reservoir intersection marks in wells;
- Structure map along the reservoir roof;
- Average values of well-specific porosity factors and permeability indices;
- Well-specific effective thickness values.

Structure basis was formed using Converged interpolation method. At that, tolerance of ≤ 0.4 m for well "setting" upon structural surface was among main requirements. This requirement is mandatory, because even insignificant vertical shifts of wells can lead to essential inconsistency between maps and hence, distortion of main estimated characteristics of the model, such as effective thickness, porosity maps, and gas saturation maps.

Among quantitative criteria of accuracy for the obtained reconstruction is evaluation of gas storage capacity based on the model data, which had less than 1% deviation.

The created geological model of the Object UGS "X" has X and Y axes cell sizes varying from 0.1975 m to 14.35 m. Vertical cell size was 8.399 m.

When determining dimensions of filtration model, the task was solved to maintain detailed geological structure of the reservoir and achievement of optimal calculation time. As a result, X- and Y-axis enlargement of cells was performed, so that their size

Table 3. Geometrical characteristics of Fully Functional Hydrodynamic Model of the Object UGS "X"

Parameters	Values
X axis cell quantity, pcs.	124
Y axis cell quantity, pcs.	303
Z axis cell quantity, pcs.	22
Total active cell quantity, pcs.	826,584

became 20 m × 20 m. Geometrical characteristics of fully functional hydrodynamic model of the Object UGS "X" are presented in Table 3.

In order to solve the task of gas filtration at UGS "X", Eclipse 100 "Black oil" model was used. This model type enables accounting for changes of reservoir fluid properties and phase transitions that would occur when reservoir pressure decreases, and at the same time, has optimal computing time parameter.

Adaptation of fully functional hydrodynamic model was performed based on factual operation features of wells, and included correction of absolute permeability parameter and production features of wells using results of gas dynamics studies.

Figure 3 shows porosity distribution in reservoir hydrodynamic model of the Object UGS "X".

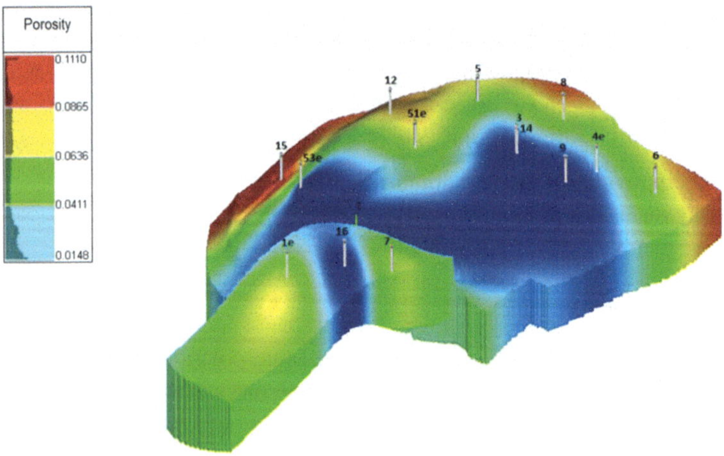

Fig. 3. Porosity distribution in hydrodynamic model of the Object UGS "X"

Initial distribution of gas saturation parameter in the model was calculated by applying a procedure of scaling end points of relative phase permeability functions [5]. Average value of gas saturation in the model was 0.75.

3.2 Proxy Model of Reservoir Part of the Object

At present, there are many studies on possible applications of simplified hydrodynamic models for calculation of reservoir parameters in oil and gas fields and UGS. Main types of such simplified models, according to Mohaghegh et al. [6] and Zangl et al. [7], are as follows:

- Surrogate models, which employ statistical approaches;
- Proxy models, which employ fluid balance calculation between areas of computational grid.

In available studies, however, capabilities of proposed theoretical models were not verified at any real object, which makes it difficult to assess their practical value.

Mathematical model for reservoir proxy model

Approach to proxy model of the reservoir system is based on the theory proposed by Geresh et al. [4], which employs decomposing the reservoir part by some areas (calculation cells).

In order to correctly account for draining areas, two cells are defined for each well – namely, well area (A) and peripheral well area (B), as shown on Fig. 4. Rectangular shape of cells is a conventional picture, for cells in the proxy model essentially repeat real shape of a reservoir, while their main properties are pore volume V_{por_i} and interaction coefficients γ_{ij}, where i is cell number of calculated cell, and j is number of adjacent cell. Adjacent cells are those, which have one common side facet.

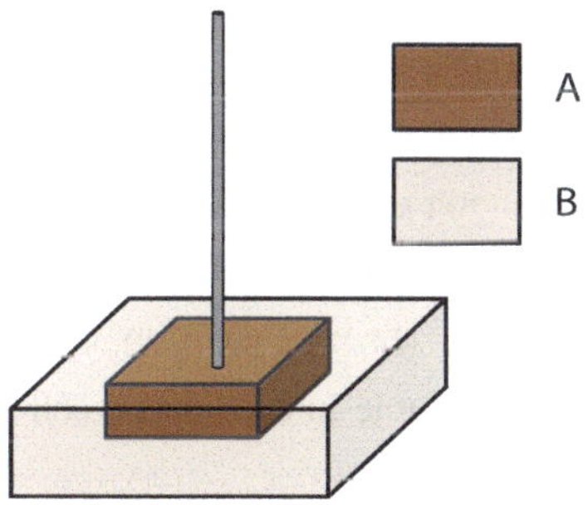

Fig. 4. Dividing well area by cells (A – well area; B – peripheral well area).

Computational grid is formed as a result of the above approach to reservoir decomposition by areas, as shows Fig. 5.

Initial data for calculations are value massifs of temperature T_i, reservoir pressure for current calculation step $P_{well_{i(n-1)}}$, and well gas flow rate Q_{well_i}.

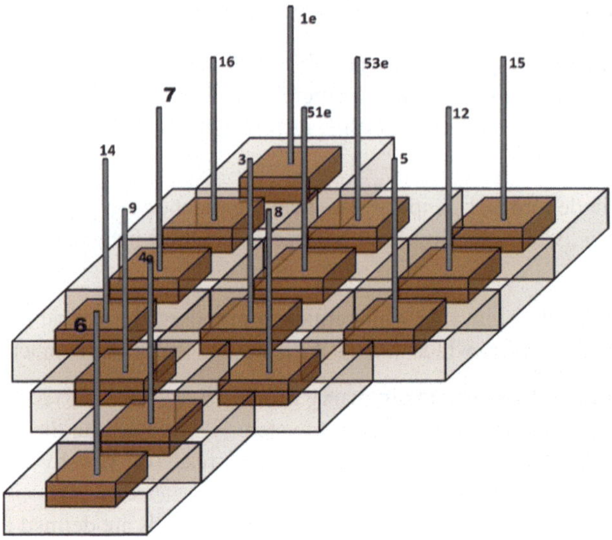

Fig. 5. Computational grid of the object UGS "X"

After obtaining initial information, reservoir pressure calculation in well area is made for the end of current calculation step (the start of next step) by the below equation, based on material balance equation for each area:

$$P_{well_{i(n)}} = \left(-\frac{V_{por_i} \cdot C_0}{z\left(P_{well_{i(n-1)}}\right)} + \sqrt{\left(\frac{V_{por_i} \cdot C_0}{z\left(P_{well_{i(n-1)}}\right)}\right)^2 - 4 \cdot \sum_j \gamma_{ij} \cdot \left(\mu_i \cdot Q_{well_i} + \frac{P_{well_{i(n-1)}} \cdot V_{por_i} \cdot C_0}{z\left(P_{well_{i(n-1)}}\right)} + \sum P^2_{well_{j(n-1)}} \cdot \gamma_{ij} \right)} \right) / 2 \cdot \sum_j \gamma_{ij} \quad (1)$$

where $C_0 = T_{SC}/T_i$, and $T_{SC} = 293\,K$ temperature in standard conditions.

Subsequently, transition to bottom-hole pressure calculation for wells is made $P_{bhp_{well_i}}\left(P_{well_{i(n)}}; a_i; b_i\right)$, where $a_i; b_i$ are well gas flow coefficients $Well_i$, the interflow between areas $\left(Q_{well_{ij}}\right)$, and gas volume in calculation cell (V_i) by equation:

$$Q_{well_{ij}} = \sum_j \left(P^2_{well_{i(n)}} - P^2_{well_{j(n)}}\right) \gamma_{ij} / \mu_i \quad (2)$$

where μ_i is fluid viscosity in area of i–th well.

$$V_i = \frac{P_{well_{i(n)}} \cdot V_{por_i} \cdot C_0}{P_{SC} \cdot z\left(P_{well_{i(n)}}\right)} \quad (3)$$

where P_{SC} pressure in standard conditions.

The result of algorithm operation is value massif $\left(P_{well_{i(n)}}; Q_{well_{ij}}; V_i; P_{bhp_{well_i}}\right)$ for each well of the reservoir part in the Object UGS "X".

Construction of reservoir proxy model of the Object UGS "X"

Creation procedure of the reservoir proxy model is conventionally divided by 3 steps:

- decomposition of the reservoir system by calculation areas (cells) via an expert decision;
- defining the matrix for fluid interflow coefficients between computational grid cells, with account for geological structure of the reservoir system;
- adaptation of the reservoir proxy model by historical data.

Decomposition of the reservoir by calculation areas was made based on factors as follows:

- draining area of each well;
- distribution of porosity and permeability parameters around the reservoir;
- presence of geological peculiarities, such as impermeable fractures and conducting cracks systems.

Draining areas of wells were defined applying mathematical tools for plotting Voronoy's diagrams [8] and with account for geological peculiarities of the object, such as distribution of porosity and permeability, and presence of impermeable fractures.

Results of construction of wells draining areas are shown on Fig. 6. Also, Fig. 6 shows possible directions of interflows, which should be examined during adaptation of proxy model to factual data.

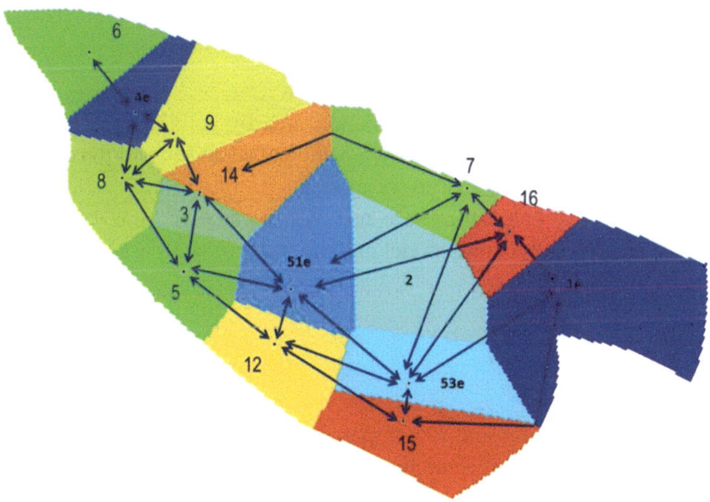

Fig. 6. Wells drainage areas

According to data, obtained during field development and subsequent UGS operation, the reservoir has a complication, as semi-conductive fracture. This fracture divides spaces of wells "107 + 116 + 001" and "114 + 51 + 53".

In order to find a matrix for interflow coefficients between areas, an iteration procedure of gradient slope was applied, consisting in minimization of mean square divergences between calculated and factual well-specific reservoir pressures. Resulting therein interaction coefficients for well areas and peripheral areas (shown in parentheses) are presented on Fig. 7. Also, Fig. 7 shows interaction coefficients between peripheral areas (shown close to arrows).

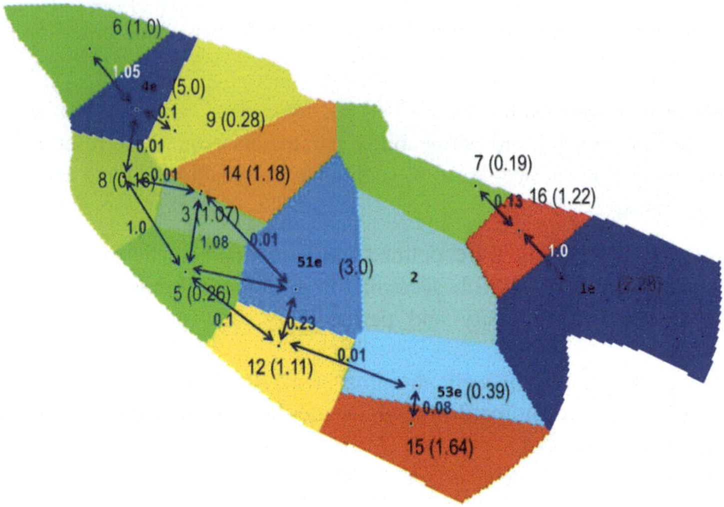

Fig. 7. Map of interflows and interaction coefficients

After making decomposition of calculation area by cells and defining interaction coefficients between cells, factual gas flow coefficients *a* and *b*, and pore volume values for each cell were entered into the proxy model.

Results of predictive calculations for reservoir parameters that were performed in the fully functional hydrodynamic model and the proxy model, are considered in detail in the below section.

4 Results

In order to assess correctness and to compare the two proposed methods for reservoir system modelling, predictive calculations for 7 month period were performed with predefined total intended gas ratio. Factual and estimated reservoir pressure values based on fully functional hydrodynamic model and the proxy model were then compared.

4.1 Reservoir Parameters

Dynamics of reservoir pressure values obtained using calculations based on fully functional hydrodynamic model and the reservoir proxy model are shown in comparison with factual values on Fig. 8.

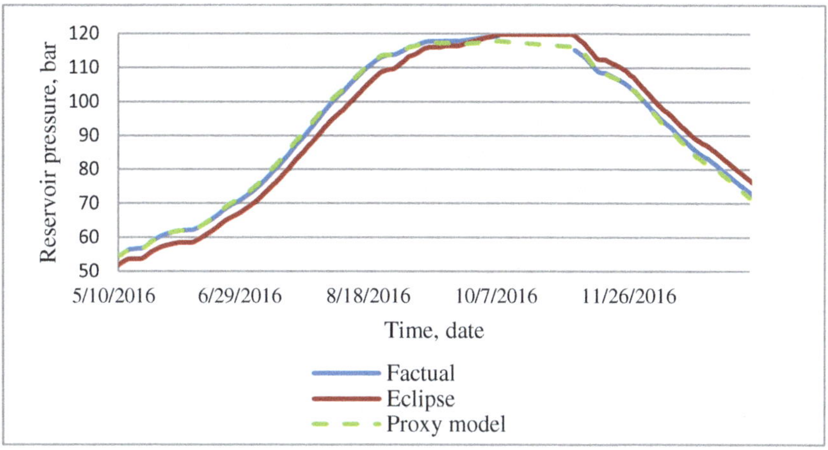

Fig. 8. Dynamics of reservoir pressure values (calculated and factual)

Figure 9 shows dynamics of deviation values between factual and calculated reservoir pressure values for fully functional hydrodynamic model and the reservoir proxy model.

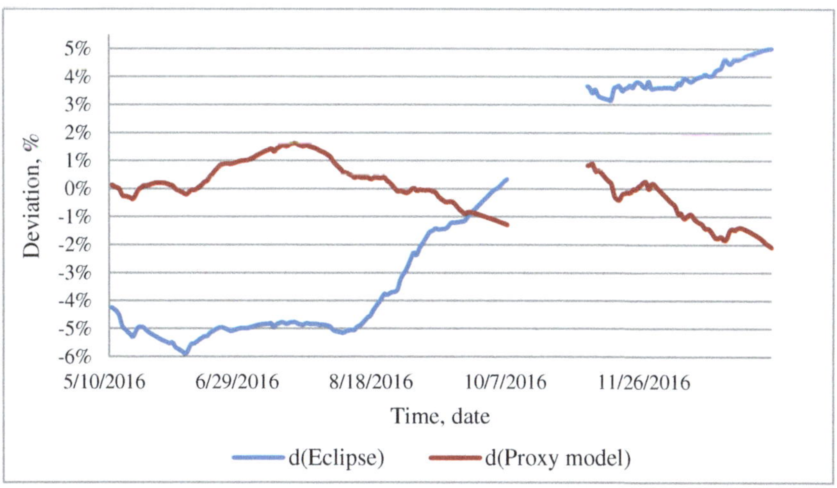

Fig. 9. Dynamics of deviation values between factual and calculated reservoir pressure values

Based on obtained results, it may be concluded that the reservoir proxy model, created according to the method proposed satisfies main requirements for accuracy of calculations. Maximum deviation by module in reservoir pressure parameter was 2.08% for the proxy model, versus 5.9% for fully functional hydrodynamic model. Depth of adaptation of fully functional hydrodynamic model was limited by the meaning of its high detalization.

4.2 Computing Time

Based on obtained results of calculations, computing time requirements for both fully functional hydrodynamic model and the reservoir proxy model were found out. It took 7.306 s to calculate one iteration in fully functional hydrodynamic model, whereas in the proxy model, calculation of one step took 0.0257 s. Thus, total times for predictive calculation of reservoir part of the Object UGS “X” operation for 7 months were:

- in fully functional hydrodynamic model – 10 h 25 min 24 s (37,524 s);
- in the proxy model – 2 min 12 s (132 s).

The Object UGS is a complex system, and methods for its management are based on automated analysis and selection of options, and imply multi-variant calculations within a series [6]. Therefore, when fully functional hydrodynamic model is used as calculation core for a system, selection period for any control action may take a few days, which can make difficult its implementation in operative management. At the same time, when the reservoir proxy model is used, computing time will be approximately 1 h, which enables employing this system even for responding to emergency situations when decision-making period required to be us short as it`s possible.

5 Discussion

5.1 Creation of Geo-Technological Model Using a Reservoir Proxy Model

Obtained results support possibility of using the reservoir proxy model as a component part of geo-technological model for the Object UGS “X”, and its subsequent application as a component part of intelligent management system for gas fields and UGS.

Main blocks of the developed algorithm for geo-technological modelling of the Object UGS “X” using the reservoir proxy model, are:

- obtaining initial data for calculation;
- initialisation of calculation procedure;
- calculation of technological parameters of pipelines operation mode;
- distribution of flows between objects and calculation of node pressure values around pipeline objects;
- calculation of technological parameters of compressor operations;

– calculation of fluid filtration parameters in the reservoir and caverns;
– calculation of interflow values and reservoir pressure distribution using the reservoir proxy model;
– obtaining results of the calculation.

General view of algorithm for geo-technological modelling of the Object UGS “X” is shown on Fig. 10.

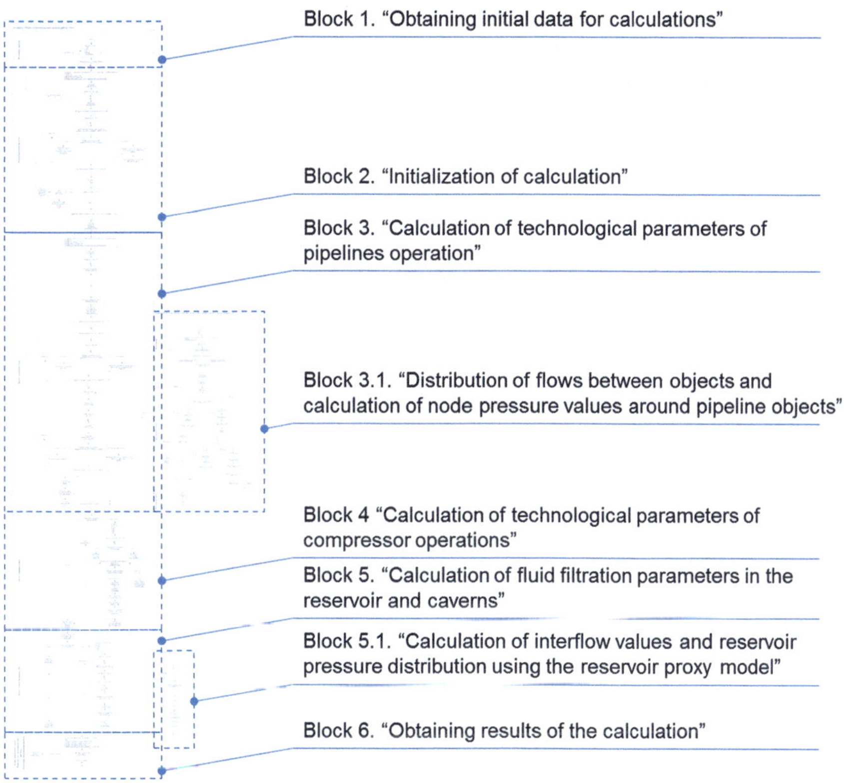

Fig. 10. Algorithm for geo-technological modelling of the object UGS “X”

Initial data for this algorithm are total gas extraction/injection for the Object UGS “X”, pressures in main gas pipelines, initial well head pressures and technological scheme of the object operation. As a result, the algorithm presents to user a massif of values for each calculation step, including values as follows:

– reservoir pressure;
– well head and bottom-hole pressures;
– interflow values between areas;

- well gas flow rates;
- well group-specific gas pressure and flow rate at the manifold;
- amount of electric power, consumed by compressor units, and compressor-specific rotation speeds;
- gas flow rates for compressor units.

Based on this algorithm, a software “Calculation of operating mode of UGS” has been created, which has main features as follows:

- predictive calculation of reservoir and cavern parts, and technological parameters of operating mode based on either defined total gas flow rate or pressures in main gas pipelines of technological equipment;
- visualisation of the object operation (Fig. 11);
- graphical presentation of calculation results (Fig. 12);
- generation of reporting documents.

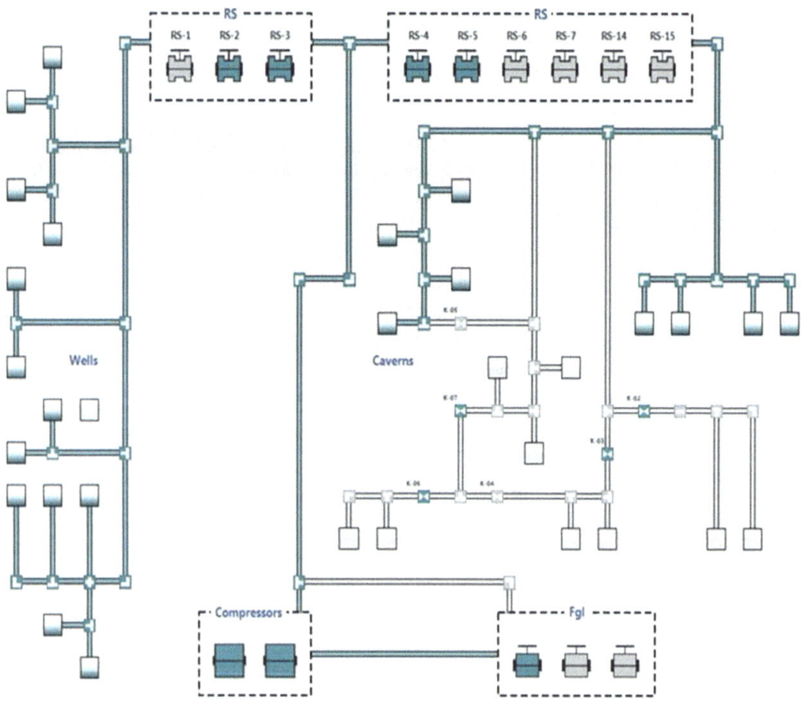

Fig. 11. Visualisation of the object operation

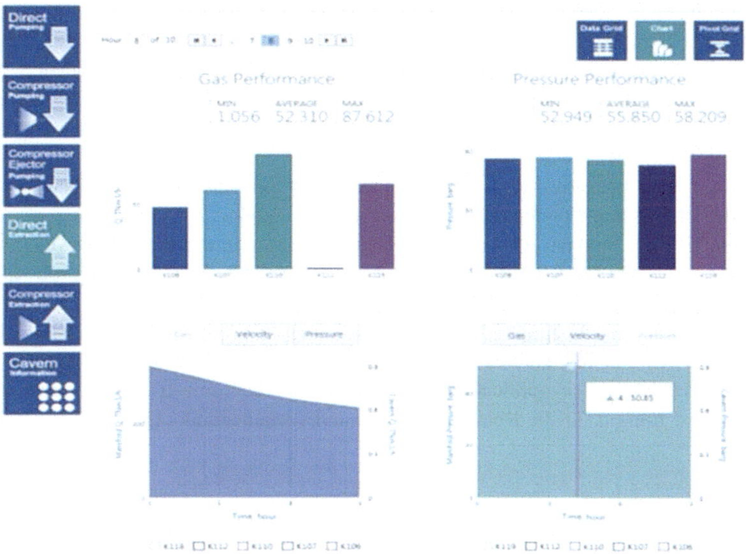

Fig. 12. Graphical presentation of calculation results

6 Conclusions

Conclusions as follows can be made as a result of the present study:

- within operative range of prediction, the reservoir proxy model can correctly implement calculations of the reservoir parameters;
- according to simplicity of mathematical tools proposed, period for calculation time of the reservoir proxy model is essentially shorter than period for similar calculation in fully functional hydrodynamic model;
- it is possible to apply proposed reservoir proxy model as a component part of geo-technological model for gas fields and UGS;
- geo-technological model of an object, based on the reservoir proxy model, is the most effective approach to a calculation core for intelligent management system for reservoir development, when implementing prediction of processes for short-time period (1 year or less) due to its high accuracy, short computing time, and low requirements to computing power.

References

1. Zinchenko, I.A., Lugay, D.V., Vasilyev, Y.N., Chudin, Ya. S., Fedorov, I.A.: A concept of intelligent management system for field development. Vesti Gazovoy Nauki **2**(26), 4–9 (2016). (in Russian)
2. Kirsanov, S.A., Pishukhin, V.M., Chudin, Y.S., Fedorov, I.A.: Special features in forming intelligent management system for field development. Automation, Telemechanization and Communication in the Oil Industry **5**, 13–16 (2016). (in Russian)

3. Russian Standart 15971-90: Information Processing Systems. Terms and Definitions, Moscow, p. 22 (1991). (in Russian)
4. Geresh, P.A., Gatsolaev, A.S., Kuznetsova, G.M., Semenova, L.N.: Application of Areal Modelling in Analysis and Prediction of Gas Reservoir Development in West Siberia, by the Example of Senoman Reservoir, Urengoy Field, Moscow, VNIIE Gazprom, vol. 13, p. 36 (1988). (in Russian)
5. Gurevich, G.R., Brusilovsky, A.I.: Reference Book on Calculation of Phase State and Properties of Gas-and-Condensate Mixtures, Moscow, Nedra (1984). (in Russian)
6. Mohaghegh, S.D., Modavi, A., Hafez, H., Haajizadeh, M., Guruswamy, S.: Development of a Surrogate Reservoir Model (SRM) for fast track analysis of a complex reservoir. Int. J. Oil Gas Coal Technol. **2**(1), 2–23 (2009)
7. Zangl, G., Graf, T., Al-Kinani, A.: Proxy modeling in production optimization. In: SPE Europec/EAGE Annual Conference and Exhibition, Vienna, Austria (2006)
8. Voronoi, G.: Nouvelles applications des paramètres continus à la théorie des formes quadratiques. Journal für die Reine und Angewandte Mathematik. **133**, 97–178 (1908)

Understanding of Rock Material Behavior Under Dynamic Loadings Based on Incubation Time Criteria Approach

A. N. Martemyanov[1(✉)] and Yu. V. Petrov[1,2]

[1] St.-Petersburg State University, 7/9 Universitetskaya nab., St. Petersburg 199034, Russia
st021081@student.spbu.ru
[2] IPME RAS, V.O., 61 Bolshoj Prospect, St. Petersburg 199178, Russia

Abstract. Different rock material dynamic laboratory tests have been analyzed with the help of incubation time criteria approach. As a result of made calculations incubation times for such rocks like granite, marble, limestone, sandstone and traverline have been estimated. This parameter according to criteria theory rule material behavior under high rate loadings. Effects of effective porosity and anisotropy on incubation time value have been considered. Experiments under high rate loadings with different saturation and temperatures have been demonstrated influence of physical conditions changes on incubation time parameter.

Keywords: Incubation time · Rock strength · Dynamic loadings

1 Introduction

Rock material strength behavior prediction plays an important role for nowadays engineering, seismic and geoscience works. Numerous investigations including laboratory, theoretical and numeric researches dedicated to studying of rock mechanical properties. Such analysis is always affected by complex inhomogeneous inner material structure, anisotropy and different external influences. Variety of rock material specific properties may be expanded by phenomenon of fast strength increasing with rising of loading rate [2–5, 9–12, 33].

Despite obvious differences between compression and tensile strengths material behavior under dynamic conditions is the same. Limiting stress value that elastic medium will withstand before macrodamage appears becomes bigger if impulse rate increase. It is natural to assume that underlying mechanism of such phenomenon related with viscous-elastic character of material which can be negligible in static case.

For the goal to describe the material behavior under high rate loadings some semi-empirical criteria were proposed (look, for example, [20]). Every such equation represents correlation between stress or strain rate and observed tensile strength data and doesn't take into account incident wave parameters which may affects the experiment results. This work is dedicated to quasi-static and dynamic strength test results analysis published in literature [1, 6–10, 13–19] based on incubation time criteria approach which supposes new material parameters to explain high rate behavior

V. Karev et al. (Eds.): PMMEEP 2017, SPRINGERGEOL, pp. 233–248, 2018.
https://doi.org/10.1007/978-3-319-77788-7_25

features [21–27]. It demonstrates well agreement with experimental results for wide range of materials like rocks, composite concretes, steels and etc.

2 Dynamic Experimental Techniques

Nowadays many different ways of strength measurements are using to obtain material characteristics and to figure out its behavior in various cases. Laboratory tests allow to receive studying medium parameters depending on temperature changes, fluid saturation, sonic or any other physical field presence, complex loading configuration and, specifically, loading rate. Moreover, deep understanding of material destruction underlying principles requires additional sample investigations like x-ray based monitoring during mechanical test for direct crack propagating detecting or tomography for inner material structure representation improving and etc.

Returning to strength measurements ultimate stress under compressive loadings often estimates from uniaxial test without any confining pressure. In tensile case strength may be detected by several techniques including direct tension test, indirect Brazilian disk, flexural (3 points) notched test and spalling. Classically external forces were produced by servo-drive machines which provide strain rates 10^{-6}–10^{-4} s^{-1}. Using of pneumatic-hydraulic and drop weight mechanisms allows to expand this area on the range of 10^{-2}–10^{1} s^{-1}. Next step of impulse velocity increasing is standard test modification by split Hopkinson bar technique. Application of such loading method makes it possible to carry out experiments in the field of 10^{2}–10^{4} s^{-1} strain rates.

According to split Hopkinson bar technique testing sample is placed between two metal bars with well-known mechanical properties. One of them calls incident bar and other one transmitter bar. Gas-driven striker acting on incident bar produces stress wave which partially goes through studying material and partially reflects. All waves are recording by computer system with the help of piezoelectric elements placed on bars.

Mechanical properties and geometrical sizes of experimental equipment are selected for the goal to make homogeneous stress field inside testing sample. The validity of this condition can be easily verified by restoring stress history on both sides of studying media. At the same time its applicability makes analyses quite simple (Fig. 1).

3 Incubation Time Criterion

Incubation time fracture criteria designed to explain features observed under high rate loadings was formulated by Petrov and Morozov [22, 24]. According to the specified one destruction appears in the point x of material if stress behavior at given point satisfies to the following condition

$$\frac{1}{d}\int_{x-d}^{x}\frac{1}{\tau}\int_{t-\tau}^{t}\sigma(y,s)\,ds\,dy \le \sigma_c, \tag{1}$$

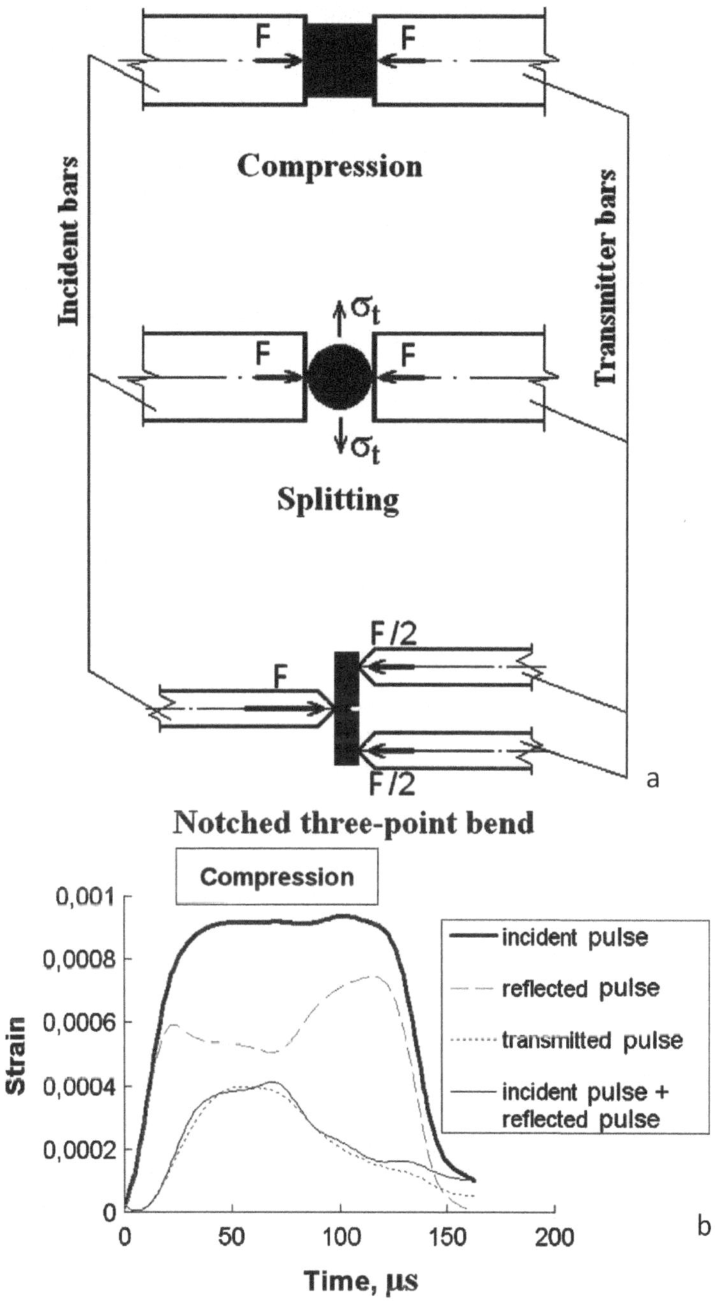

Fig. 1. (a) - Scheme of dynamic tests based on split Hopkinson bar technique; (b) - Typical waves produced during experiment.

here τ is incubation time – material constant characterizing its response to applied loadings (for given material one doesn't depend on sample geometry, load pulse shape or way of loading application), d is another material constant which reflects representative sizes of fracture process zone and thus demonstrates model spatial scale. Function $\sigma(x, t)$ represents normal stress value which may vary in space and time and σ_c is its quasi-static critical meaning (corresponding to tensile or compressive case).

Presented above fracture criterion reveals itself like rather effective tool for understanding of material reaction to dynamic impact [21–27] and relating mechanical problems solving [27, 29]. In particularly, effect of material strength growth under high rate loadings observed in numerous experiments may be successfully explained using the incubation time criteria.

Sizes of rock material samples always limited by need to satisfy the condition of homogeneity and laboratory test machine. Their length most of the time is within range from millimeters up to several centimeters and rarely up to tens of cm. In this regard, it is natural to assume that all measurements made on test sample related with its dimensions like with representative volume. Incubation time criteria (1) can be simplified and rewritten as

$$\frac{1}{\tau}\int_{t-\tau}^{t}\frac{\sigma(s)}{\sigma_c}\,ds \leq 1 \tag{2}$$

Easy to see that in the case of quasi-static force application condition (2) is equivalent to classical Irwin's damage criterion [2, 3]. For better understanding of incubation time role in the case of impulses let us consider next example. According to the classical theory of strength, the local force field instantaneously decreases to zero after achievement of a critical value σ_c because of crack. Considering described process, related with macro-fracture event, in terms of the micro-scale level kinetics, one can be interpret as a temporal process of transition from defect-free state to a completely broken state at the moment of fracture. The macro-parameter of fracture process, identified as an incubation time, is equal to the duration of known temporal process on the given scale level. The incubation time is related to the relaxation process of growth of microdefects in the structure of material, which provides its non-reversible deformation. In this case the characteristic time of relaxation can be considered as the incubation time [21]. Thus, the incubation time is a constant of material, unrelated with the geometry of the test specimen, the way the applied loading and characterize dynamic effects of the fracture process on a given scale level.

4 Interpretation Analyses

Current work based on proposing of approximately linear stress behavior acting on rock sample. Moreover stress related with strains by Hooke's law (Fig. 2).

$$\sigma(t) = E\,\dot{\varepsilon}\,t\,H(t) = \dot{\sigma}\,t\,H(t). \tag{3}$$

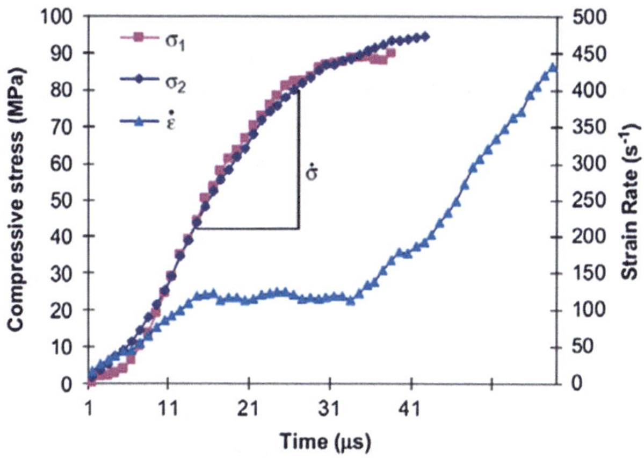

Fig. 2. Coincident of stresses detected from incident and transmitter bar approving stress homogeneity suggestion and strain rate curve.

Here, E is the Young's modulus, $\dot{\varepsilon}$ and $\dot{\sigma}$ are correspondingly loading strain and stress rates. Critical stress value $\sigma(t_*)$ may be observed from Eq. (3) using the moment t_* of fracture initiation as variable. Substituting expression Eq. (3) to incubation time damage criteria Eq. (2) in the rupture moment, ultimate stress dependence on stress/strain rate can be achieved:

$$\sigma_{dyn}(\dot{\sigma}) = \begin{cases} \sqrt{2\sigma_{st}\dot{\sigma}\tau}, & t_* < \tau; \\ \sigma_{st} + \frac{1}{2}\dot{\sigma}\tau, & t_* \geq \tau. \end{cases} \quad (4)$$

Note that calculated fracture stress is separated by two cases. The lower expression in the right-hand part of Eq. (4) describes slow processes, in which the fracture time is comparable or higher than the incubation time τ. The upper expression corresponds to the opposite case the fast dynamic loading when the rupture time is shorter that τ. Thus, the fracture stress under the quasi-static and dynamic loading is predicted based on three macroscopic parameters: the Young's modulus, the critical value of strength under quasi-static loading (static strength) and the incubation time.

The incubation time is defined by the least square method using experimental data $\sigma_{dyn}(\dot{\varepsilon})$. It is important noted, that the introduced constant of temporal parameter of criterion (2) is sensitive to changes of the inner structures of brittle material and invariant to any one impact history.

Analysis of rock test data was related with luck of information about lithological, mineralogical and filtration-capacitive material properties which makes comparisons difficult. Sample shapes, sizes and experimental techniques varies with respect to article and that makes its contribution too.

It should be noted that most of the time we perceive σ_{st} as known and reliable, but if happens that set of dynamic measurements is accompanied by the only static test or no information about elastic properties we resort to minimization procedure by both corresponding parameters.

5 Results

5.1 Experimental Technique Influence

Due to its definition incubation time value for given material doesn't depend on load application method. As good example investigations of Laurentian granite tensile strength can be seen. It has been studied under dynamic loadings by different techniques: Brazilian Disc method in [9], by flexural (3-points) notched test in [10] and spall based method in [16]. Results of laboratory tests shown on Fig. 3.

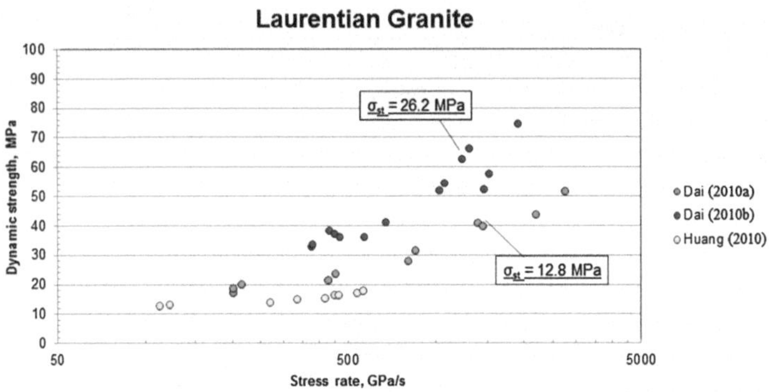

Fig. 3. Laurentian granite dynamic tensile tests results.

There is no detailed information about rock samples but in general Laurentian granite may be characterized as homogeneous medium with average density 2.63 g/cm^3, porosity 0.64% and void ratio 0.006. Mechanical properties are following: Young modulus is 92 GPa, Poisson ratio is 0.21, static tensile strength fixed is 12.8 MPa and static compressive strength is 220 MPa.

During experiments cylindrical samples were used: 40 mm in diameter and 16 mm into thickness, same geometry was used in flexural test but in last case sample was bisected and notched. For spalling cylinders had 22 mm diameter and 38 mm of longitudinal size. All samples have close dimensions and can be compared.

Application of incubation time criterion gives next values for key parameters in the case of Brazilian disc experiments: $\tau = 40 \pm 5\,\mu s$, σst=12.8 MPa. Analogical calculations for Flexural tests shown that $\tau = 50.5 \pm 10\,\mu s$, σst=26.2 MPa. Obtained results are consistent with recently received by Smirnov and coauthors data [28] (Fig. 4).

Finally dynamic properties of united set were computed. Incubation time for considered Laurentian granite studies has rough value is 38 ± 20 μs (Fig. 5).

Process of different techniques results comparison may be developed for deep understanding of static strength and incubation time role in the material destruction under dynamic conditions. This problem requires detailed understanding of stress history at given point before fracture appears.

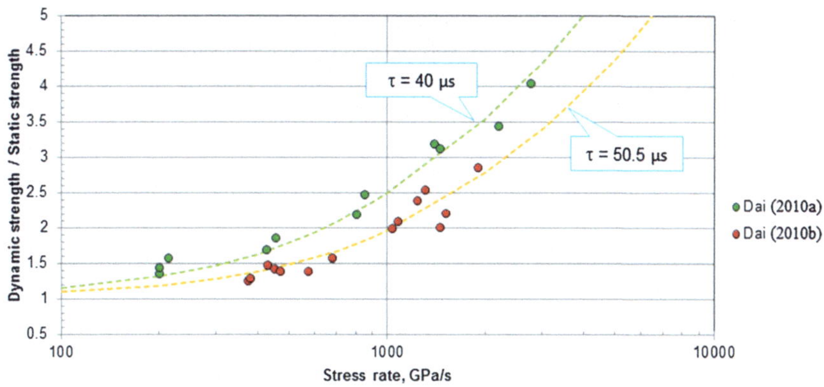

Fig. 4. Incubation time criterion analysis for Laurentian granite samples tests by different methods.

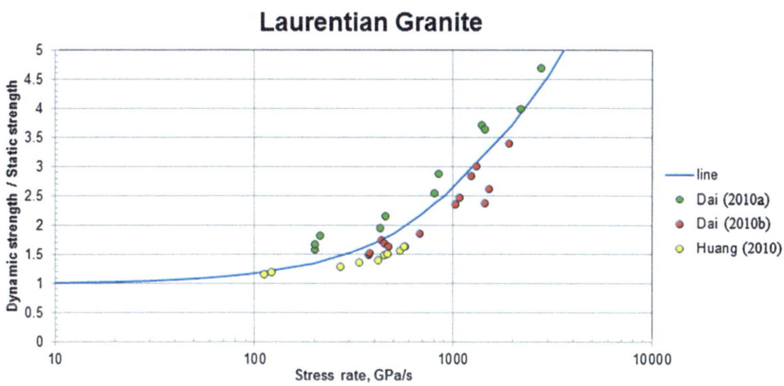

Fig. 5. Incubation time theoretical curve constructed for united set of experimental data.

5.2 Anisotropic Rock Case

It is well known that Barre granite demonstrate anisotropic mechanical properties related with presence of microcracks inside medium. Acoustic wave propagation observations reveal power of mechanical properties anisotropy in three preferential directions. Compressional wave velocity has 3.57, 4.00 and 4.75 km/s values along orthogonal axes. From mineralogical point of view Barre granite is a fine grained rock with mineral grain sizes ranging from 0.25 to 3 mm: quartz volume ratio makes up 25% of this rock and has an average grain size of 0.9 mm; feldspar is the dominant mineral (65%) and has an average grain size of 0.83 mm. The microcracks are of either the intragranular or intergranular type and are found in quartz and feldspar grains, and along cleavage planes of biotite grains [12].

Barre granite tensile strength was measured with the help of Brazilian test in both static and dynamic cases [12]. Test samples were cut from 40 mm diameter core with thickness of 16 mm and divided to six groups with different orientation. For whole data set bordering incubation time values was estimated, experimental points lay between curves corresponding τ from 30 μs to 100 μs witch indicates anisotropy.

Summarized results of data interpretation based on incubation time approach represent in Table 1. Anisotropic coefficients obtained for anisotropy estimation.

Table 1. Barre granite anisotropy parameters.

Orientation	σ_{st}, MPa	τ, μs	α_{stress}, %	α_{time}, %
ZY	16.5	41.8	16.61	3.69
YZ	11.8	45		
ZX	17.1	43.2	32.05	7.30
XZ	8.8	50		
YX	13	45.5	15.56	7.89
XY	9.5	53.3		

As follows from calculations made anisotropic behavior of given material is caused by static strength parameters and related with incubation time weakly. All obtained τ-values are typical for granites. Brittle character of rock material destruction depends on microcracks arisen from previous loadings. Based on considered case one may conclude that inner damages influence on incubation time slightly (Fig. 6).

As another example of anisotropic media Yule marble may be considered, corresponding test results contain in [15]. It is pure calcite material with grain sizes 0.3–0.5 mm and porosity 0.15% and density 2.81 g/cm^3. Young modulus depends on orientation and vary in range from 19 GPa up to 61 GPa. Samples were plugged in two orthogonal directions. Its geometrical sizes are 19 mm in diameter and 15 cm in longitudinal direction. Static tensile strength demonstrates significant difference: material two times stronger in Y-direction than another. Based on given values dynamic characteristics were obtained. Incubation time of Z-oriented samples is 21 μs, same parameter for Y-oriented cylinders takes on value 28 μs (Fig. 7).

As in granite case anisotropic characteristics of marble doesn't affect on incubation time and caused by static strength.

5.3 Incubation Time Dependence on Porosity

Not much works dedicated to investigation of dynamic strength properties and its relation with porosity. Our computations based on two articles [35, 36] describing compressive dynamic tests have been made [35] on carbonate type rocks. All analyzed information was divided by three groups for better understanding.

First one include two types of traverlines studied: T1 characterizes by 2.5 g/cm^3 density and 4.58% effective porosity. Compressional wave velocity for this rock is 5075 m/s, static strength has a value of 84 MPa in compression and 6.41 MPa in tension. Another tuff T2 has density 2.4 g/cm^3, 2.6% of effective porosity. Its dynamic elastic response demonstrate wave speed of 4056 m/s and brittle limiting stresses are 36 MPa and 3.36 MPa for compression and tension respectively.

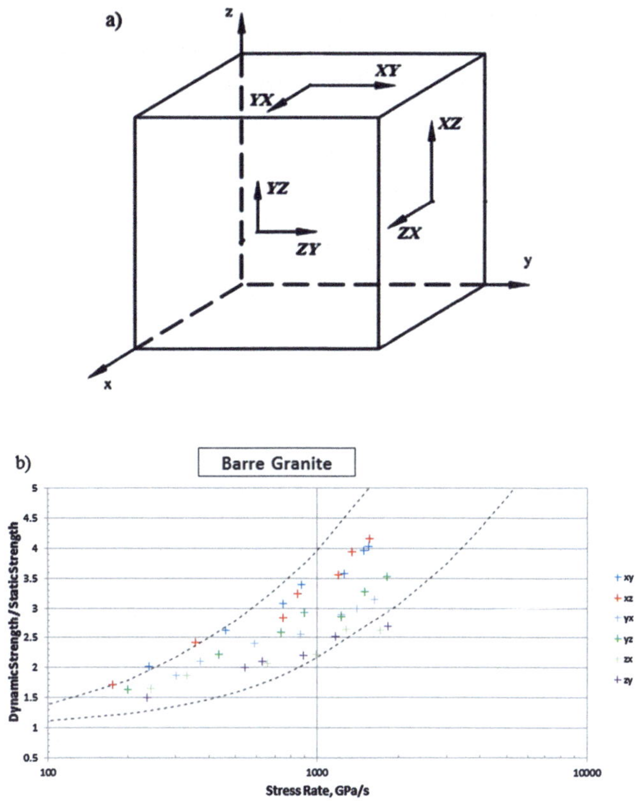

Fig. 6. (a) - Scheme of sample plugs orientation; (b) - Barre granite dynamic tensile strength dependence on sample orientation.

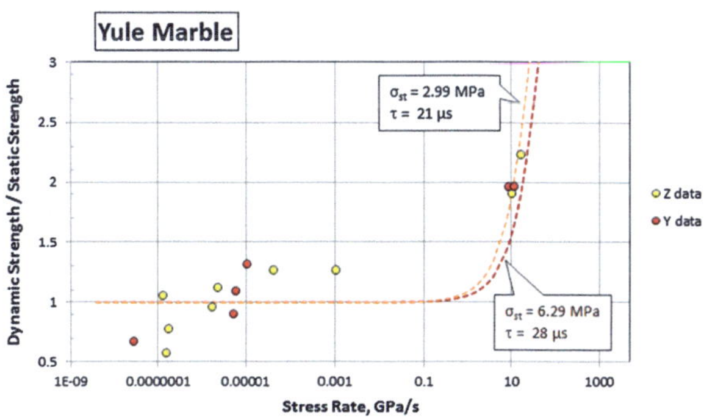

Fig. 7. Incubation time anisotropy analysis of Yule marble.

Estimation of incubation times for this rocks and their further comparison gave τ equal 77 μs for T1 and 75 μs for T2. Obtained calculations quite close and don't represent porosity influences on dynamic damage accumulation processes (Fig. 8).

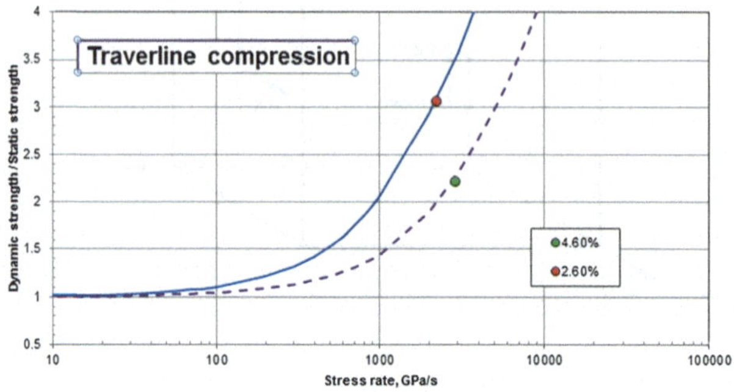

Fig. 8. Dynamic strength properties comparison of traverlines with different porosity based on incubation time approach.

Second group contains dynamic compressional studies made for 5 limestone types. Their physical properties shown in Table 2. To figure out strength features depending on porosity limestones with this parameter below 1% analyzed together.

Table 2. Mechanical characteristics of limestones.

Rock code	Bulk density (g/cm^3)	P-wave velocity (m/s)	Effective porosity (%)	Quasi-static compressive strength (MPa)	Quasi-static tensile strength (MPa)	Schmidt hardness
L1	2.69	6218	0.386	126	8.23	52
L2	2.689	6022	0.342	119	7	49
L3	2.675	5803	0.575	100	8.01	47
L4	2.635	5900	1.842	91	5.73	45
L5	2.377	4270	9.32	43	3.9	34

Calculation of incubation times gave following results: 37.3 μs for rock with effective porosity less than 1%, 45 μs for L4 and 85 μs with one's value close to 10% (Fig. 9).

Most of experimental data demonstrate close results. Significant increasing of incubation time for L5 may be related with nonlinear character of porosity influence on strength. But it is probable that such behavior caused by another reasons. To fully understand pore volume effect standard set of static-dynamic experiments should be supplemented by measurements of total and effective porosity values.

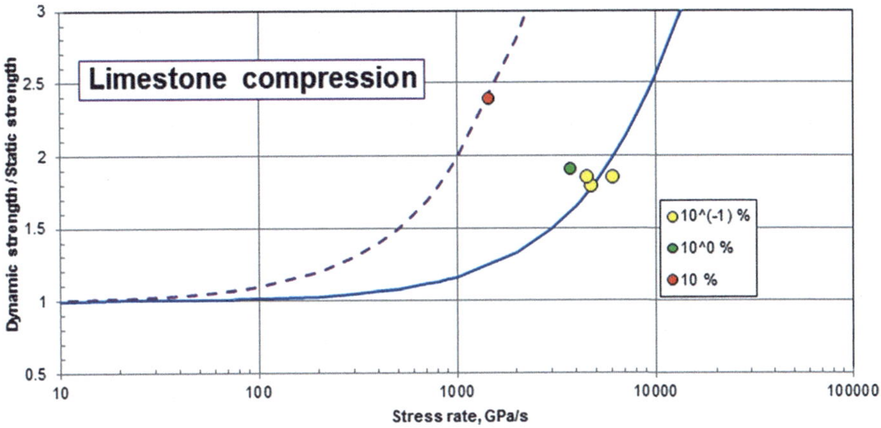

Fig. 9. Limestones with different effective porosity dynamic strength estimation based on incubation time criteria.

Third group consider results observed by Demirdag [36]. Cylindrical specimens 18 mm in diameter and 11 mm long have been used for compressive testing. Physical properties of studying materials and incubation time calculations may be seen in Table 3 below. Experimental points and fitting theoretical curve shown in Fig. 10.

Table 3. Rock material characteristics and obtained incubation time calculation results for limestones

Rock name	Unit volume weight (kg/m^3)	Porosity (%)	Schmidt hardness	Quasi-static compressive strength (MPa)	τ (μs)
Beige	2660	0.44	59	110.8	13.1
Travertine (light)	2458	2.14	46	68.13	21.1
Travertine (dark)	2410	2.3	43	65.2	24.8
Lymra	2382	8.98	41	60.5	22.1

It should be noted that remarkably different structural-temporal parameters computed in the case of Beige limestone related with its quasi-static strength and thus may be connected with inappropriate representation volume. The same reason underlies the previous result.

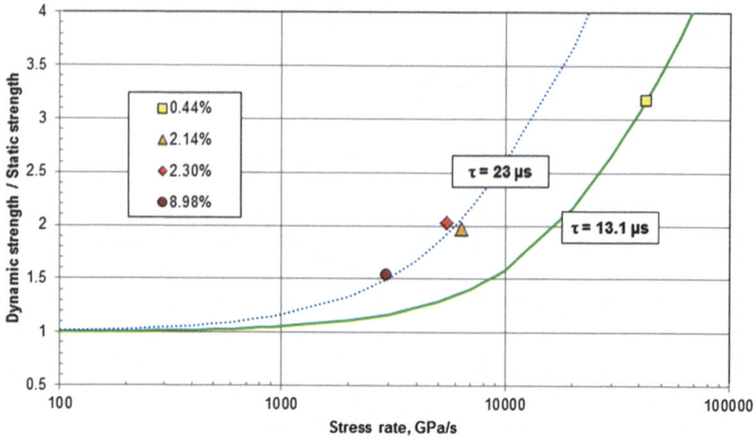

Fig. 10. Compression tests data for several types of limestone rocks interpretation based on incubation time approach.

5.4 Incubation Time Dependence on Saturation

In natural conditions rocks is always fully or partially saturated by fluids of different types. And this affect on its strength characterization notably: limiting stress under tensile or compressive conditions reduces several times while saturation changes for couple percent. Strength dependence on fluid saturation for Longyou sandstone in both dynamic and quasi-static cases was studied in [17]. In addition to mechanical experiments medium was inspected using microscope and X-ray diffraction technique. This let one to obtain mineral composition and establish basic properties. Dry rock density is 2.15 g/cm^3 (for saturated case it is 2.33 g/cm^3), porosity has a value as 17%, compressional wave velocity is 1600 m/s (correspondingly for saturated rock 1560 m/s). General part of test material constitutes quartz (55% of weight) and feldspar (33% of weight) grains which sizes vary in range 0.02-0.35 mm.

Cylindrical samples with 44 mm in diameter and 20 mm in thickness were used for tensile strength tests. Ten of them have been kept into water for 7 days to achieve full saturation (Fig. 11).

Incubation time criteria analyses allow demonstrating changes in material temporal response behavior with saturation. For dry case calculation gives $\sigma_{st} = 3 \pm 0.1\,\mathrm{MPa}$ and $\tau = 58.5 \pm 20\,\mu\mathrm{s}$; saturated samples in its turn characterized by $\sigma_{st} = 0.39 \pm 0.01\,\mathrm{MPa}$ and $\tau = 196.8 \pm 50\,\mu\mathrm{s}$. With standard deviation is not more than 0.5 MPa.

5.5 Incubation Time Dependence on Temperature

To learn how material strength properties change with temperature variation two different types of rock material have been studied [13]: Barre granite has been described early and Italian sandstone which characterizes like homogeneous medium with major grain size is 0.22 mm distributed uniformly. Both rocks were tested by Brazilian disk method. Disk-shaped test specimens were cut from 46.23 mm-diameter core into

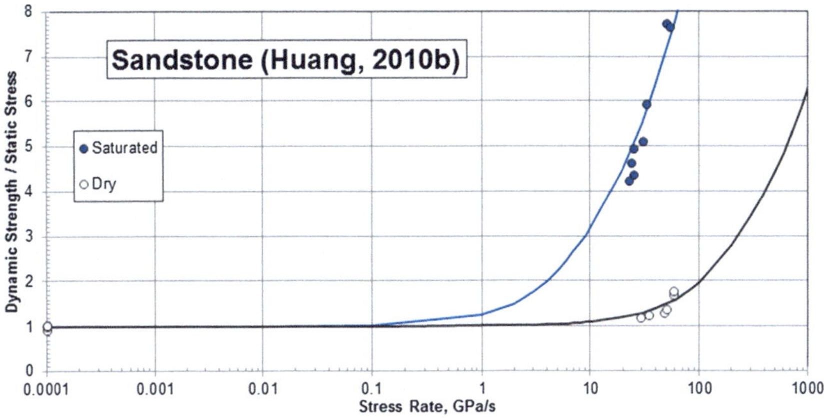

Fig. 11. Incubation time curves correlation with experiment results for dry and saturated sandstone rocks

19.81 mm thickness. Static-dynamic set of experiments were conducted from one side at room temperature (+24 °C) and at low temperatures (−30 °C for static and −40 °C for dynamic) from another side (Table 4).

Incubation time calculation demonstrates τ-independency on temperature for studied rocks. Resulting theoretical curves with analyzed data shown in Fig. 12.

Table 4. Incubation time properties estimated under different temperature conditions.

Rock, temperature conditions	σ_{st}, MPa	τ, μs
Barre granite, +24 °C	11.76±0.72	14.1±1
Barre granite, −40 °C	13.48±0.57	12.2±1
Italian limestone, +24 °C	5.41±0.37	34.9±1
Italian limestone, −40 °C	5.79±0.50	41.7±1

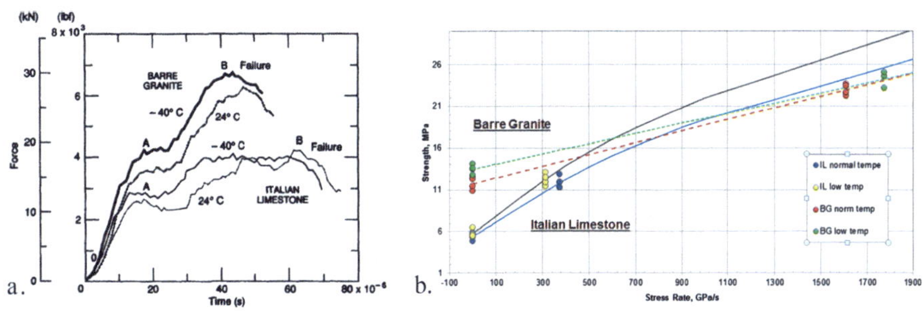

Fig. 12. a. Applied force-time curves recorded in tests; b. Incubation time approach analyses for Barre granite and Italian limestone in case of normal and low temperatures

5.6 Stress Invention Effect

"Stress inversion" effect has clearly dynamical nature and presents itself the change of the dominant strength between the two rock materials [4]. A material, which has relatively lower strength in quasi-static case, can have greater strength under dynamic loading. Such effect has been observed empirically and describes by incubation time criteria well.

Results of previous subsection demonstrate that effect: as follows from calculations performed limestone which static properties significantly lower in comparison with granite has a stronger behavior in dynamic conditions.

6 Conclusions

Dynamic properties of rock materials are governed by incubation time value. Consideration of published experimental results allowed to trace its variation due to additional factors. Analyzed data let us formulate and approve next conclusions:

- Anisotropic rock properties most of the time caused by presence of microcracks and haven't got much influence to incubation time.
- Incubation time for given material doesn't depend on external load application method and may be used for critical strength prediction in wide range of cases.
- Effective porosity apparently haven't got a significant influence on incubation time. But to make a reliable conclusion standard set of static-dynamic experiments should be supplemented by measurements of total and effective porosity values.
- Incubation time critically affected by saturation of rock matrix. Its value increases with saturation.
- From the other hand decreasing of temperature influences on τ negligible.
- Stress inversion effect may have a crucial meaning for planning and realization of engineering operations including high rates impacts.

Acknowledgements. The work was supported by the Russian Science Foundation (grant 17-11-01053).

References

1. Asprone, D., Cadoni, E., Prota, A., Manfredi, G.: Dynamic behavior of a Mediterranean natural stone under tensile loading. Int. J. Rock Mech. Min. Sci. **46**, 514–520 (2009)
2. Bragov, A.M., Petrov, Yu.V., Karihaloo, B.L., Konstantinov, A.Yu., Lamzin, D.A., Lomunov, A.K., Smirnov, I.V.: Dynamic strengths and toughness of an ultra-high performance fibre reinforced concrete. Eng. Fract. Mech. **110**, 477–488 (2013)
3. Bragov, A.M., Karihaloo, B.L., Petrov, Yu.V., Konstantinov, A.Yu., Lamzin, D.A., Lomunov, A.K., Smirnov, I.V.: High-rate deformation and fracture of fiber reinforced concrete. J. Appl. Mech. Tech. Phys. **53**(6), 926–933 (2012)

4. Bragov, A.M., Konstantinov, A.Yu., Petrov, Yu.V., Evstifeev, A.D.: Structural-temporal approach for dynamic strength characterization of rock. Mater. Phys. Mech. **23**, 61–65 (2015)
5. Bratov, V.A., Gruzdkov, A.A., Krivosheev, S.I., Petrov, Yu.V.: Energy balance in the crack growth initiation under pulsed-load conditions. Dokl. Phys. **49**(5), 338–341 (2004)
6. Cai, M., Kaiser, P.K., Suorineni, F., Su, K.: A study on the dynamic behavior of the Meuse/Haute-Marne argillite. Phys. Chem. Earth **32**, 907–916 (2007)
7. Cadoni, E.: Dynamic characterization of orthogneiss rock subjected to intermediate and high strain rates in tension. Rock Mech. Rock Eng. **43**, 667–676 (2010)
8. Cho, S.H., Ogata, Y., Kaneko, K.: Strain-rate dependency of the dynamic tensile strength of rock. Int. J. Rock Mech. Min. Sci. **40**, 763–777 (2003)
9. Dai, F., Huang, S., Xia, K., Tan, Z.: Some fundamental issues in dynamic compression and tension tests of rocks using split Hopkinson pressure bar. Rock Mech. Rock Eng. **43**, 657–666 (2010)
10. Dai, F., Xia, K., Tang, L.: Rate dependence of the flexural tensile strength of Laurentian granite. Int. J. Rock Mech. Min. Sci. **47**(3), 469–475 (2010)
11. Dai, F., Chen, R., Iqbal, M.J., Xia, K.: Dynamic cracked chevron notched Brazilian disc method for measuring rock fracture parameters. Int. J. Rock Mech. Min. Sci. **47**, 606–613 (2010)
12. Dai, F., Xia, K.: Loading rate dependence of tensile strength anisotropy of Barre granite. Pure. appl. Geophys. **167**, 1419–1432 (2010)
13. Dufta, P.K., Kim, K.O.: High-strain- rate tensile behavior of sedimentary and igneous rocks at low temperatures. CRREL Report, 93-16 (1993)
14. Goldsmith, W., Sackman, J.L., Ewert, C.: Static and dynamic fracture strength of Barre granite. Int. J. Rock Mech. Lin. Sci. Geomech. Abstr. **13**, 303–309 (1976)
15. Howe, S.P., Goldsmith, W., Sackman, J.L.: Macroscopic static and of Yule marble. Exp. Mech. **8**, 337–346 (1974)
16. Huang, S., Chen, R., Xia, K.W.: Quantification of dynamic tensile parameters of rocks using a modified Kolsky tension bar apparatus. J. Rock Mech. Geotechn. Eng. **2**(2), 162–168 (2010)
17. Huang, S., Xia, K., Yan, F., Feng, X.: An experimental study of the rate dependence of tensile strength softening of Longyou sandstone. Rock Mech. Rock Eng. **43**, 677–683 (2010)
18. Khan, A.S., Irani, F.K.: An experimental study of stress wave transmission at a metallic-rock interface and dynamic tensile failure of sandstone, limestone, and granite. Mech. Mater. **6**, 285–292 (1987)
19. Kubota, S., Ogata, Y., Wada, Y., Simangunsong, G., Shimada, H., Matsui, K.: Estimation of dynamic tensile strength of sandstone. Int. J. Rock Mech. Min. Sci. **45**, 397–406 (2008)
20. Lu, D., Wang, G., Du, X., Wang, Y.: A nonlinear dynamic uniaxial strength criterion that considers the ultimate dynamic strength of concrete. Int. J. Impact Eng. **103**, 124–137 (2017)
21. Morozov, N.F., Petrov, Yu.V.: Dynamics of fracture. Springer, Berlin (2000)
22. Petrov, Yu.V., Utkin, A.A.: Dependence of the dynamic strength on loading rate. Mater. Sci. **25**, 153–156 (1989)
23. Petrov, Yu.V.: On the “quantum” nature of dynamic fracture in brittle solids. Sov. Phys. Dokl. **36**, 802–804 (1991)
24. Petrov, Yu.V., Morozov, N.F.: On the modeling of fracture of brittle solids. J. Appl. Mech. **61**(3), 710–712 (1994)
25. Petrov, Yu.V., Morozov, N.F., Smirnov, V.I.: Structural macromechanics approach in dynamics of fracture. Fatigue Fract. Eng. Mater. Struct. **26**, 363–372 (2003)

26. Petrov, Yu.V.: Incubation time criterion and the pulsed strength of continua: fracture, cavitation, and electrical breakdown. Dokl. Phys. **49**, 246–249 (2004)
27. Petrov, Yu.V., Karihaloo, B.L., Bratov, V.V., Bragov, A.M.: Multi-scale dynamic fracture model for quasi-brittle materials. Int. J. Eng. Sci. **61**, 3–9 (2012)
28. Smirnov, I., Konstantinov, A.Yu., Lomunov, A., Bragov, A., Petrov, Yu.V.: The structural temporal approach to dynamic and quasi-static strengthof rocks and concrete (2017)
29. Volkov, G.A., Bratov, V.A., Gruzdkov, A.A., Babitsky, V.I., Petrov, Yu.V., Silberschmidt, V.V.: Energy based analysis of ultrasonically assisted turning. Shock Vib. **18**, 333–341 (2011)
30. Wang, Q.Z., Li, W., Xie, H.P.: Dynamic split tensile test of Flattened Brazilian Disc of rock with SHPB setup. Mech. Mater. **41**, 252–260 (2009)
31. Yan, F., Feng, X., Chen, R., Xia, K., Jin, C.: Dynamic tensile failure of the rock interface between tuff and basalt. Rock Mech. Rock Eng. **45**, 341–348 (2012)
32. Zhang, Q.B., Zhao, J.: A review of dynamic experimental techniques and mechanical behaviour of rock materials. Rock Mech. Rock Eng. **47**(4), 1411–1478 (2013)
33. Zhang, Q.B., Zhao, J.: Determination of mechanical properties and full-field strain measurements of rock material under dynamic loads. Int. J. Rock Mech. Min. Sci. **60**, 423–439 (2013)
34. Zhao, J., Li, H.B.: Experimental determination of dynamic tensile properties of a granite. Int. J. Rock Mech. Min. Sci. **37**, 861–866 (2000)
35. Yavuz, H., Tufekci, K., Kayacan, R., Cevizci, H.: Predicting the dynamic compressive strength of carbonate rocks from quasi-static properties. Exp. Mech. **53**, 367–376 (2013)
36. Demirdag, S., Tufekci, K., Kayacan, R., Yavuz, H., Altindag, R.: Dynamic mechanical behavior of some carbonate rocks. Int. J. Rock Mech. Min. Sci. **47**, 307–312 (2010)

Analytical Research of Character of Relative Permeability Function Under Unsteady Two-Phase Filtration

D. U. Semiglasov(✉) and V. M. Maximov

Institute of Oil and Gas Problems of the Russian Academy of Sciences, Moscow, Russia
samuell@bk.ru

Abstract. The article discusses methodological contradictions and their effects that occur when interpreting the results of researches of relative permeability by the method of non-stationary displacement. For the case of mutual displacement of liquids with equal viscosities on the simplest structural model of a porous medium, an important aspect is shown: the discrepancy between the functions of the average filtration rate and the flow of the displacing phase in the exit section of the sample from the saturation of the cross section. This aspect changes the concept of the nature of the processes of two-phase displacement of liquids in an inhomogeneous porous medium.

Keywords: Filtration · Analytical research · Inhomogeneous porous medium

1 Relevance

Currently the relative permeability function is the determining concept in modern subsurface hydromechanics and is introduced as an equilibrium parameter that determines the force interactions of the porous medium and the filtering phases by A. Darsy's law. The scalar form of A. Darsy's law, neglecting capillary forces, looks like this:

$$w_i = -\frac{k}{\mu_i} k_i^*(s_i) \frac{\partial p}{\partial x}, \; i = 1, 2 \tag{1}$$

where $i-$ phase number; $k-$ coefficient of absolute permeability; μ_i- phase viscosity; $k_i^*(s_i)-$ dependence of relative permeability on phase saturation; $\partial p/\partial x-$ pressure gradient.

The modern theory of two-phase filtration uses relative permeability function for the motion characteristics of the displacement agent in the oil reservoir, and is used for the prediction the development parameters of oil fields in the implementation of waterflooding, as well as for the activities planning aimed at increasing the production capabilities of the producing well stock and ultimate oil recovery factor of the field. The generalization of the materials of modern scientific research devoted to the identification of the main regularities of the form and nature of the relative permeability

V. Karev et al. (Eds.): PMMEEP 2017, SPRINGERGEOL, pp. 249–258, 2018.
https://doi.org/10.1007/978-3-319-77788-7_26

function is an actual task of developing modern theoretical concepts on the technical aspects of the development and operation of oil fields.

The relative permeability function are usually determined from the data of laboratory research of core material on two-phase measuring installations, are less often estimated by petrophysical methods according to the data on the core material. Attempts to generalize experimental data on the nature of the relative permeability function are difficult because of the absence of theoretical presentations, models of the process of displacement at the microlevel, and also qualitative, analytical solutions of the problems of averaging flow characteristics in inhomogeneous medium. However, a large amount of experimental data [1–5] allows us to make some generalizations. Qualitatively, according to the results of the vast majority of laboratory studies, it is possible to distinguish two characteristic types of relative permeability functions (Fig. 1): the first type is functions concave to the axis of the corresponding phase saturation, the second type is functions with the presence of an inflection point in range the mobile saturation of phase.

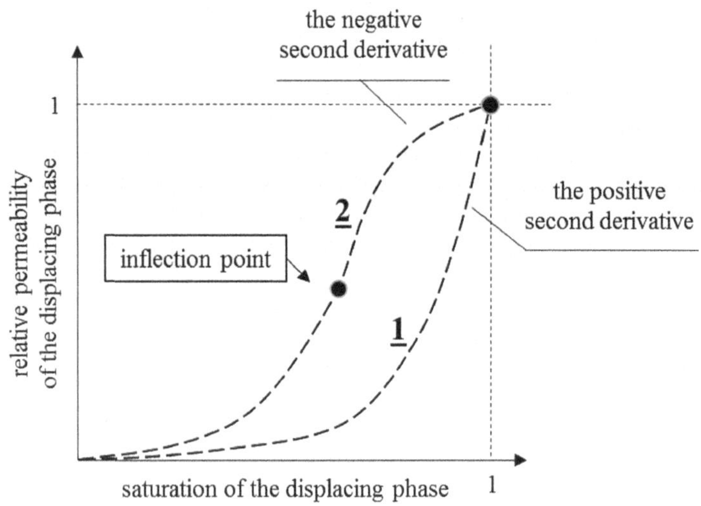

Fig. 1. The result of a qualitative generalization of experimental researches (1 – first type relative permeability function - with the maximum derivative at the saturation point; 2 – second type relative permeability function, containing point of inflection).

Another possible criterion is the value of the second derivative to the corresponding saturation at the point of irreducible saturation. If you classify the relative permeability functions by this characteristic, the first type has a positive second derivative at the specified point, for the second type, the corresponding second derivative is negative.

2 The Mathematical Model of the Process

To identify the reasons of such typification relative permeability functions in the process of two-phase experimental studies on core material, we use the mathematical Stiles model [6]. Model Stiles describes the process of two-phase filtration of liquids of equal viscosities in an inhomogeneous porous medium consisting of independent flows, in each of which a "piston" displacement character is realized at a constant pressure drop. The filtration properties of the inhomogeneous soil are set through the statistical frequency function of the dimensionless permeability $f(b)$ within the Stiles model for $b \in [b_{\min}, 1]$, which determines the distribution of the volume of the porous medium between the independent elements, each of which is characterized by a dimensionless permeability b (Fig. 2). If we apply the principles of the Styles model to describe A. Darsi law in the gallery for an inhomogeneous porous medium, we obtain:

$$w_L^* = \frac{1}{\mu} b_{\max} \int\limits_{b_{\min}}^{1} bf(b)db \left[\frac{\int_{b_\alpha}^{1} bf(b)db}{\int_{b_{\min}}^{1} bf(b)db} \right] \frac{\Delta p}{L} \tag{2}$$

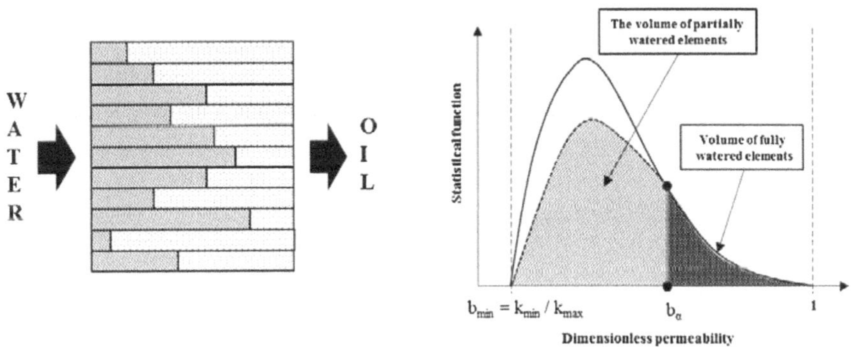

Fig. 2. Principled approach of description the two-phase displacement of the Stiles model.

where w_L^* – velocity of the displacing phase in the gallery L; μ – phase viscosity; b_α – dimensionless permeability of the element«breakthrough»; $b_{\max}$ – maximum permeability of the specimen; $\Delta p / L$ – pressure gradient on the specimen.

The expression for the flow rate of the displacing phase in the gallery, following Stiles model takes the form:

$$q_L^* = \frac{1}{\mu} b_{\max} \int_{b_{\min}}^{1} bf(b)y(b)db \left[\frac{\int_{b_\alpha}^{1} bf(b)y(b)db}{\int_{b_{\min}}^{1} bf(b)y(b)db} \right] F \frac{\Delta p}{L} \tag{3}$$

where q_L^*- flow rate of the displacing phase in the gallery L; $F-$ sectional area of the specimen; $y(b)-$ proportion area fraction with dimensionless permeability b in the total cross section of the sample.

The expression presented in Eq. (2) in square brackets corresponds to the classical definition of relative permeability function (further k_L^w) in A. Darsi law and serves as a coefficient of proportionality between the average filtration rate and the pressure drop across the sample:

$$k_L^w(b_\alpha) = \frac{\int_{b_\alpha}^{1} bf(b)db}{\int_{b_{\min}}^{1} bf(b)db} \tag{4}$$

The expression presented in Eq. (3) in square brackets characterizes the relative permeability pseudofunction (further k_L^q) and serves as a proportionality coefficient between the flowrate of the displacing phase and the pressure drop across the sample:

$$k_L^q(b_\alpha) = \frac{\int_{b_\alpha}^{1} bf(b)y(b)db}{\int_{b_{\min}}^{1} bf(b)y(b)db} \tag{5}$$

Saturation s in the gallery L can be represented in accordance with the Stiles model in the form:

$$s_L(b_\alpha) = \int_{b_\alpha}^{1} f(b)db \tag{6}$$

The dependencies of expressions (4) and (5) on the saturation (6) have parametric form and connected through a parameter b_α.

Functions $k_L^w(s_L)$ and $k_L^q(s_L)$ can be expand in a Taylor series in powers of saturation in a small neighborhood of an arbitrary saturation point s_m corresponding to the condition $b = b_\alpha$ until the second term of the expansion for qualitative investigation of the nature of the dependences. The decomposition will take the form for the function $k_L^w(s_L)$:

$$k_L^w(s_L) \approx k_L^w(s_m) + \left[\frac{b_\alpha}{b_{cp}}\right](s_L - s_m) - \left[\frac{1}{b_{cp} f(b_\alpha)}\right]\frac{(s_L - s_m)^2}{2} \tag{7}$$

where $b_{cp} = \int\limits_{b_{\min}}^{1} bf(b)db \equiv const.$

The coefficients of the decomposition depend not only on the form of the function $f(b)$, as follows from the expression (7), but also directly from its domain of definition, the dependence $k_L^w(s_L)$ at an arbitrary saturation point has a negative second derivative, regardless of the form of the function $f(b)$, which determines its convex shape to the saturation axis. Such a result obviously contradicts the generally accepted ideas about the form of the function $k_L^w(s_L)$ [7], since the classical approach assumes a power approximation of a function $k_L^w(s_L)$ having a positive second derivative at an arbitrary saturation point.

A similar expansion of the function $k_L^q(s_L)$ takes the form:

$$\begin{aligned} k_L^q(s_L) \approx k_L^q(s_m) + \left[\frac{b_\alpha y(b)}{b'_{cp}}\right](s_L - s_m) - \\ - \left[\frac{y(b_\alpha)}{f(b_\alpha)}\left(1 + b_\alpha \frac{y'(b_\alpha)}{y(b_\alpha)}\right)\frac{1}{b'_{cp}}\right]\frac{(s_L - s_m)^2}{2} \end{aligned} \tag{8}$$

where $b'_{cp} = \int\limits_{b_{\min}}^{1} bf(b)y(b)db \equiv const.$

In expression (8), the sign of the second derivative depends on the form of the function $y(b)$, and, therefore, the dependence $k_L^q(s_L)$ can be at an arbitrary saturation point, both convex and concave, to the saturation axis. In this case, the condition for the existence of an inflection point of a function $k_L^q(s_L)$ is determined completely by the parameters of the function $y(b)$ at an arbitrary saturation point.

3 Analytical Example

Consider the analytical example using the power of the family of statistical functions $f(b)$:

$$f(b) = \frac{(n+1)(1-b)^n}{(1-b_{\min})^{n+1}} \quad b \in [b_{\min}; 1] \tag{9}$$

where $n > 0$ arbitrary constant.

The proposed form of the statistical function can be considered as a particular case of a more general probability function of the gamma distribution, with the exclusion of a certain fraction of the low-permeability volume from consideration at the previously valid critical value of the minimum permeability of the reservoir.

From the additional condition of the analogy of functional dependencies $y(b) = f(b)$, excluding the parameter b_α from expressions (4), (5), (6), the functions take the analytical form $k_L^w(s_L)$ and $k_L^q(s_L)$:

$$k_L^w(s_L) = \frac{s_L\left(1 + (n+1)\left[1 - (1 - b_{\min})s_L^{\frac{1}{n+1}}\right]\right)}{1 + (n+1)b_{\min}} \tag{10}$$

$$k_L^q(s_L) = \frac{s_L^{\frac{2n+1}{n+1}}\left(1 + (2n+1)\left[1 - (1 - b_{\min})s_L^{\frac{1}{n+1}}\right]\right)}{1 + (2n+1)b_{\min}} \tag{11}$$

For the examples in Figs. 3 and 4, the function graphs are given for the cases $b_{\min} = 0,05$ and $b_{\min} = 0,5$.

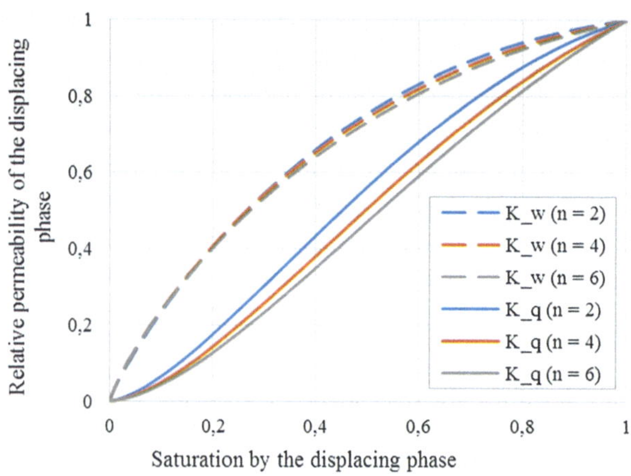

Fig. 3. Functions $k_L^w(s_L)$ and $k_L^q(s_L)$ for $b_{\min} = 0,05$.

As can be seen from the graphs, the parameters $b_{\min}$ and n of the distribution density function $k_L^w(s_L)$ and $k_L^q(s_L)$ in different degrees affect the form of the functions. The parameter $b_{\min}$ determining the width of the permeability spectrum of the sample has the greatest influence on the form of the functions, in contrast to the degree n that influence a weak effect on the shape of the curves $k_L^w(s_L)$ and $k_L^q(s_L)$.

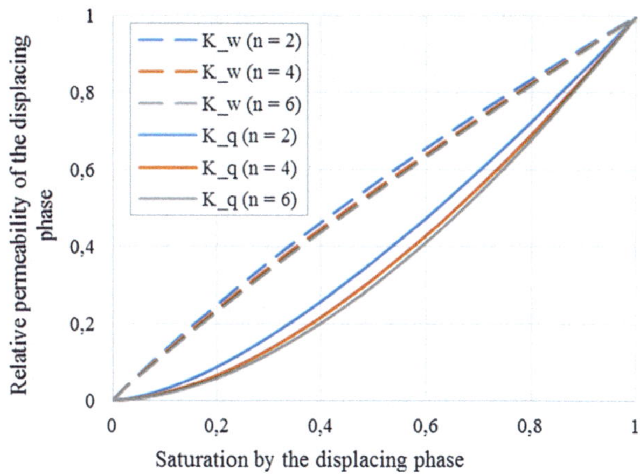

Fig. 4. Functions $k_L^w(s_L)$ and $k_L^q(s_L)$ for $b_{\min} = 0,5$.

Expressions of the first and second derivatives $k_L^w(s_L)$, according to the saturation s_L, take the form:

$$\frac{\partial k_L^w}{\partial s_L} = (n+2)\frac{1-(1-b_{\min})s_L^{\frac{1}{n+1}}}{1+(n+1)b_{\min}} \quad \frac{\partial^2 k_L^w}{\partial s_L^2} = -\left(\frac{n+2}{n+1}\right)\frac{(1-b_{\min})s_L^{-\frac{n}{n+1}}}{1+(n+1)b_{\min}} \tag{12}$$

The analysis of expressions (12) shows that on the interval $s_L \in [0; 1]$ the function $k_L^w(s_L)$ doesn't contain extreme values, and also inflection points. The negative second derivative indicates the convex character of the function to the saturation axis.

Similarly for the function $k_L^q(s_L)$:

$$\begin{aligned} \frac{\partial k_L^q}{\partial s_L} &= 2(2n+1)\cdot\frac{s_L^{\frac{n}{n+1}} - s_L(1-b_{\min})}{1+(2n+1)b_{\min}} \\ \frac{\partial^2 k_L^q}{\partial s_L^2} &= 2\left(\frac{2n+1}{n+1}\right)\cdot\frac{ns_L^{-\frac{1}{n+1}} - (n+1)(1-b_{\min})}{1+(2n+1)b_{\min}} \end{aligned} \tag{13}$$

Equation (13) show that function $k_L^q(s_L)$ on the interval $s_L \in [0; 1]$ as well as the function $k_L^w(s_L)$ doesn't have extreme but, unlike $k_L^w(s_L)$ function can have an inflection point, which is defined as:

$$s_f = \left[\left(\frac{n+1}{n}\right)(1-b_{\min})\right]^{-(n+1)} \quad s_f \in [0; 1] \tag{14}$$

For the examples considered above for $b_{\min} = 0,05$ and $n = 2,4,6$ expression (14) is $s_f = 0,35; 0,42; 0,49$. The condition for the presence of an inflection point on the interval $s_L \in [0; 1]$ has the form:

$$b_{\min} \leq \frac{1}{n+1} \tag{15}$$

For the examples considered for $b_{\min} = 0,5$, the inflection point of the function $k_L^q(s_L)$ doesn't exist on the specified saturation interval for $n > 1$ - the function $k_L^q(s_L)$ has a concave shape to the saturation axis.

Another important consequence of expression (15) is the case of a maximally inhomogeneous sample ($b_{\min} = 0$), in which the inflection point of the function $k_L^q(s_L)$ exists for any value of the parameter n.

4 Conclusion

A qualitative analysis of the form of the dependences $k_L^w(s_L)$ and $k_L^q(s_L)$ shows that within the framework of the considered model of liquids of equal viscosities, the following conclusions can be formulated:

- the character of the function $k_L^w(s_L)$ of the displacing phase doesn't depend on the form of the function $f(b)$, and is a convex function to the saturation axis by a phase whose shape depends only on the width of the permeability spectrum of the sample;
- the character of the function $k_L^q(s_L)$ of the function depends on the form of the function $y(b)$ and can have an arbitrary number of inflection points, depending on the complexity of the latter;
- the expression (3) is used in interpreting two-phase filtration studies and allows one to calculate the function $k_L^q(s_L)$ from the known phase flow rate and the pressure drop across the sample. Hereinafter the function $k_L^q(s_L)$ is used to construct a macroscopic profile of two-phase displacement using the well-known, solution problem J-function obtained for a pseudo-homogeneous porous medium with averaged characteristics, and having the form [7]:

$$\frac{d\varepsilon}{dt} = F'(s_L) = [k_L^q(s_L)]' \tag{16}$$

where $d\varepsilon/dt-$ dimensionless filtration velocity, $F(s_L)-$ J-function.

Since in the context of the problem under consideration the dependence $k_L^q(s_L)$ is similar to the J-function, the further use of the function $k_L^q(s_L)$ in the construction of the two-phase extrusion profile leads to the formation of an incorrect displacement profile. Thus, for the example considered earlier, using the first derivative of the function from (13), expression (16) takes the form (on condition $\varepsilon(0) = 0$):

$$\varepsilon(t) = \left[2(2n+1) \cdot \frac{s_L^{\frac{n}{n+1}} - s_L(1 - b_{\min})}{1 + (2n+1)b_{\min}}\right]\Delta t \tag{17}$$

The correct displacement profile is characterized by a function $k_L^w(s_L)$ whose use in macroscopic modeling, unlike the function $k_L^q(s_L)$, for any parameters of the function $f(b)$ makes it possible to obtain a smooth (without rupture) saturation profile corresponding to the physics of the problem under consideration (Figs. 5 and 6). For the example considered, the correct displacement profile must be determined from expression (16) with regard to (12) as:

$$\varepsilon(t) = \left[(n+2) \cdot \frac{1 - (1 - b_{\min})s_L^{\frac{1}{n+1}}}{1 + (n+1)b_{\min}}\right]\Delta t \tag{18}$$

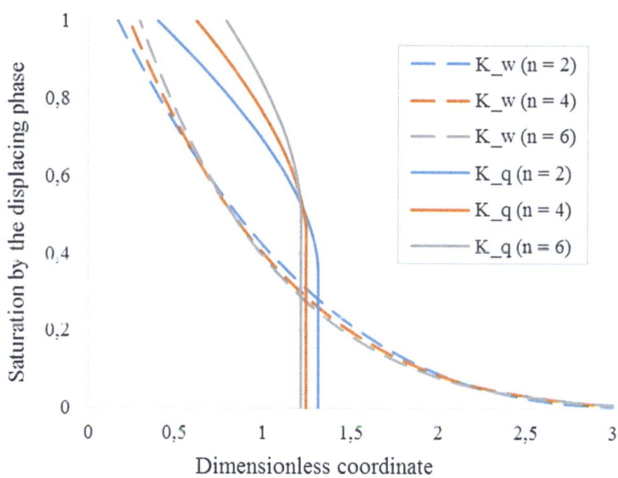

Fig. 5. Displacement profile for the case of $b_{\min} = 0,05$ for $t = 1$.

Within the framework of the physical representations expounded, the direct results of two-phase experimental studies used for mathematical modeling of the displacement process are incorrect and don't allow us to reflect the patterns of saturation distribution under nonstationary two-phase displacement.

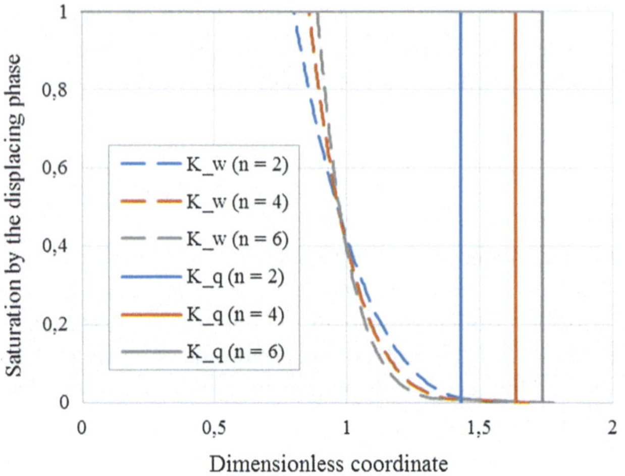

Fig. 6. Displacement profile for the case of $b_{\min} = 0,5$ for $t = 1$.

References

1. Helset, H.M., Nordtvedt, J.E., Skjaeveland, S.M., Virnovsky, G.A.: Relative permeabilities from displacement experiments with full account for capillary pressure. SPE Reservoir Eval. Eng. **1**(2), 92–98 (1998)
2. Mott, R.E., Cable, A.S., Spearing, M.C.: Measurements of relative permeabilities for calculating gas-condensate well deliverability. SPE Reservoir Eval. Eng. **3**(6), 473–479 (2000)
3. Robert, M.C., Alan, P.B.: Relative permeability in tight gas sandstone reservoirs – the "Permeability Jail" model. In: SPWLA 51st Annual Logging Symposium, 19–23 June 2010
4. Zubkov, M.Y., Semenov, V.V., Mikulina, O.I., Pushin, A.V.: The results of studies of the relative phase permeabilities of the different-age productive deposits of the Krasnoleninsk field. Vestnik nedropolzovatelya KhMAD **19**, 16–21 (2008)
5. Orlov, D.M., Ryzhov, A.E., Savchenko, N.V., Perunova, T.A.: Complex experimental investigation of two-phase flow in the reservoirs of the Chayandinskoye oil and gas condensate field and development of a methodology for quantifying the effect of filtration conditions on relative phase permeabilities. Collection of scientific articles "Topical issues of hydrocarbon field deposits research", Moscow (2010)
6. Orlov, V.S.: Design and Analysis of the Development of Oil Fields Under the Regime of Oil Displacement by Water. Nedra, Moscow (1973)
7. Basniev, K.S., Kochina, I.N., Maksimov, V.M.: Underground Hydromechanics. Nedra, Moscow (1993)

Estimation of the Hydraulic Fracture Propagation Rate in the Laboratory Experiment

M. Trimonova[1(✉)], E. Zenchenko[1], N. Baryshnikov[1], S. Turuntaev[1,2], P. Zenchenko[1], and A. Aigozhieva[2]

[1] Institute of Geosphere Dynamics of Russian Academy of Sciences, Moscow, Russia
Trimonova.ma@gmail.com
[2] Moscow Institute of Physics and Technology, Moscow, Russia

Abstract. In this paper we present the results of the laboratory experiment aimed at the research of the hydraulic fracture propagation rate. The laboratory experiment was carried out according to similarity criteria. The fracture growth rate was determined by one direct method and two indirect methods. The direct method was based on registering the formation of the fracture by conductive strips. The first indirect method was based on the variations of the pressure in the well. The second one was based on a numerical solution of the problem of hydraulic fracture propagation.

Keywords: Hydraulic fracturing · Pore pressure · Laboratory experiment Fracture growth · Fluid injection · Numerical model

1 Introduction

Hydraulic fracturing is one of the most effective methods for increasing the productivity of hydrocarbon field development. Since the 1980s, Russia has produced more than 10 thousand hydraulic fracturing operations. The method consists in a highly conductive crack creation by pumping fluid into the well under pressure exceeding the strength of the rock. The rock breaks in a direction perpendicular to the minimum tectonic stress. Due to the continuous fluid injection, the crack increases in size [1]. The studies of hydraulic fracturing in hydrocarbon fields are expensive, therefore the results of such studies are few and not always available. So, conducting experiments on small-scale fracturing in controlled laboratory conditions is a common practice [2, 3].

2 Method of Laboratory Experiments

2.1 Selecting of Sample Material

Laboratory experiments were carried out in accordance with similarity criteria derived for the case of radial crack propagation (1) [4, 5].

V. Karev et al. (Eds.): PMMEEP 2017, SPRINGERGEOL, pp. 259–268, 2018.
https://doi.org/10.1007/978-3-319-77788-7_27

$$N_t = \frac{ti}{r_w^3} \quad N_{\overline{E}} = \frac{\overline{E} r_w^3}{\overline{\mu} i} \quad N_{K_l} = K_l \sqrt{\frac{r_w}{i}}$$
$$N_\sigma = \frac{\sigma}{\overline{E}} \quad N_{K_1} = \frac{K_{1c}^2}{4 r_w \overline{E}^2} \quad N_\lambda = \frac{\sigma_{max}}{\sigma_{min}}, \tag{1}$$

where t - time, i - injection rate, r_w - wellbore radius, $\overline{E} = E/4/(1-\upsilon^2)$, E - Young's modulus, υ - Poisson ratio, $\overline{\mu} = 12\mu$, μ - fluid viscosity, K_l - leakoff coefficient, K_{1c} - critical stress intensity factor, $\sigma = \sigma_{min}$, $\sigma_{min}, \sigma_{max}$ - minimum and maximum principal stresses.

According to the dimensionless complexes and to preserve the similarity of the model and nature, a sample material was chosen. It consists of gypsum and cement mixture in the ratio of 9:1. Water was added to the mixture. The water mass was 0.45 of the mass of the mixture. Gypsum was mixed with lemon acid to reduce the rate of solidification. The gypsum mixture was dried during 2–3 days.

To determine the properties of the material, the preliminary studies were conducted to find the modulus of elasticity and permeability of the sample.

2.2 Estimation of the Sample Permeability

To determine the permeability, the samples were placed between two parallel plates (Fig. 1a).

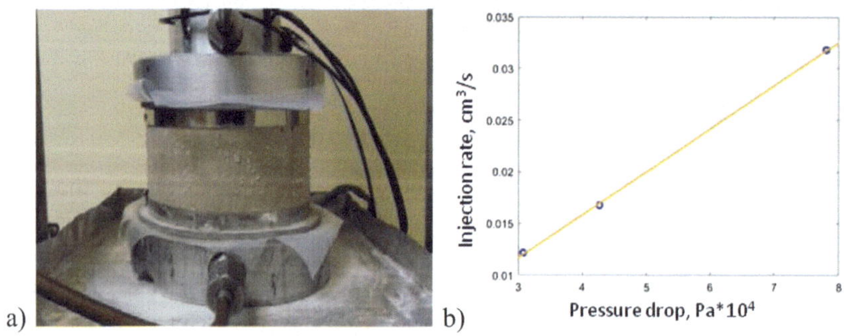

Fig. 1. Setup for measuring the permeability of the sample (a), the dependence of the flow rate of water on the pressure (b).

Fluid was injected into the central hole in the sample with the constant flow rate. The pressure in the center of the sample was measured. The dependence of the water flow on the pressure is shown in Fig. 1b. The flow rate Q through the cylindrical sample in the radial direction depends on the differential pressure according to formula (2) [6, 7]:

$$Q = \frac{2\pi kh(p_B - p_H)}{\mu \ln R_H/R_B} \tag{2}$$

k – sample's permeability, h – sample's height, p_{B} – pressure in the central hole; p_{H} – boundary pressure, μ – fluid viscosity, R_{H} – sample's external radius; R_{B} – central hole radius.

The measurements were made at several flow rates during two days. It should be noted that the permeability of the model material decreased with time. The range of changes was from 2.7–2.1 mD to 1.7–1.0 mD.

2.3 Determination of the Elastic Properties of the Model Material

The tests were carried out under normal conditions. Six cylindrical samples were tested. The sample was deformed at a constant rate until the failure. During the sample loading, ultrasonic scanning was performed and the velocities of longitudinal and transverse waves were measured.

The dependence of deformations on the applied stress occurred to be non-linear for all samples.

This is probably due to the considerable porosity of the samples. As a result of the analysis of the stress-strain curves, the static Poisson and Young's moduli were calculated for all the tested samples, the values of the moduli varied from 0.18 to 0.21 and 3.4 to 4 GPa, respectively.

To determine the dynamic moduli of elasticity, the ultrasonic records were examined. The velocities of longitudinal and transverse waves in the samples were calculated by the time of the signal arrival to the ultrasound receiver.

Dynamic Young's moduli and Poisson's coefficients were calculated from the relations for longitudinal and transverse elastic waves velocities.

$$
\begin{aligned}
c_p &= \sqrt{\frac{E_{dyn}(1-\nu_{dyn})}{\rho(1+\nu_{dyn})(1-2\nu_{dyn})}} \\
c_s &= \sqrt{\frac{E_{dyn}}{2\rho(1+\nu_{dyn})}}
\end{aligned}
\tag{3}
$$

where ρ is the medium density. As a result of the calculations, the value of the Young's modulus determined in the range from 7 to 8 GPa; the value of the Poisson's ratio ranged from 0.22 to 0.28. It can be seen that the values of the Poisson's ratios measured by both methods are similar. Young's moduli determined dynamically are approximately twice as large as those obtained by deformation measuring. The dynamic elastic moduli are calculated from the velocities of longitudinal and transverse waves under the condition of absolute homogeneity of the medium in all directions although the sample material is heterogeneous. That's why the deformation characteristics of rocks determined during static measurements are more reliable and reflect the real properties of both homogeneous and heterogeneous anisotropic rock.

3 Description of the Experimental Setup

The setup consists of the upper and lower covers, between which is a ring with an internal diameter of 43 cm and a height of 6.5 cm (Fig. 2a).

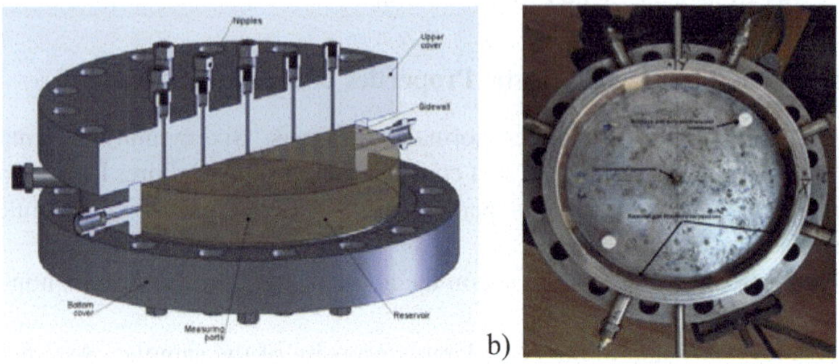

a) b)

Fig. 2. Schematic view of the setup (a), top view of the lower cover of the installation (b)

The top cover is separated from the sample by a rubber membrane. A gap is created between the membrane and the cover. This gap is filled with a water under pressure, which allows us to model the lithostatic pressure in the reservoir model. To create horizontal stresses in the sample, the model is horizontally loaded using four sealed chambers located on the inner side surface (Fig. 2b). The lateral loading is produced by injecting gas or fluid into the opposite chambers.

Holes with a diameter of 6 mm are drilled in both covers and sidewall. These holes are used for both sensors and fluid pumping out or injection.

The fluid injection into the sample can be carried out at constant pressure or at constant flow rate depending on the task. During the experiment it is possible to monitor the change in the sample pore pressure by the pressure sensors located in the holes of the lower cover of the setup.

4 Experiment for the Fracture Propagation Rate Estimation

4.1 Stages of the Experiment

The experiment consisted of the following stages: (a) preparation of the setup; (b) filling of the setup by gypsum (solidification about 2 days); (c) saturation of the sample with a water solution of gypsum and creation of stationary pore pressure distribution in the sample through technological wells (about 1 h) (d) applying horizontal and vertical loads; (e) pumping fracturing fluid into the central well (about 200 s) with simultaneous recording of the pore pressure in the center.

4.2 Experiment Description

To detect and register the fracture propagation, the conductive strips of low-melting alloy with graphite coating were placed on the surface of the sample. The resistance of the strips was between 0.3 and 3 kΩ. The strips were connected to the power supply and their resistivity was registered. The sample was covered with teflon film to protect the sensors.

The experiment was carried out at a vertical load of 4.8 MPa and a horizontal load of 2 MPa. The fracturing pressure was approximately 5 MPa. It is depicted in Fig. 3.

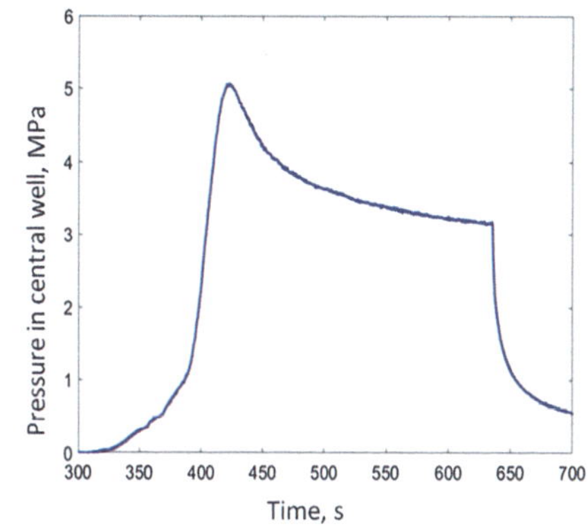

Fig. 3. Well pressure vs. time.

As a result of hydraulic fracturing, the fracture was formed in the sample. It propagated in both directions from the central hole with the wing length about 13 cm. The sample photo after fracturing is shown in Fig. 4.

Fig. 4. Photo of the sample with the fracture after the experiment.

The fracture formed and propagated in the direction of the main compressive stress.

One branch of the crack reached the technological well and stopped, another one slightly deviated from the well. It was possible to estimate the fracture propagation with the help of the strip resistivity measurements. Values were taken for the time and distance of the fracture input-output from the strips. Time was estimated by measuring the voltage drop on the strips.

In the considered experiment, four strips with numbers 4, 5, 6, 7 reacted to the growth of the fracture. The dependence of the fracture length on the time constructed according to the data from the reacted strips is shown in Fig. 5.

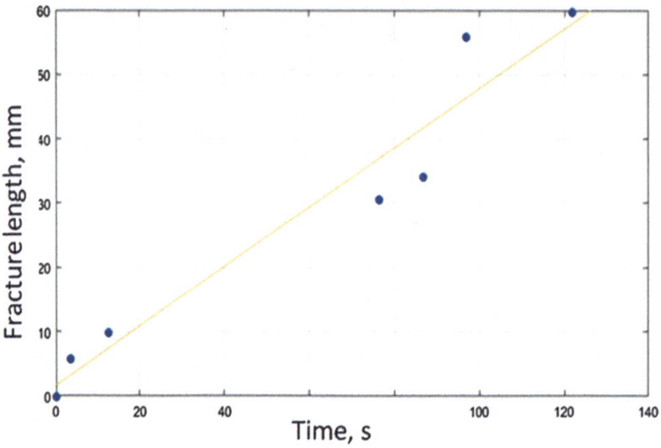

Fig. 5. Dependence of the fracture length on the entry and exit time of strips.

From the dependence, it was possible to estimate the average value of the velocity, which was found to be 0.45 mm/s ± 0.07 mm/s.

The average fracture rate was also estimated as the ratio of the fracture length to the time of its growth, measured from the pressure plot in the well (the time from the moment of the pressure maximum to the end of injection). The resulting velocity value was 0.6 mm/s, which approximately corresponds to the value of the velocity estimated by the conductive strips.

5 Mathematical Model of the Hydraulic Fracture Propagation

The solution of the fracture propagation problem is a step-by-step solution of several subtasks. First, the flow of fluid in the fracture is considered. The result of this calculation is the pressure field in the fracture. Then the piezoconductivity equation is solved for the two-phase flow in the sample; the pressure in the fracture is used as boundary condition. Thus, the pressure distribution in the sample becomes known. According to the known pressure in the fracture and in its small vicinity, the leaks are estimated, which are used in the next step to solve the equation in the fracture. The stress intensity factor is calculated after the final calculation of the pressure field in the fracture. The growth of the fracture is determined by this factor. When the factor reaches a value greater than the known critical value, the fracture begins to grow: a new element is added to the fracture length. The size of this element is determined by the size of the grid. The described algorithm is then reproduced for a new fracture length. In this paper, the problem is considered in the following assumptions:

1. The problem is assumed to be two-dimensional.
2. The medium is assumed to be homogeneous and isotropic in its properties.
3. The flow is assumed to be two-phase. First, the sample is saturated with water, and then vacuum oil is injected into the central well. The viscosity of this oil is much greater than the viscosity of the water. This fact allows us to use a simplified model of two-phase filtration: the frontal drive simplification.

5.1 Fluid Flow in the Fracture

A simultaneous consideration of the continuity equation and the equation of motion for the case of the two-dimensional Perkins-Kern fracture model leads to the solution of the following differential equation [6, 7]:

$$\frac{\partial w}{\partial t} - \frac{1}{12\mu w_0} \frac{\partial \left(w^3 \frac{\partial w}{\partial r} \right)}{\partial r} = -2q_L. \tag{4}$$

This equation is solved with the following boundary conditions:

$$\begin{aligned} w(r = c, t) &= w_0 (p_{well} - \sigma) \\ w(r = l(t), t) &= 0 \end{aligned} \tag{5}$$

5.2 Fluid Filtration in thc Sample

The fluid filtration in the sample is described by the piezoconductivity equation in cylindrical coordinates for a two-phase process in the frontal drive simplification [6, 7].

$$\kappa_{o,w} \frac{\partial P_{o,w}}{\partial t} = \frac{1}{r} \left(\frac{\partial}{\partial r} \left(r \frac{\partial P_{o,w}}{\partial r} \right) \right) + \frac{1}{r^2} \frac{\partial^2 P_{o,w}}{\partial \phi^2} \tag{6}$$

Here $P_{o,w}(x, y, t)$ is the fluid (oil or water respectively) pressure inside the sample; $\kappa_{o,w}$ – is the corresponding piezoconductivity coefficient.
On the surface between the phases the following condition has to be valid:

$$P_o = P_w, \quad V_0 = V_w. \tag{7}$$

Equations 6 and 7 are solved with the following boundary and initial conditions:

$$\begin{aligned}&\frac{\partial P(r=r_s,\varphi,t)}{\partial r}=0\\&P(r=r_{fr},\varphi=\varphi_{fr},t)=P_{fr}(r_{fr},t)\\&P(r=r_{well(i)},\varphi=\varphi_{well(i)},t)=P_{well}(i)\\&P(x,y,t=0)=P_{\infty}\end{aligned}\tag{8}$$

5.3 Fracture Growth

It is assumed that the fracture is a normal cut. The fracture propagation condition is used in the following form [8]:

$$K_{\rm I}=\frac{1}{\sqrt{\pi L}}\int\limits_{-L}^{L}\left(p_f(x,t)-\sigma_i\right)\sqrt{\frac{L+x}{L-x}}d\xi=K_{\rm Ic}\tag{9}$$

The medium is considered to be homogeneous with constant characteristics: minimum horizontal stress and fracture toughness (σ and K_{Ic}). The integral in Eq. 9 is simplified and calculated analytically in each cell. Stress intensity factor is calculated at each time step, then it is compared with K_{Ic}, and a conclusion is made about the growth of the fracture:

$K_I > K_{Ic}$ - elongation of the boundary points;
$0 \le K_I \le K_{Ic}$ - no elongation.

A detailed numerical solution of Eqs. 4–9 is presented in the works of the authors [9, 10].

5.4 Results of Numerical Simulation

In numerical simulation, the initial length of the fracture was set equal to the length of the initiators in the borehole. Their orientations determine the initial direction of the fracture.

The fracture length, pressure in the fracture after 500 s of the experiment are shown in Figs. 6 and 7.

Pressure in the sample and compared view of the fractures received from the experiment and numerical simulation is shown in Fig. 8.

It can be seen that the fracture ceased to grow after 330 s. It is possible to determine the average fracture growth rate by dividing the fracture length increase by the time of its growth. Thus, the average fracture growth rate was 0.54 mm/s, which coincides approximately with the rate determined by direct experiment.

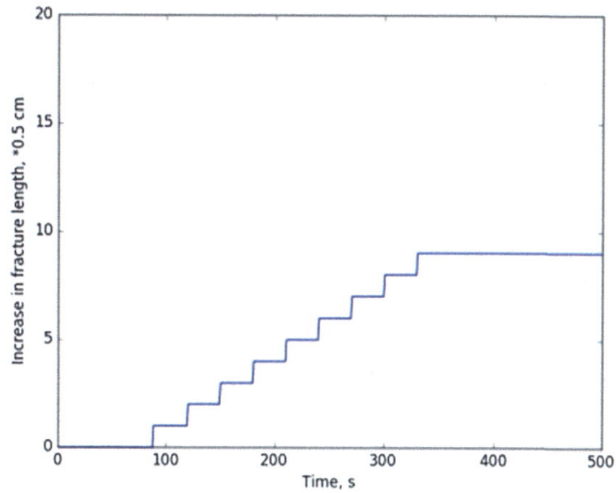

Fig. 6. Fracture length vs. time after 500 s of the experiment.

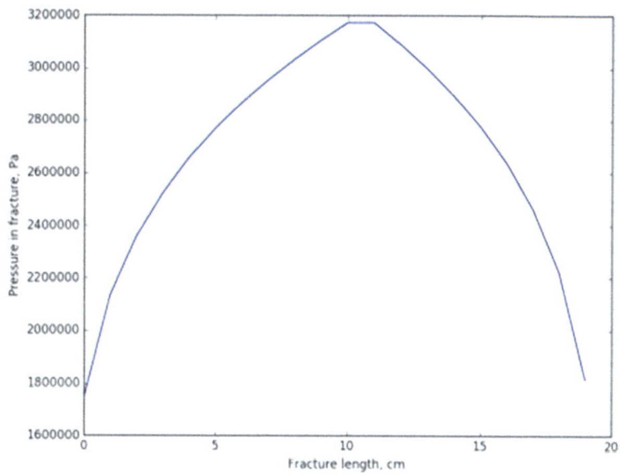

Fig. 7. Pressure in the fracture after 500 s of the experiment.

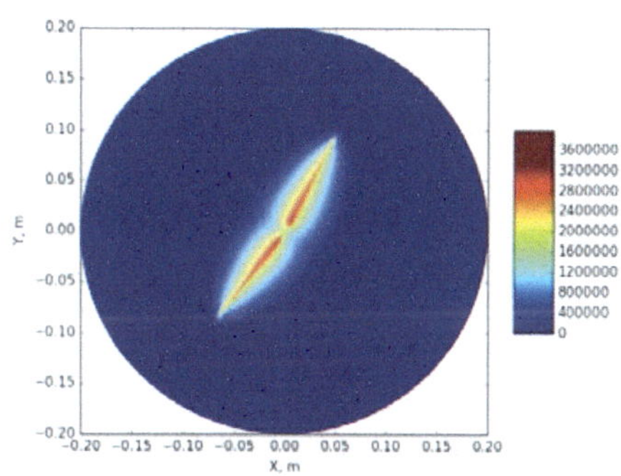

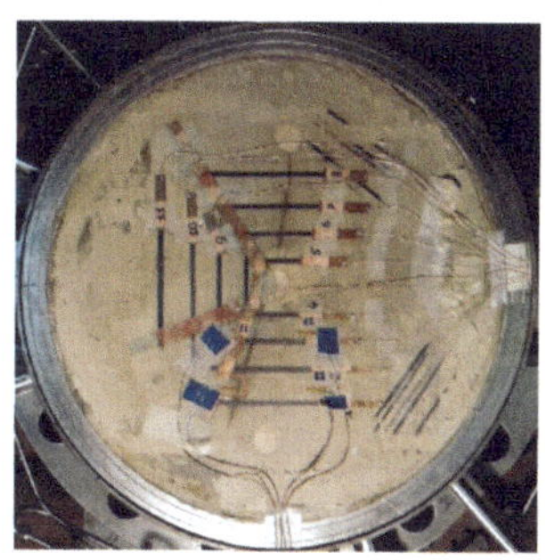

Fig. 8. Pressure in the sample after 500 s of the experiment and compared view of the fractures.

6 Conclusions

The fracture growth rate was measured in three different ways and was approximately 0.5 ± 0.1 mm/s in each of the methods. This fact is a confirmation that numerical simulation reflects fairly accurate hydraulic fracture propagation process. It means that in the numerical modeling of the real oil field and real hydraulic fracturing we can estimate the rate of the hydraulic fracture propagation too. Also, we can achieve the necessary velocity of the fracture propagation depending on the injection pressure or injection rate

The work was carried out according to the State task (№ 0146-2014-0012), with the financial support of LLC "RN-UfaNIPIneft" and RFBR (project No. 16-05-00869).

References

1. Economides, M.J., Nolte, K.G.: Reservoir Stimulation, 3rd edn. Willey, New York (2000)
2. Zoback, M.D., Rummel, F., Jung, R., Raleigh, C.B.: Laboratory hydraulic fracturing experiments in intact and pre-fractured rock. Int. J. Rock Mech. Min. Sci. Geomech. Abstr. **14**, 49–58 (1977)
3. Medlin, W.L., Masse, L.: Laboratory experiments in fracture propagation. Soc. Pet. Eng. **24**(3) (1984)
4. de Pater, C.J., Cleary, M.P., Quinn, T.S.: Experimental verification of dimensional analysis for hydraulic fracturing. SPE Prod. Facil. **9**, 230–238 (1994)
5. de Pater, C.J., Weijers, L., Savic, M.: Experimental Study of Nonlinear Effects in Hydraulic Fracture Propagation. SPE Prod. Facil. **9**, 239–249 (1994)
6. Aziz, K., Settari, A.: Petroleum Reservoir Simulation. Institute of Computer Research, Moscow (2004)
7. Basniev, K., Kochina, I., Maximov, V.: Subsurface Hydromechanics. Nedra, Moscow (1993)
8. Cherepanov, G.: Mechanics of brittle fracture. Nauka, Moscow (1974)
9. Trimonova, M., Dubinya, N.: The main tendencies of water-induced hydraulic fracture propagation. SPE-176638-MS (2015)
10. Trimonova, M., Baryshnikov, N., Zenchenko, E, Zenchenko, P., Turuntaev, S.: The study of the unstable fracture propagation in the injection well: numerical and laboratory modeling. SPE-187822-MS (2017)

Paleomagnetism of Some Basalts Samples from the Red Sea Rift Zone

V. I. Maksimochkin(✉) and L. R. Preobrazhenskii

Lomonosov Moscow State University, Moscow, Russian Federation
maxvi@physics.msu.ru

Abstract. Based on the results of a study of the magnetic properties of basalt samples dredged from the Red Sea rift zone at a latitude of 18° N, and of geomagnetic field paleointensity determination by Thellier–Coe method, it was concluded that the studied basalts are related to three different stages of crust formation in this region. It is determined that the geomagnetic field paleointensity at the time of formation of the studied basalts samples was 1.1–1.9 times higher than current geomagnetic field intensity in this region. Obtained results indicate that the geomagnetic field intensity in this region have been decreasing in the last 100 thousand years.

Keywords: Paleomagnetism · Red Sea · Rift zone · Basalts

1 Introduction

According to modern ideas, the formation of new earth crust occurs in oceanic rift zones. An important for understanding the formation of the earth's crust in the context of the concept of tectonics of lithospheric plates is the study of the rift zone of the Red Sea, where, according to modern concepts, the young ocean begins with the formation and development of the oceanic crust.

So, in paper [1] it was shown that the growth of the bottom of the Red Sea for Brunhes and Matuyama epochs occurs asymmetrically. According to the calculations given in this work, the rate of movement of the African plate is 7 mm/year for Brunhes epoch, and rate Arabian plate -6 mm/year. The magnetic characteristics of the basalts of the bottom of the Red Sea are high different from those, for example, for the Mid-Atlantic Ridge. The basalts of the Red Sea have a NRM 1.5-2 times higher than the basalts of the Reykjanes Ridge (North Atlantic) or the South of the MAR [2, 3], and high value of the Koenigsberger parameter ($Q_n > 150$). The titanium content in the titanomagnetite of the basalts of the Red Sea as a whole was lower than in the MAR basalts, and the Curie points is higher correspondingly. Basalts with signs of the oxyexsolution of titanomagnetite were discovered. The reason for such features, undoubtedly due to the peculiarities of the formation of basalts, is still unclear.

Based on geochemical studies in paper [4], it was shown that the formation of basalts of the rift zone of the Red Sea occurred in two stages. These stages differ by the rate of magma supply and, accordingly, different oxidation-reduction conditions, which is probably was reflected in the distribution of the anomalous geomagnetic field in the area under study.

V. Karev et al. (Eds.): PMMEEP 2017, SPRINGERGEOL, pp. 269–281, 2018.
https://doi.org/10.1007/978-3-319-77788-7_28

Currently geotectonic constructions based on geomagnetic data are based on the assumption that the source of magnetic anomalies, which characteristic for rift zones, are rocks of a magnetoactive layer, most of magnetization of which is irreversible and formed in ancient magnetic field. Determining the paleointensity geomagnetic field on the NRM of basalts of the rift zone of the Red Sea, we can formulate the stages of lava eruption and stages of crust formation in this region, using data on the time variation of the Earth magnetic field [5, 6]. According to the physical and chemical properties of the titanomagnetites - carriers of the remanent magnetization of basalts, we can draw conclusions about the thermodynamic condition of their formation. In [7] attempt was made to determine the paleointensity of the geomagnetic field (Hpl) in the Red Sea region by NRM of basalts of rift zone. It was shown that Hpl, determined by the NRM of some samples, is 2.5 times higher than its current value. The reason for this discrepancy is still unclear.

In this work, for improvement the periods of volcanic activity in the rift zone of the Red Sea and to determine the paleointensity of the geomagnetic field of time of the formation of basalts, paleomagnetic studies of samples raised from the bottom of the Red Sea have been carried out.

2 Samples, Techniques and Methods of Experiment

2.1 Samples

Paleomagnetic studies in this work were carried out on basalt samples from 9 sites of rift of the Red Sea: №№ 53-4, 59-2, 62-1, 62-2, 57, 61-1, 67, 68-2, 71-3. Samples were raised by the R/V "Aquanaut" in area of 17° 58' N., 40° 04' E. and kindly provided to us by the leading scientific staff of the IO RAS, professor A. A. Shreider. The age of the rocks of the seabed according to the geological structure of the rift zone of the Red Sea in region of 18° N. less than 100 thousand years [8]. In paper [4], two periods of lava overflow were revealed from the petrochemical and magnetic data of the basalts of this region of the rift zone of the Red Sea. The conclusion is drawn that basalts №№ 61-1, 57, 68-2, 71-3 are older than basalts №№ 53-3, 59-2, 62-2.

2.2 Techniques and Methods of Experiment

Measurement of the magnitude and direction of the remanent magnetization of the samples at room temperature was carried out on JR-6 magnetometer (AGICO company). The magnetic susceptibility (k) of the samples at room temperature was measured on the IMBO-M instrument.

The Curie point was determined from the temperature dependence of the magnetic susceptibility k(T) on the MFK1-A instrument, also on the dependence of the magnetization I(T) in the field B = 0.24 T on the made by us vibration magnetometer VM-1. On this magnetometer, a program for control and collecting data was implemented using the ZETLab 220 module [9].

The saturation magnetization (Ms) was determined by the magnetic flux meter (F-191) according to the Weiss and Forrer method. Hysteresis characteristics: remanent saturation magnetization (Mrs), coercive force (Hc) and residual coercive force Hcr were measured on a VMA-1 vibration magnetometer [10]. The paleointensity of the geomagnetic field was determined by the Thellier method in the Coe modification [11]. The magnetic field paleointensity was calculated by the formula:

$$H_{pl} = KH_{lab} \tag{1}$$

K is coefficient of linear approximation, obtained from the Arai–Nagata diagram, Hlab-laboratory field acting on the sample at the time of formation PTRM in the Thellier cycles.

For estimate the reliability of the Hpl determination, the q parameter was calculated – the quality index of Arai–Nagata diagram in interval T1, T2, combining and taken into account four factors simultaneously [12].

$$q = \frac{|K|fg}{\sigma_K} \tag{2}$$

g (gap factor) is factor of the uniformity of the distribution of points along the y-axis on the Arai–Nagata diagram in the temperature range T1 to T2.

$$g = 1 - \left[\frac{\sum_{i=1}^{i=N-1} (NRM_i - NRM_{i+1})^2}{\left(\sum_{i=1}^{i=N-1} (TRM_i - TRM_{i+1})^2 \right)^2} \right] \tag{3}$$

N is the number of points on the Arai–Nagata diagram in the interval T1–T2.
f is the fraction of NRM that falls on the interval T1, T2, if Hpl is determined on the interval T0–Tc, then f = 1.

$$f = \frac{NRM_{T_0}}{NRM_{T_1} - NRM_{T_2}} \tag{4}$$

K is the tangent of the slope of the approximating line in the Arai–Nagata diagram in the interval T1, T2. NRMm and TRMm are arithmetic mean values in the interval T1, T2.

$$K = -\left[\frac{\sum_{i=1}^{i=N} (NRM_i - NRM_m)^2}{\sum_{i=1}^{i=N} (TRM_i - TRM_m)^2} \right]^{-0.5} \tag{5}$$

σ_K is mean square error of the arithmetic mean K

$$\sigma_K = \left[\frac{2\sum_{i=1}^{i=N} (NRM_i - NRM_m)^2 - 2K\sum_{i=1}^{i=N} (TRM_i - TRM_m)(NRM_i - NRM_m)}{(N-2)\sum_{i=1}^{i=N} (TRM_i - TRM_m)^2} \right]^{-0.5} \tag{6}$$

Correspondingly, the higher q, the higher the reliability of determining Hpl. According to [12], the data having the quality factor q > 5 are valid.

3 Results

3.1 Natural Magnetic Characteristics

In the work 28 samples from 9 sites were studied. The natural remanent magnetization (NRM) and magnetic susceptibility in the field 300 A/m were measured. The results are shown in Fig. 1. The values of the magnetic susceptibility and natural remanent magnetization of samples from different sites varied within wide limits.

NRM = 4–80 A/m (<NRM> = 45 A/m), k = (0.204–1,94)*10^{-2} units SI (<k> = 1,01*10^{-2} units SI). Variations of magnetization and magnetic susceptibility were much smaller for samples of each site (Table 1, Fig. 1)

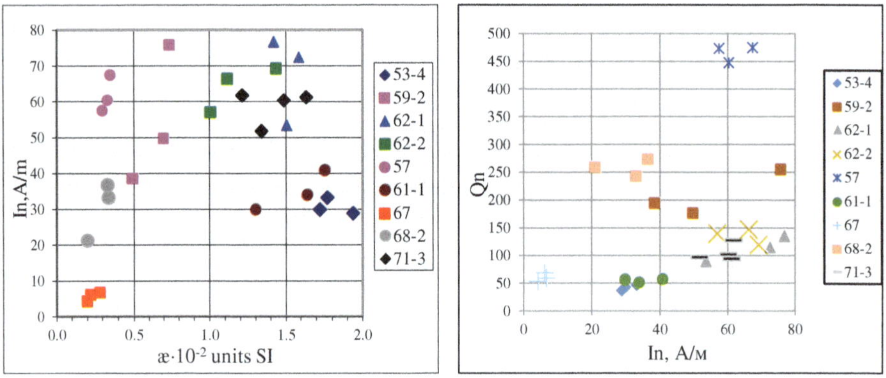

a) b)

Fig. 1. (a) Natural remanent magnetization and magnetic susceptibility of a collection of samples of basalts of the Red Sea. (b) Koenigsberger parameter and natural remanent magnetization.

Since the values of the NRM and magnetic susceptibility k of the rock are proportional to the concentration of ferrimagnetic grains in it, such a strong spread is probably associated with a variation concentration of ferrimagnetic grains in the sample. Therefore, neither on the magnitude of the natural remanent magnetization nor on the value of the magnetic susceptibility can conclusions be drawn about the preservation of the natural magnetic state of the sample. The Koenigsberger parameter does not depend on the concentration, but depends on the type of magnetization and the structural state of the ferrimagnetic grains that included in the rock, and also characterizes the preservation of the primary remanent magnetization.

Table 1. Natural magnetic characteristics basalts samples of the Red Sea

Sample	NRM, A/m	$k \cdot 10^{-2}$ units SI	Q
53-4(1)	29,99	1,727	43
53-4(2)	28,98	1,942	37
53-4(3)	33,32	1,773	47
59-2(1)	75,85	0,741	256
59-2(2)	49,78	0,702	177
59-2(3)	38,61	0,496	195
62-1(1)	76,79	1,424	135
62-1(2)	72,54	1,590	114
62-1(3)	53,47	1,508	89
62-2(1)	66,41	1,119	148
62-2(2)	69,35	1,437	121
62-2(3)	57,01	1,009	141
57(1)	60,48	0,338	448
57(2)	67,45	0,355	475
57(3)	57,51	0,304	473
61-1(1)	40,88	1,759	58
61-1(2)	34,04	1,646	52
61-1(3)	29,92	1,304	57
67(1)	6,295	0,227	69
67(2)	6,845	0,288	59
67(3)	4,392	0,205	53
68-2(1)	36,62	0,334	274
68-2(2)	21,16	0,204	259
68-2(3)	33,11	0,340	243
71-3(1)	61,11	1,636	93
71-3(2)	51,64	1,339	96
71-3(3)	60,25	1,488	101
71-3(4)	61,69	1,213	127

$$Q_n = \frac{NRM}{Hk} \tag{7}$$

H = 40 A/m. The values of the Koenigsberger parameter for the studied samples:

$$\mathrm{Qn} = 37\text{–}475(<\mathrm{Qn}> = 158).$$

The high values of the parameter Qn indicate a high potential paleoinformativity of NRM of basalts samples.

3.2 Hysteresis Characteristics

To determine the structural state of ferrimagnetic grains, the hysteresis characteristics of the samples were measured: saturation magnetization (Ms), residual saturation magnetization (Mrs), coercive force (Hc), residual-coercive force (Hcr). Based on this results, we constructed the Day's diagram [13] (Fig. 2).

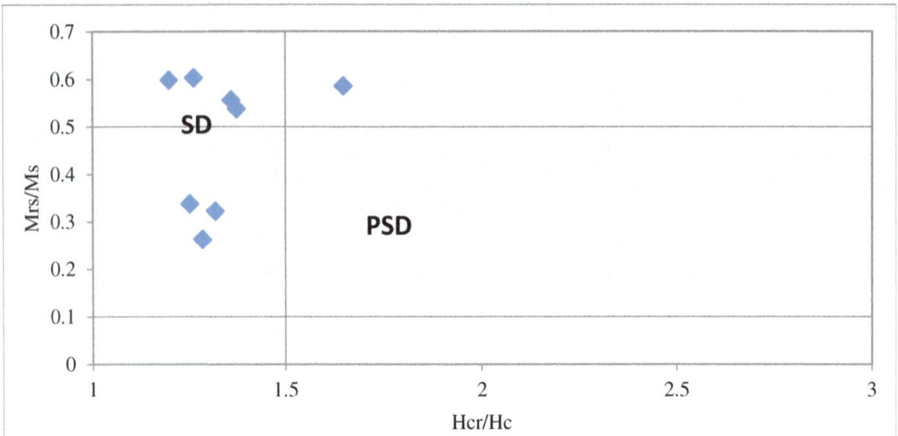

Fig. 2. Day's diagram.

According to Day's criterion, the ferrimagnetic fraction of the samples is represented by single-domain and pseudo-single-domain grains. Therefore, for determine the paleointensity of the geomagnetic field we can use the Thellier method.

3.3 The Results of Thermomagnetic Analysis

For the study the phase state of ferrimagnetic grains, the magnetic susceptibility was measured as a function of temperature (see Fig. 3) and the magnetization in the field B = 0.24T was measured as a function of temperature (see Fig. 4). The results of thermomagnetic analysis show that in samples 53-4, 59-2, 62-1, 62-2, 57, 61-1, 71-3 dominated a single-phase composition. The Curie temperature of this phase, determined from the k(T) dependence, is in the range from 234 °C to 431 °C. The Curie temperature, determined from the Ms(T) dependence, turned out to be 10–15 lower. (see Table 2). In some samples, a small amount of phase with a Curie temperature equal to Tc of magnetite was found. Sample 67 has a two-phase composition (Tc1 = 412 °C и Tc2 = 580 °C), the magnetization of one of those components has chemical nature probably. In sample 68-2, two phases with Curie temperatures Tc1 = 280 °C and Tc2 = 500 °C were also found from the Is(T) dependence.

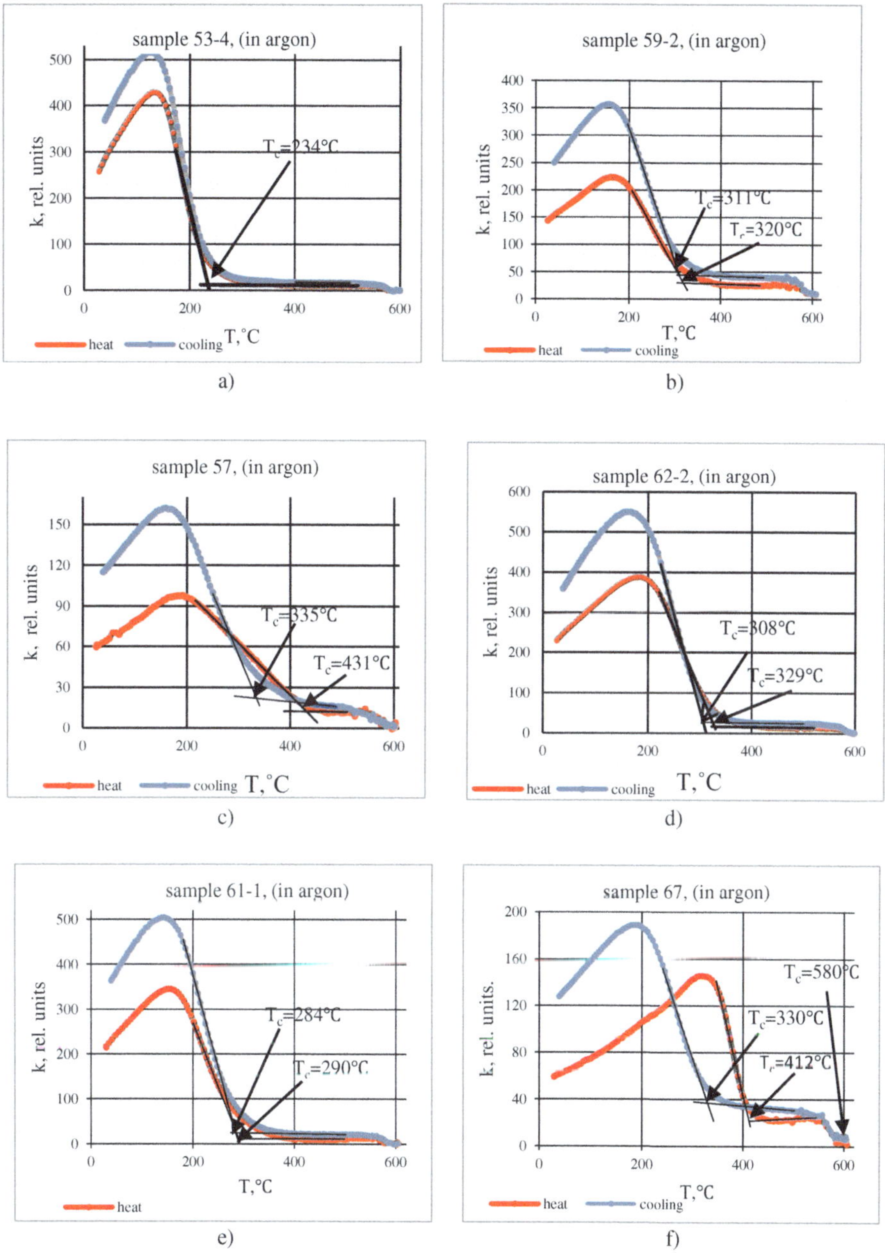

Fig. 3. Thermomagnetic analysis of basalts sample of the Red Sea (red curve – heating, blue curve – cooling)

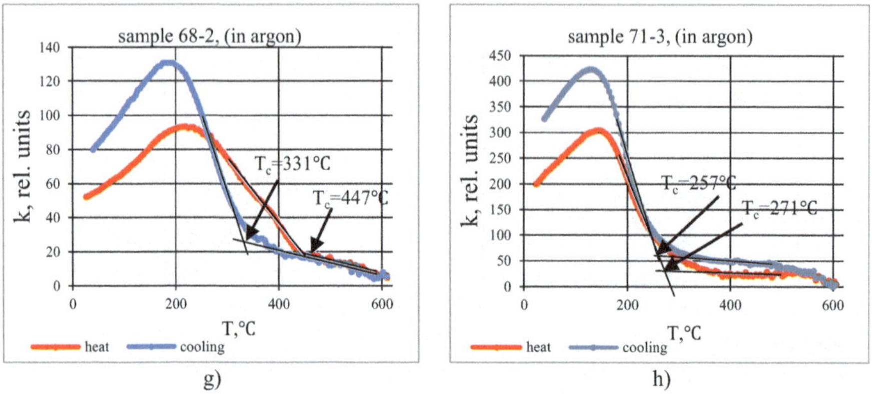

Fig. 3. (*continued*)

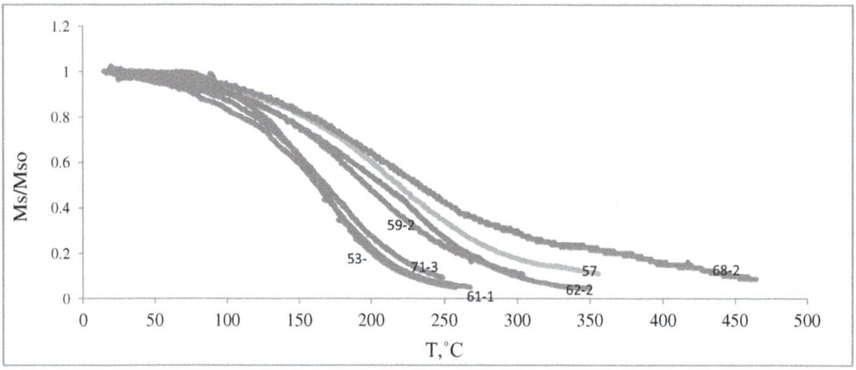

Fig. 4. Change in magnetization in the magnetic field of 0.24 T when heating basalt samples. (The numbers near the curves are the numbers of the basalts samples)

Table 2. Results

Sample	53-4	71-3	59-2	61-1	62-2	57	68-2	67
T_c, °C{k(T)}	234	271	320	290	329	431	447	412/500
T_c, °C{k(T)} (after heating)	234	257	310	284	308	335	331	330
T_c, °C {Ms(T)}	225	240	280	230	305	300	280/500	-
Mrs/Ms	0.21	0.26	0.34	0.22	0.28	0.38	0.32	0.29
Hcr/Hc	1.37	1.29	1.25	1.36	1.26	1.2	1.32	1.65
K	−0.809	−0.996	−1.130	−1.247	−1.184	−1.501	−1.344	−0.442
g	0.9879	0.9969	0.9967	0.9885	0.9979	0.9975	0.9921	0.50
f	0.9092	0.8906	0.9129	0.9746	0.8373	0.8162	0.8129	0.3253
σ_b	0.0147	0.0235	0.0550	0.0447	0.0418	0.0341	0.0196	0.0620
q	49	38	19	27	24	36	55	1
Hlab, A/m	40	40	40	40	40	40	40	40
Hpl, A/m	32.4 ± 0.6	39.8 ± 0.9	45.5 ± 2.2	49.9 ± 1.8	47.4 ± 1.7	60.0 ± 1.4	53.8 ± 0.8	18 ± 2
VADM, A·m^2	6.42E + 22	7.88E + 22	9.01E + 22	9.88E + 22	9.39E + 22	11.9E + 22	10.7E + 22	-

3.4 Results of Determining of the Geomagnetic Field Paleointensity

The results of the NRM study of basalt samples by the Thellier–Coe method in the representation of the Arai–Nagata and Zijderveld [14] diagram are shown in Fig. 5. Diagrams of Zijderveld show that the natural magnetization of these samples is practically one-component. According to these regularities and the linear relationship between NRM and PTRM in the Arai–Nagata diagram for samples 53-4, 59-2, 62-1, 62-2, 57, 61-1, 68-2, 71-3, we can state that the primary magnetization of these samples is preserved and has a thermally remanent nature. From the Arai–Nagata diagram for sample 67, it can be seen that the remanent magnetization has a two-component structure. It is also seen that after heating the sample to 482 °C without magnetic field only about 30% of the NRM collapses, i.e. the main contribution to the magnetization of the sample is made be a component with Curie temperature close to Curie temperature of magnetite, and most of NRM is probably of chemical origin. For each sample, the paleointensity of the ancient magnetic field Hpl was determined from the slope of the approximation line of data in the Arai–Nagata diagram. Also, for each sample, the coefficients characterizing the quality of the Hpl determination were determined (see Table 2). From the getting value of paleointensity, the dipole magnetic moment of Earth (VADM) was calculated under the assumption of axial dipole according to the formula:

$$VADM = \frac{4\pi H_{pl} R_E^3}{\sqrt{1 + 3cos^2\Theta}} \qquad (8)$$

Θ – polar angle (for the studied region $\Theta = 72^0$).

The calculated value of VADM varied from $6.42*10^{22}$ A*m^2 to $11.9*10^{22}$ A*m^2

3.5 The Discussion of the Results

As is known, Curie temperature of titanomagnetite of basalt depends on the content of titanium in it and the degree of oxidation, which is determined by the oxidation medium and grows with age of the rock. Thermomagnetic analysis based on dependence of the magnetic susceptibility on the temperature under heating in argon showed that titanomagnetites of basalt № 53-4 and № 71-3 (the first group) have smaller Curie points (Tc = 234 °C Tc = 271 °C, respectively), i.e. contains more titanium that titanomagnetite samples of basalt of the second group-№№ 59-2, 61-1, 62-2 (Tc = 320 °C, Tc = 290 °C, Tc = 329 °C, respectively). Moreover, it was found that after heating of the samples noted above to 600 °C in argon, the Curie temperature decreased insignificantly, maximum of 20 °C (see Fig. 3), which indicates a low degree of change in titanomagnetite in situ. It was also found that the paleointensity values of the geomagnetic field, determined from the NRM of basalts of the second group, are quite close to each other: Hpl = 45.5; 47.4 and 49.9 A/m (see Table 2). The magnitude of the paleointensity of the geomagnetic field, determined from samples № 53-4 and 71-3 with low Curie points, turned out to be lower: Hpl = 32.4 and 39.8 A/m.

The two-phase composition of the ferrimagnetic fraction of basalts №№ 57, 68-2 and 67 with Tc1 = (412–447) °C, showed a decrease by about 80–110 °C of the Curie

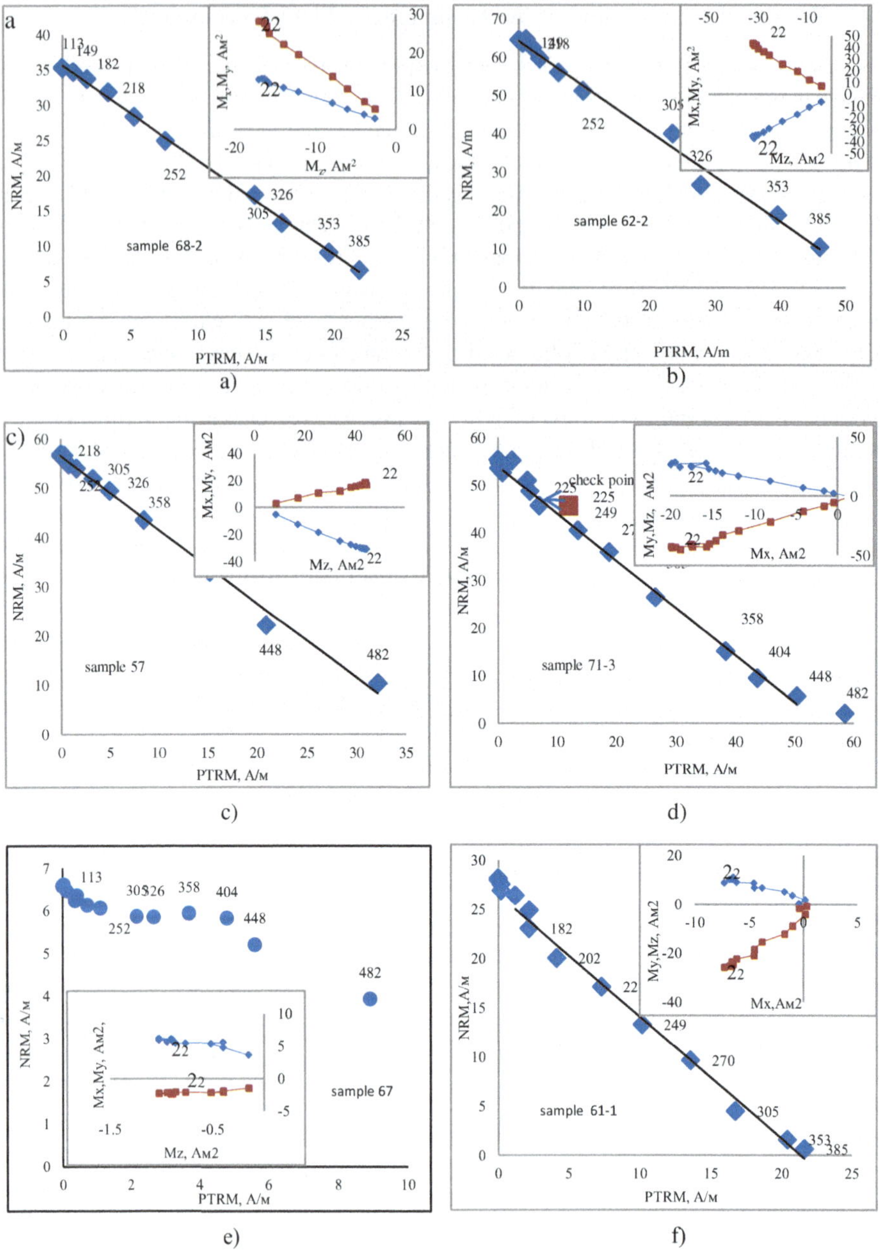

Fig. 5. Thermal demagnetization results of NRM in the Thellier–Coe cycles in the representation of the Arai–Nagata and Zijderveld diagrams (the number near the points are the maximum value of the temperature in the Thellier cycle)

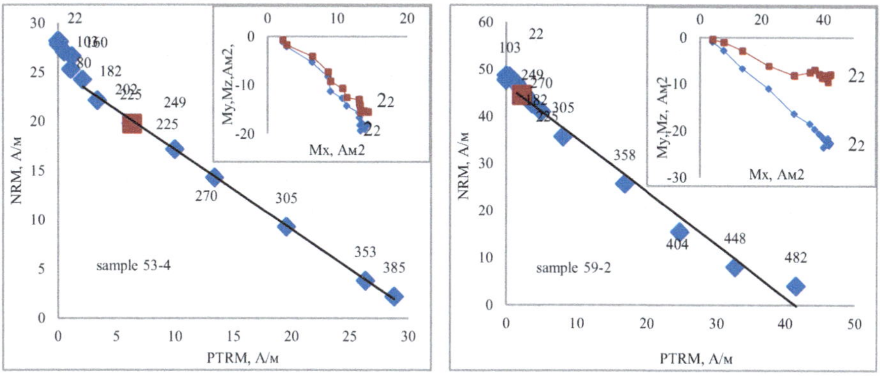

g) h)

Fig. 5. (*continued*)

temperature when the samples were heated to 600 °C in argon (see Fig. 3), which indicates that the titanomagnetite of these basalts has a greater degree of non-stoichiometry [15], than the titanomagnetite of the basalt samples seen above. Consequently, the condition for their formation are significantly different, and they are older in age than basalts №№ 59-2, 61-1, 62-2. The geomagnetic field paleointensity, determined from NRM of samples № 68-2 and № 57, were higher: Hpl = 53.4 and 60 A/m, respectively.

The geomagnetic field paleointensity Hpl, determined by us, turned out to be 1.1–1.95 times higher than the geomagnetic field calculated by the IGRF12 model for

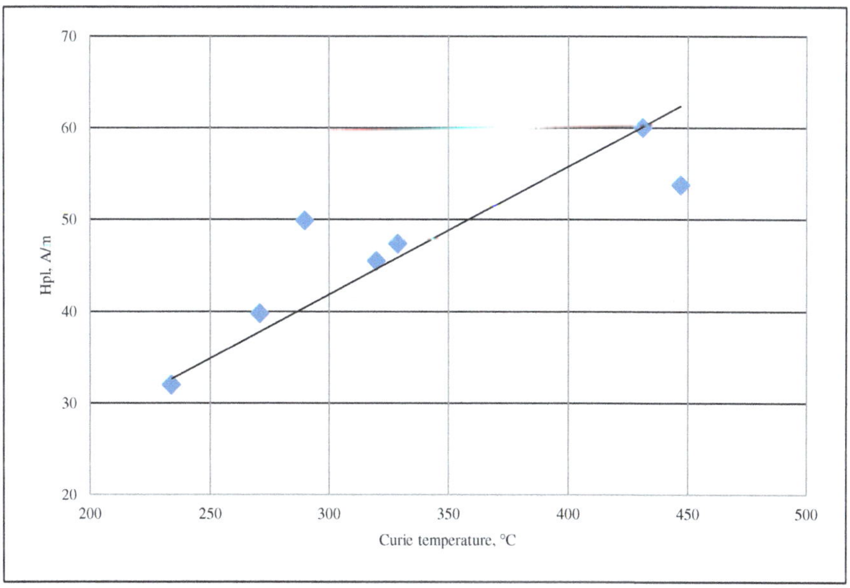

Fig. 6. The relationship between the Curie temperature and geomagnetic field paleointensity

the 2015 epoch for the region of sampling Hmain = 30.7 A/m. The growth tendency of geomagnetic field paleointensity and VADM is observed with increasing Curie temperature of basalt titanomagnetite (Table 2, Fig. 6).

If we assume that basalts with a larger Curie point are of a larger age, then we can conclude that geomagnetic field intensity has been decreasing over the last 100 thousand years.

Thus, the result of thermomagnetic analysis and the paleointensity values of the geomagnetic field determined by us indicate that the investigated samples of basalts refer to three stages of lava outflow.

4 Conclusions

1. It is established that the natural magnetization of most of the basalt samples studied has thermo-remanent nature and has a high paleoinformativity. Paleoinformativity of natural magnetization of sample № 67 is low, as one of its components has a chemical nature.
2. The values of the hysteresis characteristics indicate the predominance of magneticgrains with a single-domain and pseudo-single domain structure, which indicates the reliability of determining the geomagnetic field paleointensity by Thellier method.
3. The investigated samples of the basalts of the Red Sea refer to the three stages of lava overflow.
4. The geomagnetic field paleointensity, determined from the basalts of the Red Sea, was 1.1–1.9 times higher than the current field for the region. The data of thermomagnetic analysis and the results of determination of Hpl indicate that the geomagnetic field intensity in this region have been decreasing in the last 100 thousand years.

Acknowledgments. This work was supported by the Russian Foundation for Basic Research, project no. 16-05-00144

References

1. Shreider, A.: Geomagnetic studies of the Indian Ocean. Science, Moscow (2001)
2. Trukhin, V., Maksimochkin, V., Zhilyaeva, V., et al.: Magnetic properties of basalts and geodynamics of the rift zone in the southern red sea. Izvestiya - Physics of the Solid Earth **42**(11), 928–941 (2006)
3. Kurochkina, E.: Comparative analysis of the magnetic properties of basalts selected in the Red Sea and other rift zones. Mosc. Univ. Phys. Bull. **62**(5), 303–307 (2007)
4. Trukhin, V., Maksimochkin, V., Zhilyaeva, V., Shreider, A., Kashintsev, T.: Magnetic and petrochemical properties of rocks of the rift zone of the Red Sea. Mosc. Univ. Phys. Bull. **62**(5), 296–302 (2007)
5. Channell, J., Xuan, C., Hodell, D.: Stacking paleointensity and oxygen isotope data for the last 1.5 Myr (PISO-1500). Earth Planet. Sci. Lett. **283**, 14–23 (2009)

6. Koronovsky, N.: Magnetic field of the geophysical past of the Earth. Soros Educ. J. (6), 69–73 (1996)
7. Maksimochkin, V., Mbele, J., Trukhin, V., Shreider, A.: Paleointensity of the geomagnetic field in the last half a million years in the regions of the Red Sea and the south of the Mid-Atlantic Ridge. Mosc. Univ. Phys. Bull. **65**(6), 531–538 (2010)
8. Monin, A., Bogdanov, Y., Zonenshain, L.: Underwater research with inhabited vehicles. Science, Moscow (1985)
9. http://www.zetlab.com
10. Maksimochkin, V., Trukhin, V., Garifullin, N., Khasanov, N.: Automated high-sensitivity vibration magnetometer. Devices and technics of experiment, Moscow (2003)
11. Thellier, O., Thellier, E.: Sur la direction du champ magnetique terrestre dans le passe historique et geologique. Annales de Geophysique. **15**, 285-376 (1959)
12. Coe, R., Gromme, C., Mankinen, E.: Geomagnetic paleointensity from radiocarbon-dated lava flows on Hawaii and the question of the Pacific nondiple low. Geophys. Res. **83**, 1740–1756 (1978)
13. Day, R., Fuller, M., Schmidt, V.: Hysteresis properties of titanomagnetites: grain-size and compositional dependence. Phys. Earth Planet. Inter. **13**, 260–268 (1977)
14. Zijderveld, J.: Demagnetization of rocks: analysis of results. In: Collinson, D., Creer, K. (eds.) Methods in Paleomagnetism, p. 254 (1967)
15. Maksimochkin, V., Tselebrovsky, A.: Influence of the chemical magnetization of oceanic basalts on the determination of the paleointensity of the geomagnetic field by the Thellier method. Mosc. Univ. Phys. Bull. **66**(6), 566–576 (2015)

Influence of Hydrodynamic Perturbations on Dispersion Characteristics of a Near-Water Aerosol

V. N. Nosov[1(✉)], S. G. Ivanov[1], V. I. Pogonin[1], V. I. Timonin[2], N. A. Zavyalov[1], E. A. Zevakin[1], and A. S. Savin[2]

[1] Institute of Geochemistry and Analytical Chemistry, V.I. Vernadsky RAS, Moscow, Russia
viktor_nosov@mail.ru, ivanovsg5167@yandex.ru
[2] Moscow State Technical University, N.E. Bauman, Moscow, Russia

Abstract. We report on the experimental studies of the influence of two different types of hydrodynamic perturbations on the characteristics of laser radiation scattering in the near-water aerosol layer. Laboratory experiment was carried out in a tank with flowing and standing aerated water. The first type of hydrodynamic perturbations was created by a streamlined obstacle at the tank bottom; the second one - by rotating propellers. Two sorts of statistical amplitude distribution of red laser light scattering in a near-water aerosol located above the perturbation region are established. It is established that various statistical distributions of the amplitudes of the laser radiation scattering pulses on the particles of the near-water aerosol and, accordingly, the different aerosol distributions in size, correspond to these various hydrodynamic perturbations. Normalized amplitude distribution histograms of light scattering are compared.

Keywords: Aerated aqueous medium · Near-water aerosol · Laser radiation
Hydrodynamic perturbations · Histograms · Registration of light scattering

1 Introduction

Hydrodynamic perturbations that arise in the sea depths under the influence of various sources can be manifested in the near-water layer of the atmosphere. Earlier [1–6] it was established that one of the mechanisms of the impact of the marine environment perturbation on the characteristics of the near-water aerosol is associated with the release of gas bubbles. Hydrodynamic perturbations of the water environment change the characteristics of the gas bubbles that reach the water surface. When the bubbles burst, small particles of water containing organic and inorganic substances are released into the atmosphere. The intensity of the particles emission into the atmosphere increases with the increasing density of the gas bubbles flow. The particle size distribution of the near-water aerosol also changes.

The establishment of connection between aerosol characteristics and hydrodynamic perturbations in the marine environment is an actual problem. By observing perturbations

V. Karev et al. (Eds.): PMMEEP 2017, SPRINGERGEOL, pp. 282–288, 2018.
https://doi.org/10.1007/978-3-319-77788-7_29

in the near-water layer of the atmosphere it is possible to indirectly obtain information about the actual hydrodynamic processes. Remote recording of the characteristics of the near-water aerosol and its particle size distribution can be carried out using laser-optical instruments [7–10]. In the context of creating new remote probing tools this line of research is promising and allows not only to register various anomalies in the water layer, but also to identify them.

We report on the laboratory experimental studies of the variability of disperse characteristics of aerosols in the near-water atmospheric layer under the influence of two types of hydrodynamic perturbations (HDP), which were based on the laser radiation elastic scattering method. In our work we used installations with a flow-through and non-flow-through tank.

2 Registration of Aerosol Laser Radiation Scattering Over the Water in a Tank, Disturbed by Propellers

Rotation of two propellers from the ship model was used to generate the HDP in the setup with a non-flow-through tank (Fig. 1).

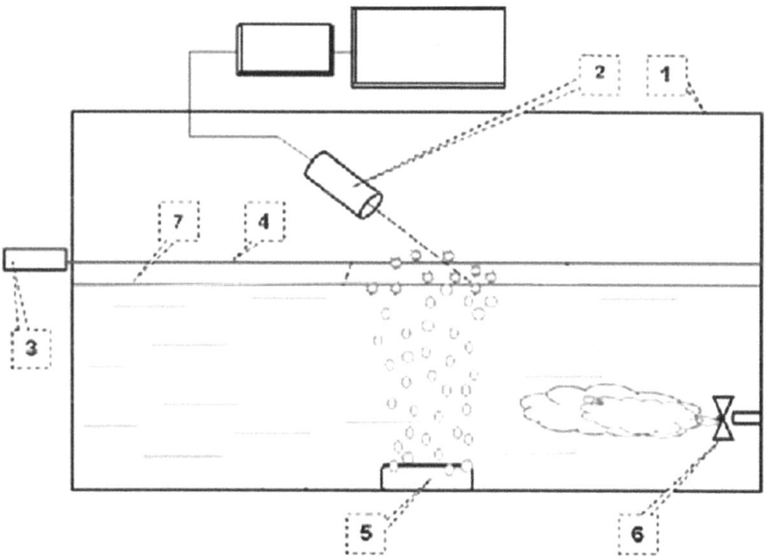

Fig. 1. Scheme of a laboratory setup with a non-flow-through tank: 1 – tank with a set of fasteners; 2 – an objective with a photodetector based on a photomultiplier; 3 – laser emitter (λ = 660 nm); 4 – laser beam; 5 – aerator; 6 – turbulent flow generator (propellers); 7 – water surface.

In order to investigate the effect of HDP on the characteristics of the laser radiation scattering on a near-water aerosol experiments were conducted by aerating water in the tank by means of air pumping through the ceramic rods lying on the bottom of the tank.

The measurements were performed with recording the light scattering amplitudes on individual aerosol particles in the LGraph2 program in the accumulation mode with a duration of 300 s. Based on the measurement results, histograms of the distribution of the aerosol scattering signals amplitudes, normalized to the total number of pulses, as well as difference histograms were constructed. For the convenience of comparison of results, all the histograms were multiplied by a factor of 10^6. Figure 2 shows normalized histograms of the aerosol scattering signals amplitudes for HDP in the form of propellers at their different rotation speeds, differing by a factor of 1, 3. The left histogram in Fig. 2 is obtained with a higher rotational speed of the propellers.

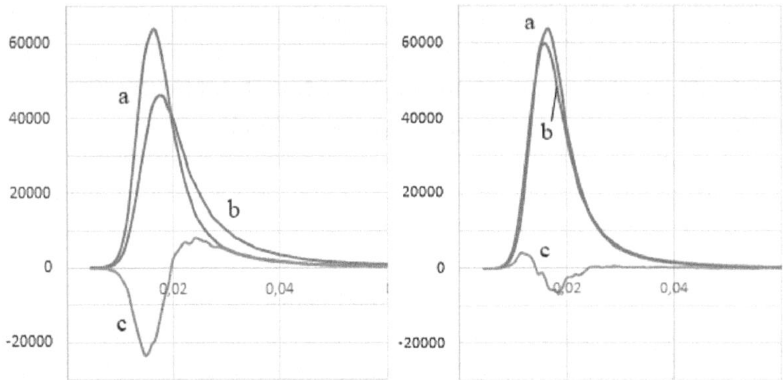

Fig. 2. Normalized histograms of the aerosol scattering signals amplitudes without HDP-influence (a), in case of water perturbation by the propellers (b) and their difference (c).

From the data obtained, it can be seen that the HDP, caused by propellers, leads to a noticeable change in the forms of the histograms of the aerosol scattering amplitudes, which indicates a change in the dispersion of aerosols in the presence of such HDP. The received distributions of the amplitudes of the detected signals also change significantly with change of the propellers rotational speed.

3 Registration of Aerosol Laser Radiation Scattering Over Water Flowing Around an Underwater Obstacle in a Tank

The experiments were carried out in a flow-through tank (Fig. 3). An obstacle in the form of a hump with a smooth profile, installed at the bottom of the tank, was used as a source of perturbations.

Water was supplied to the flow-through tank from the water pipe through a comb with holes. Aerating of water in this case was carried out by feeding air into the comb from the compressor. From the opposite end of the tank water was flowing out, which provided a constant water flow in the tank at a certain speed. To disturb the water flow, an underwater barrier was installed on the bottom of the tank. The distance from the top of the barrier to the water surface was 40–50 mm.

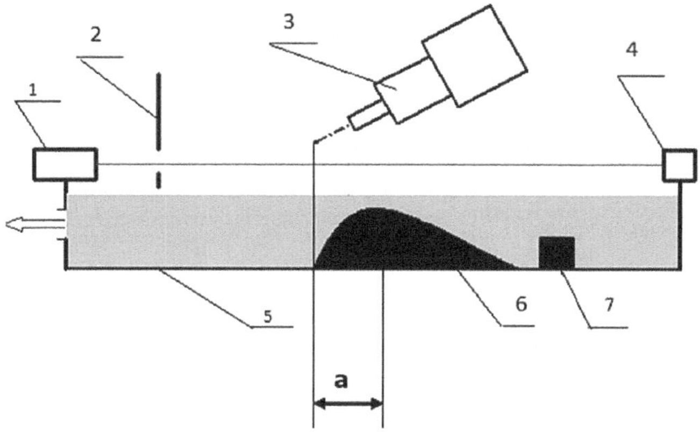

Fig. 3. Scheme of the laboratory setup with a flow-through tank: 1 – laser emitter (λ = 660 nm); 2 – screen with a hole; 3 – an objective with a photodetector device based on a photomultiplier; 4 – laser radiation trap; 5 – the tank; 6 – underwater obstacle (a – displacement of the registration area from the top of the underwater obstacle); 7 – generator – aerator of laminar flow.

Measurements of aerosol scattering in the flow-through tank were carried out at several positions of the barrier, which was moved horizontally by steps of 5 cm in the direction of the flow. At the same time, the position of the registration zone was not changed. Figure 4 shows the normalized histograms and their difference when the position of the underwater barrier has the maximum effect on the distribution of the amplitudes of the aerosol scattering impulses (a = −22 mm).

Figure 5 shows peak-to-peak values (the difference between the maximum and minimum values) of the differences in the normalized histograms obtained during experiments with the movement of the obstacle.

The largest peak-to-peak value of the normalized histograms differences corresponds to distance a = −22 mm, when the top of the barrier was "downstream" from the detection zone. The data shown in Fig. 5 illustrate that there is a significant dependence of the aerosol particles distribution on the location above the underwater obstacle.

To compare the characteristics of various types of HDP, Fig. 6 shows the normalized histograms of the amplitudes of the signals of aerosol scattering with propellers and aerosol scattering with an obstacle.

The shapes of the histograms in Fig. 6 show a significant difference in the size distribution of aerosol particles for HDP created by propellers and HDP in a flowing tank with a barrier. This difference can serve as an identification feature for these types of perturbations and can be used in solving application tasks in full-scale marine conditions.

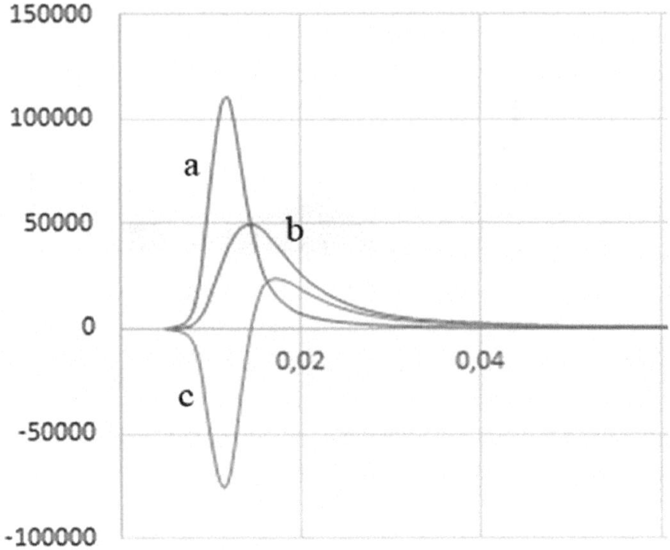

Fig. 4. Normalized histograms of the amplitudes of aerosol scattering signals without perturbation of the water flow by the barrier (a), with perturbation of the water flow by the barrier (b) and their difference (c).

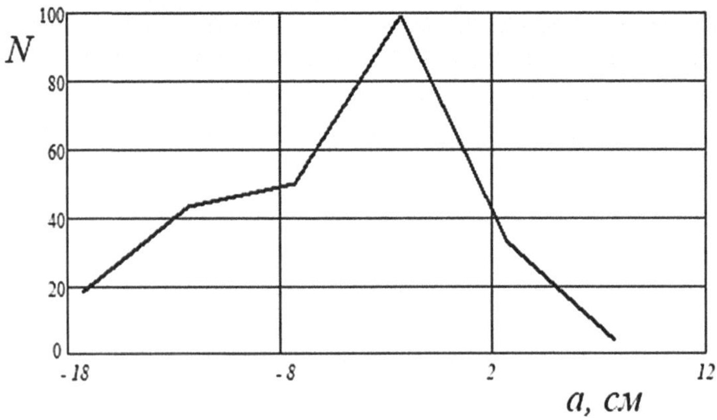

Fig. 5. Dependence of peak-to-peak values N ($\times 10^{-3}$) of the normalized difference histograms of aerosol scattering on the displacement of the barrier in the tank (distance "a" in Fig. 4).

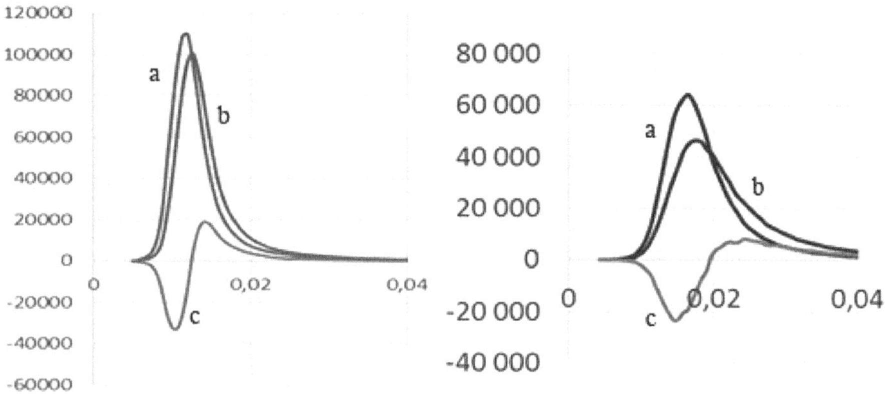

Fig. 6. Normalized histograms of amplitudes of aerosol scattering with propellers (left) and aerosol scattering with an obstruction (right); without HDP (a), in the presence of HDP (b) and their difference (c).

4 Conclusion

As a result of laboratory experiments, it has been established that the HDP of an aqueous medium change the characteristics of the laser radiation scattering on a near-water aerosol. This happens due to the change in the particle size distribution of aerosols caused by HDP. Different by nature HDP affect the size distribution of aerosols in different ways, which is manifested in the form and magnitude of differences in the histograms of the amplitudes of the aerosol scattering signals. This effect can be used not only to record the presence of HDP in the aquatic environment, but also to identify them.

References

1. Day, J.A.: Production of droplets and salt nuclei by the bursting of air-bubble films. Quart. J. Roy. Meteorol. Soc. **90**, 72–78 (1964)
2. McIntyre, F.: Flow patterns in breaking bubbles. J. Geophys. Res. **77**(27), 5211–5228 (1972)
3. Julanov, Y.V., Petryanov, I.V.: Investigation of mechanism of marine aerosol generation. Rep. RAS, vol. 253, no. 4 (1980)
4. Thorpe, S.A.: On the clouds of bubbles formed by breaking wind-waves in deep water, and their role in air-sea gas transfer. Philos. Trans. R. Soc. London Ser. A **304**, 155 (1982)
5. Nosov, V.N., Ivanov, S.G., Kaledin, S.B., Savin, A.S.: Registration of the manifestations of deep processes in the near-surface layers of sea water and the atmosphere. Process. Geomed. **2**(11), 522–528 (2017)
6. Bakhanov, V.V., Goryachkin, Y.N., Korchagin, N.N., Repina, I.A.: Local manifestations of deep processes on the sea surface and in the near-water layer of the atmosphere. Rep. RAS, vol. 414, no. 1, pp. 111–115 (2007)
7. Zielinski, A., Piskozub, J., Irczuk, M.: Lidar studies of marine aerosol in the coastal zone. In: Proceedings of SPIE, vol. 2471, pp. 428–438 (1995)

8. Piskozub, J.: Study of spatial distribution of marine aerosol over sea coast with a multifrequency lidar system. In: Proceedings of SPIE, vol. 2471, pp. 387–389 (1995)
9. Nosov, V.N., Gorelov, A.M., Kaledin, S.B., Kuznetsov, V.A., Leonov, S.O., Savin, A.S.: Laser emission scattering over a marine surface during hydrodynamic disturbances within the water mass. Dokl. Earth Sci. **433**(1), 920–921 (2010)
10. Nosov, V.N., Kaledin, S.B., Gorelov, A.M., Leonov, S.O., Kuznetsov, V.A., Pogonin, V.I., Savin, A.S.: Light scattering in the atmospheric near-water layer above areas of long-living hydrodynamic disturbances of the marine environment. Rep. Earth Sci. **442**(2), 247–248 (2012)

A Comparative Analysis of Optical Methods for Detection and Prediction of Radionuclides Migration in the Geosphere

B. P. Yakimov[1], G. S. Budylin[2], V. G. Petrov[2], V. V. Fadeev[1], S. N. Kalmykov[2], S. A. Evlashin[3], and E. A. Shirshin[1(✉)]

[1] Faculty of Physics, M.V. Lomonosov Moscow State University, Moscow, Russia
shirshin@lid.phys.msu.ru
[2] Faculty of Chemistry, M.V. Lomonosov Moscow State University, Moscow, Russia
[3] Skolkovo Institute of Science and Technology, Moscow, Russia

Abstract. Here we report on a comparative analysis of laser techniques for detection and speciation of radionuclides and its complexes in the geosphere. The application of different methods is illustrated by the example of uranium(VI) speciation in aqueous environment and detection of trace elements on the ppm level in gases, which appear as a result of reprocessing of spent nuclear fuel.

Keywords: Laser spectroscopy · Radionuclides · Speciation Time-resolved fluorimetry

1 Introduction

Radionuclides migration in geosphere is a hot spot in biogeochemistry due to its adverse impact on the human organism. Besides the fraction of nuclides, which is naturally present in the environment, recent developments of atomic energy resulted in a problem of nuclides accumulation due to the necessity of radioactive waste disposal, and also leakages accompany the functioning of nuclear plants. For instance, it is known that uranium is present in elevated concentration in the hydrosphere of certain regions in Germany [1] due to the sewage waters, which come from uranium mines. The process of radionuclides migration is also of a high interest because of the necessity to create long-term repositories for high level radioactive waste, and in this case mathematical modeling of nuclides' transport behavior is crucial. As these repositories are usually located deep underground, one needs to consider several mechanisms of nuclides migration, including complexation and transportation by mineral nanoparticles or dissolved organic matter. This fact stimulated the development of multiple methods for the assessment of interactions between radionuclides and various geologically significant ligands in solution, as well as for characterization of metal sorption on the surface of colloid nanoparticles. First of all, this requires determination of equilibrium constants and their temperature in laboratory conditions behavior as a prerequisite for modelling

V. Karev et al. (Eds.): PMMEEP 2017, SPRINGERGEOL, pp. 289–297, 2018.
https://doi.org/10.1007/978-3-319-77788-7_30

of the speciation of different metal complexes in the system. The second task is the detection of radionuclides and determination of its speciation in the geosphere. Both tasks can be solved by a set of optical methods, which allow for high sensitivity and selectivity of detection and, importantly, to perform this in real time without sampling. Moreover, the analysis can be performed remotely, that is required to control processes in a radioactive zone. This work is aimed at the discussion of the state-of-art optical methods capable of solving the described task.

Namely, we will focus on the following optical methods: time-resolved laser-induced fluorescence spectroscopy (TRLIFS), laser induced breakdown spectroscopy (LIBS) and surface enhanced Raman scattering (SERS) as the proven optical methods that meet the requirements of remote analysis.

Brief description and possible applications of these methods related to detection of radionuclides will be given in theoretical background section. Next, in the results and discussion section, the experimental results obtained by our group using these methods will be presented, after which we summarize the merits and demerits of these approaches.

2 Theoretical Background

2.1 Time-Resolved Laser-Induced Fluorescence Spectroscopy

Time-resolved laser-induced fluorescence spectroscopy (TRLIFS) was extensively used for uranium speciation in aqueous solution. Though its application is mostly motivated by uranyl analysis in drainage waters [2], this technique can also benefit the technological process of uranium mining using the leaching process and uranium disposal – e.g., this initially stimulated the developments of TRLIFS in our group [3, 4].

Usual experimental setup of TRLIFS consists of the excitation source which is usually a pulsed laser system and a spectrograph, combined with a so-called intensified CCD-camera and delay generator providing time-resolved mode [5].

At low levels of the excitation source intensity and uranyl concentrations, such that no nonlinear effects associated with the processes of collisional deactivation between two excited complexes are observed, the decay signal of the luminescence of a mixture of N fluorescent radionuclide complexes is represented in the form [3]:

$$I_{lum}(\lambda, t) = \sum_{i=1}^{N} A_i S_i(\lambda) \exp(-t/\tau_i) \quad (1)$$

where Ai, τi, Si is the contribution of the i-th form to the general luminescence signal, the lifetime of the excited state, and the spectral band shape of the i-th form at given wavelength λ.

Taking into account that the total luminescence signal is proportional to the fraction of the i-th form of uranyl complex in the solution, it becomes possible to restore the uranium speciation in the solution. It is worth noting, that for uranyl complexes the luminescence spectra of various forms differ insignificantly, while the time for deactivation of the excited state differs in between complexes by several orders of magnitude [6], which makes it possible to effectively distinguish the forms of uranyl complexes in solution using the TRLIFS method.

However, TRLIFS sometimes gives incorrect predictions [7]. Namely, we aimed at understanding the differences in the values of photophysical parameters of uranium complexes obtained by different research groups and came up to the importance of non-linear processes under laser excitation, annihilation of the excited states.

Still, the main advantages of TRLIFS are its sensitivity and capability to identify different complexes of uranyl, however, not all significant radionuclides exhibit laser-induced fluorescence. Though sometimes luminescent metals (e.g. terbium or europium) are used to mimic the speciation or interaction with nanoparticles by non-luminescent radionuclides (e.g. americium) [8], this approach obviously is not applicable for in situ sensing.

2.2 Laser-Induced Breakdown Spectroscopy

Hence, universal optical methods are of a high demand for radionuclides detection, and in this area laser-induced breakdown detection (LIBS) can be considered to be the most promising one. LIBS method is free of TRLIFS restrictions and allows to detect a signal form any radionuclide by its atomic emission in laser-induced plasma.

The LIBS method is based on the phenomenon of optical breakdown in a sample: when the intensity of the incident radiation exceeds a certain threshold value, a luminous plasma forms in the sample. The radiation spectrum of the resulting plasma is formed due to the effects of inverse bremsstrahlung of plasma electrons and due to radiation relaxation of the excited ions and atoms of the substance from which the plasma was generated. Using atomic emission spectrum one can get fraction of certain atoms in a complex compound [9].

A typical LIBS experimental setup is rather simple and includes a pulsed laser capable of creating a power density above a predetermined breakdown threshold and a spectrometer (often also with a temporal resolution, to separate the most representative analytical lines) [9].

LIBS operates in liquid, solid and gas phases, and has already been commercialized for the use in nuclear industry e.g. by the Applied Photonics (UK) company. Disregarding the general problems of LIBS such as non-linearity of signal with concentration, the influence of matrix composition on the signal etc. it can be adapted to almost any industrial process with certain compromises. Though being less sensitive that expensive laboratory methods as ICP-AES or ICP-MS, it attract the users from industry by the possibility to control the composition of media in real time without sampling, flexibility and price.

For instance, our group installed the LIBS technique to perform on-line control of wastes during the pyrochemical reprocessing of spent nuclear fuel. In the case of nuclides sorption on nanoparticles, modification of LIBS – LIBD (laser-induced breakdown detection) is known to detect ultralow quantities of colloids in solution. However, LIBS and LIBD generally do not provide for "molecular" information, i.e. speciation, thus not allowing to discriminate different forms of nuclides.

2.3 Surface Enhanced Raman Scattering

From this point of view, researchers become interested in Raman scattering, the method, which allows determining molecules by their fingerprint spectra originated from vibronic structure. Raman spectroscopy in sometimes used simultaneously with LIBS as an additional detection mode and allows to extend LIBS capabilities. At the same time, the problem of Raman scattering is its low sensitivity (almost 13 orders of magnitude lower than for fluorescence spectroscopy!). To solve this problem, specific substrates can be used, which provide for SERS – surface enhanced Raman scattering. While exhibiting vibrational spectra, SERS may in certain cases detect single molecules, however, the problem here is the development of substrates for concrete purposes, i.e. optimized for radionuclides detection. Recently we reported on the development of a carbon nanowalls-based chips coated with Au nanoparticles for SERS [10], which also demonstrated highly efficient detection of uranyl complexes. Hence, the SERS chips can become a prospective platform for radionuclides complexes express detection in liquid samples.

3 Experimental Setup

A custom-built laser fluorometer was used for TRLIFS measurements and was similar to the described earlier [3].

For the LIBS experimental setup Nd:YAG was used ($\lambda_{excitation}$ = 1064 nm, pulse duration ~ 10 ns, pulse energy ~ 100 mJ). The detector was combination of monochromator "M266-IV" ("Solar", Belorussia) and intensified CCD-camera "Nanoscan-24" ("Nanoscan", Russia). For flow control, flowmeters of Bronkhorst High-Tech BV (Netherlands) were used.

4 Materials

All liquid samples were prepared in deionized Millipore water. Samples for detection of complexation processes were prepared with total concentration of uranium 10^{-4} M, pH = 3.0, and ionic strength 0.1 M that were controlled by addition of known amount of $HClO_4$, NaF and $NaClO_4$.

Samples for demonstration of dependence of luminescence decay rate on intensity of excitation source were prepared with concentration of UO_2^{2+} 0.01 M, H_3PO_4 0.1 M and $HClO_4$ 0.1 M. Measurements were performed at ambient temperature (25 ± 1) °C.

For experiments with gas, gases of argon and hydrogen were used with a purity of 1 ppm.

5 Results and Discussion

First, we demonstrate the determination of the species distribution of uranyl fluoride complexes as a function of the fluorine concentration obtained by TRLIFS method. On Fig. 1 typical luminescence decay curve observed in experiment presented.

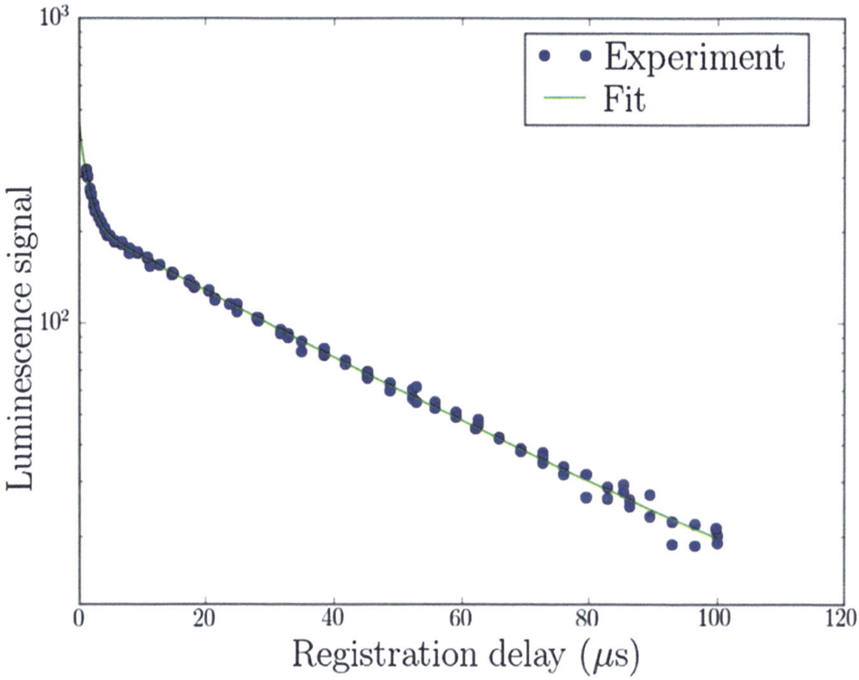

Fig. 1. Typical kinetics of the luminescence decay of uranyl fluoride complex in aqueous solution.

Using (1), from the kinetics of luminescence decay for solutions of uranyl fluoride with different fluorine concentrations, a distribution of uranyl complexes with a different number of attached fluorine ligands was obtained. Figure 2 shows the concentrations of uranyl fluoride complexes in solution recovered from the TRLIFS-data.

As noted earlier, our group found the fact that the rate of deactivation of the excited state of uranyl depends on the intensity of the incident radiation. In addition, in [3] a model was proposed that explains this dependence by the interaction between two excited uranyl ions. The characteristic rate of such interaction, according to the assumptions [3], would depend on the concentration of uranyl in the solution, the diffusion constant of the complex and the radius of the bimolecular interaction.

We estimated this rate of non-linear deactivation of uranyl for the uranyl phosphate complex. Using the values of the radius and the diffusion constant, as in [3], we estimated indirectly the concentration of uranyl in the solution, it turned out to be

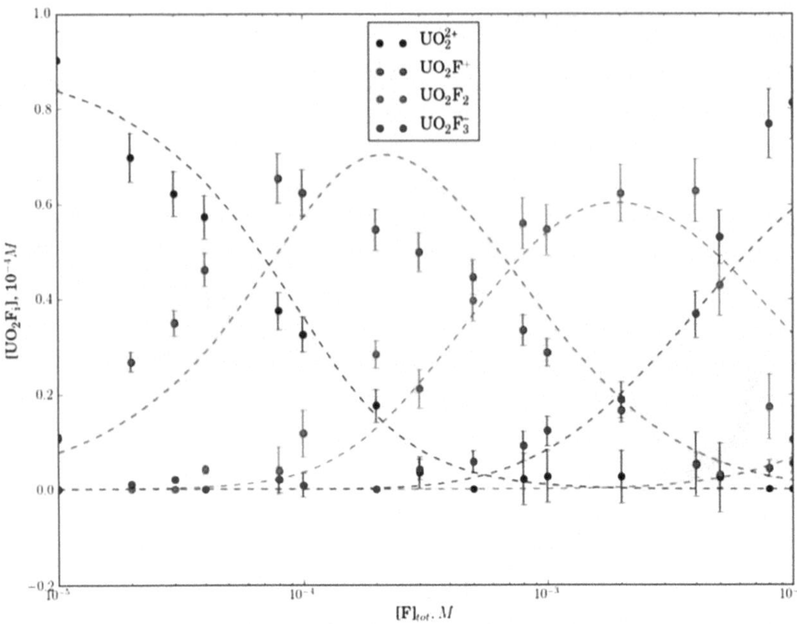

Fig. 2. Speciation diagram for uranyl fluoride complexes. Conditions were $[UO_2^{2+}] = 10^{-4}$ M, pH = 3.0. The dashed line indicates theoretical distributions of uranyl fluoride complexes in solution with specified parameters calculated using the MEDUSA-HYDRA software [11] on the complexation constants from [12]

0.05 ± 0.04 M. Despite the relatively small accuracy (true concentration of uranyl was 0.01 M), it can be assumed that the method of measuring the dependence the decay rate of luminescence on the intensity of the incident radiation, can be used as a calibration-free method for determining the concentration of uranyl in inorganic solutions.

Figure 3 shows the kinetics of the luminescence decay at various intensities of excitation radiation. The difference in kinetics for different intensity levels allows us to tell about the presence of an additional decay channel, which is related to the interaction between excited uranyl complexes.

Further we present some of results that can be obtained using other mentioned optical method – LIBS. In the process of reprocessing spent nuclear fuel using pyrochemical methods, it becomes important to control the waste gas component.

To simulate the atmosphere of the pyrochemical chamber, the experimental setup presented in Fig. 4 was created. A custom designed gas cuvette was connected to gas cylinders filled with gases, which are known to be present in the atmosphere of a pyrochemical chamber. The gas inlet from these cylinders occurred under the control of the flowmeters, and the pressure was controlled by a manometer. Thus, it was possible to obtain a calibration curve for the LIBS device, as well as determine its detection limit.

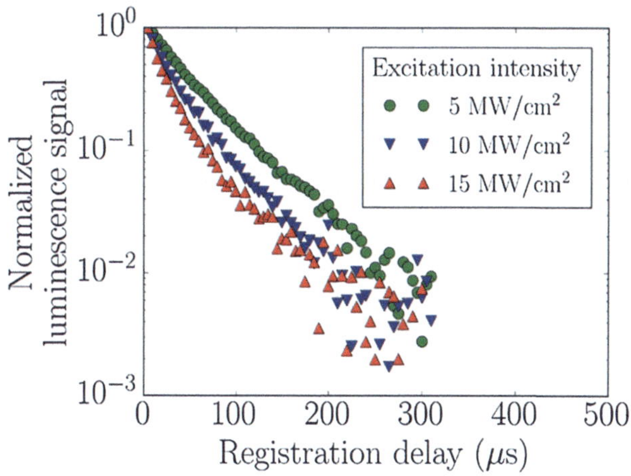

Fig. 3. Kinetics of luminescence decay of uranyl phosphate complexes at various intensities of excitation radiation.

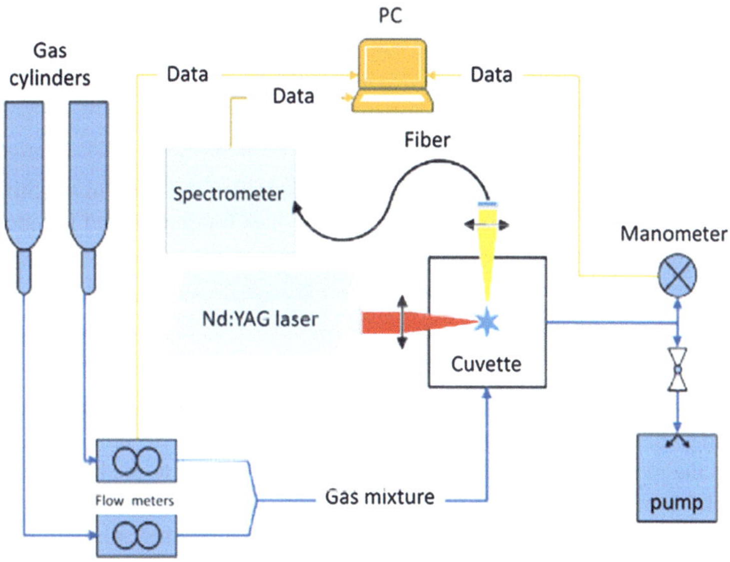

Fig. 4. The scheme of the experimental setup for modeling the determination of the gas composition in the pyrochemical chamber by the LIBS method.

The obtained kinetics of the hydrogen inlet in argon atmosphere, which is determined as a ratio of the intensities of the hydrogen spectral lines at 656 nm and the argon at 696 nm, is shown in Fig. 5. As we see, this experimental setup made it possible to easily determine the elemental composition of gas atmosphere at the level of 100 ppm. The estimated limit of detection for various gases was about below 1 ppm.

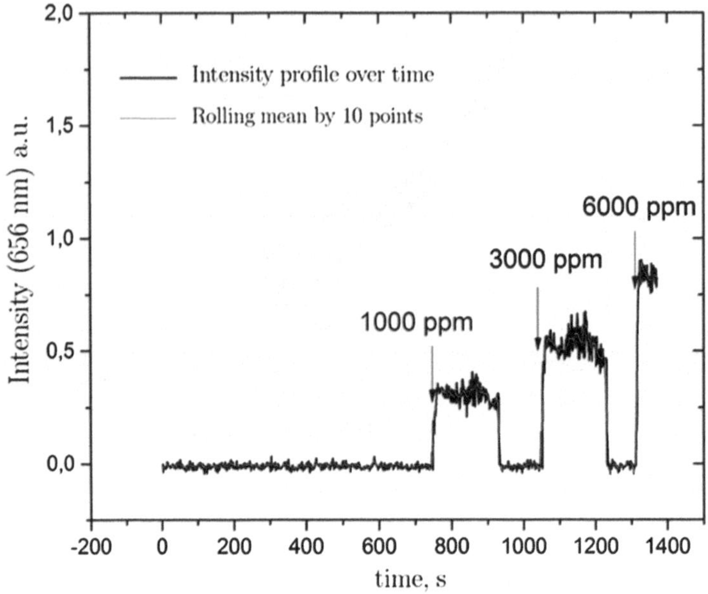

Fig. 5. The kinetics of the hydrogen inlet in argon atmosphere, which is determined as a ratio of the intensities of the hydrogen spectral lines at 656 nm and the argon at 696 nm. Concentration of hydrogen is indicated by the arrows.

The experiments demonstrated that it is possible to detect 1–10 pm sensitivities for nitrogen, oxygen and inert gases, while the sensitivities for carbon and iodode were in the >100 ppm range.

6 Conclusion

Here we report on the comparative study of several laser techniques which allow for radionuclides detection and speciation. While in the case of luminescent species TRLIFS is the most advantageous method due to its sensitivity to the local environment of the metal ion, in case of technical applications, such as control of technological processes, where real-time analysis is required, LIBS is a reliable solution. Namely, in the case of gas analysis, LIBS allows for detection of ~ 1 ppm concentrations of the major components of spent nuclear fuel, e.g. during the pyroprocessing.

Acknowledgments. The work was supported by the Russian Federation President's grant (MK-9394.2016.2) and grant № 8719GU/2015 from 16.12.2015 (code 0019175), competition UMNIK 15-11.

References

1. Meinrath, A., Schneider, P., Meinrath, G.: Uranium ores and depleted uranium in the environment, with a reference to uranium in the biosphere from the Erzgebirge/Sachsen Germany. J. Environ. Radioact. **64**(2), 175–193 (2003)
2. Arnold, T., Baumann, N., Krawczyk-Bärsch, E., Brockmann, S., Zimmermann, U., Jenk, U., Weiß, S.: Identification of the uranium speciation in an underground acid mine drainage environment. Geochim. Cosmochim. Acta **75**(8), 2200–2212 (2011)
3. Budylin, G., Shirshin, E., Fadeev, V., Petrov, V., Kalmykov, S.: Laser-induced fluorescence of uranyl complexes in aqueous solutions: the role of diffusion-controlled excited states annihilation. Opt. Express **21**(18), 20517–20528 (2013)
4. Batuk, D.N., Shiryaev, A.A., Kalmykov, S.N., Batuk, O.N., Romanchuk, A.Y., Shirshin, E. A., Zubavichus, Y.V.: Sorption and speciation of uranium on silica colloids. In: Actinide Nanoparticle Research, pp. 315–332. Springer Berlin, Heidelberg (2011)
5. Billard, I., Geipel, G.: Luminescence analysis of actinides: instrumentation, applications, quantification, future trends, and quality assurance. In: Standardization and Quality Assurance in Fluorescence Measurements I, pp. 465–492 (2008)
6. Geipel, G., et al.: Uranium (VI) sulfate complexation studied by time-resolved laser-induced fluorescence spectroscopy (TRLFS). Radiochim. Acta **75**(4), 199–204 (1996)
7. Billard, I., et al.: Aqueous solutions of uranium (VI) as studied by time-resolved emission spectroscopy: a round-robin test. Appl. Spectrosc. **57**(8), 1027–1038 (2003)
8. Gracheva, N.N., Romanchuk, A.Y., Smirnov, E.A., Meledina, M.A., Garshev, A.V., Shirshin, E.A., Kalmykov, S.N.: Am (III) sorption onto TiO 2 samples with different crystallinity and varying pore size distributions. Appl. Geochem. **42**, 69–76 (2014)
9. Noll, R.: Laser-induced breakdown spectroscopy, pp. 7–15. Springer, Heidelberg (2012)
10. Dyakonov, P., Mironovich, K., Svyakhovsky, S., Voloshina, O., Dagesyan, S., Panchishin, A., Suetin, N., Bagratashvili, V., Timashev, P., Shirshin, E., Evlashin, S.: Carbon nanowalls as a platform for biological SERS studies. Sci. Rep. 7(1), 13352 (2017)
11. Guillaumont, R., Fanghänel, T., Fuger, J., Grenthe, J., Neck, V., Palmer, D., Rand, M.: Update on the Chemical Thermodynamics of Uranium, Neptunium, Plutonium, Americium and Technetium. Elsevier Science Publishers, Amsterdam (2003)
12. Puigdomenech, I.: Chemical Equilibrium Diagrams. https://sites.google.com/site/chemdiagr/

Advanced Procedure for Estimation of Phytoplankton Fluorescence Quantum Yield Using Remote Sensing Data: A Comparative Study of the Amundsen Sea Polynyas

Elena E. Nikonova[1], Evgeny A. Shirshin[1], Victor V. Fadeev[1(✉)], and Maxim Y. Gorbunov[2]

[1] Department of Physics, M.V. Lomonosov Moscow State University, 119991 Leninskie Gory 1/2, Moscow, Russia
victor_fadeev@mail.ru
[2] Department of Marine and Coastal Sciences, Rutgers University, New Brunswick, USA

Abstract. The algorithm for estimation of the quantum yield of phytoplankton fluorescence from the remote sensing satellite of the MODIS tool is discussed in the paper. There is an example of manifestation of the iron limitation in the Amundsen Sea. Amundsen Sea encloses two polynyas: Fe-limited ASP (Amundsen Sea Polynya) and Fe-replete PIP (Pine Island Polynya). We present a procedure for comparing the mean values of the quantum yield of phytoplankton fluorescence in these regions and the requirements for it. To meet these requirements, the data of two satellite systems were analyzed and compared: MODIS and AVHRR. Analysis of the data made it possible to observe the differences in the mean values of the quantum yield of phytoplankton fluorescence in these two regions during the bloom period of the Amundsen Sea phytoplankton in 2012.

Keywords: Remote sensing · MODIS · AVHRR · Phytoplankton Quantum yield · Amundsen sea

1 Introduction

Phytoplankton is a natural marker that characterizes the ecological state of the ocean. Monitoring of phytoplankton physiology by remote sensing methods is a key task of research of global biogeochemical cycles in the ocean. Polar regions are most interesting as they are areas with highest values of primary production [1, 2]. The efficiency of CO2 absorption by phytoplankton cells in the polar regions largely determines its concentration in the global atmosphere [3]. Also in these areas there are significant changes associated with the melting of glaciers, manifesting themselves, including in the physiology of phytoplankton [4].

The most informative for this purpose is the satellite instruments of high spatial resolution. However, for interpretation obtained data, especially for the estimation to phytoplankton parameters characterizing its physiological state, it is necessary to use

V. Karev et al. (Eds.): PMMEEP 2017, SPRINGERGEOL, pp. 298–304, 2018.
https://doi.org/10.1007/978-3-319-77788-7_31

complex models and their regional correction using in situ data. Also in these areas there are problems of atmospheric correction associated with high region cloudiness, low angles of incident solar radiation and ice drift. Therefore, it is necessary to compare the sensing data in the visible range (ocean color) with the results of sounding in the microwave range to determine atmospheric correction errors and estimation sea ice concentrations.

The object of this study is Amundsen Sea – the most productive region (per unit area) in the Antarctic. The Amundsen Sea harbors two particularly productive polynyas, in each of which there is phytoplankton bloom: ASP (Amundsen Sea Polynya) of about 27000 km^2 and PIP (Pine Island Polynya), 18000 km^2 [5]. The phytoplankton bloom is controlled by the flow of iron entering the water during the melting of glaciers, as well as by high-nutrient flows of circumpolar deep waters. Oceanographic studies revealed strong Fe stress in the ASP, whereas the PIP showed virtually no signatures of Fe limitation [6, 7]. Other factors that determine the vital activity of phytoplankton (the surface flux, the amount of nutrients, the temperature) in these regions are the same, which makes them an interesting object from the point of view of remote sensing. Thus, a comparison of these areas allows us to consider the manifestation of the effect of iron limitation in satellite remote sensing data.

Iron is a necessary element for the functioning of the electronic transport chain in phytoplankton cells. It is contained in the photosystem II reaction systems [8, 9]. The variable fluorescence signal of photosystem II (PSII) is sensitive to iron limitation [10, 11] and can be used to determine the physiological state of phytoplankton. The fluorescence quantum yield is used as a parameter of the iron stress in its cells. Here and below, under the quantum yield of fluorescence, we will mean the quantum yield of chlorophyll a fluorescence in phytoplankton cells in vivo, that is the ratio of the fluorescence photons of chlorophyll-a emitted throughout the fluorescence band to photons absorbed by all cellular phytoplankton pigments. It is known that the fluorescence quantum yield of phytoplankton tends to increase with iron stress [12], which is associated with a decreasing efficiency of photochemical processes. Thus, in this paper, the task was to investigate the possibility of remote detection of differences in the fluorescence quantum yield in ASP and PIP as an indicator of the saturation of these regions with iron.

For this reason, we corrected the procedure for estimation of the fluorescence quantum yield from satellite data. Further the distribution of the fluorescence quantum yield calculated from satellite data was compared with the results of in situ measurements of the 2012 fluorescence quantum yield [13].

2 Methods

Level 3 monthly and 8-days composite regional maps of chlorophyll-a, normalized fluorescence line height and diffusive attenuation coefficient from Moderate-Imaging Spectroradiometer (MODIS) Aqua were obtained from the Goddard Space Flight Center. We used algorithm Hout for estimation values of quantum yield and Python programming language. We analyzed data of 2012 year 8-days data sets, with the spatial resolution of approximately 9 km.

We also used AVHRR microwave remote sensing data, retrieved from the NOAA [14] for sea ice coverage during the study.

3 Results and Discussion

In the paper, we will rely on in situ data obtained in the 2012 expedition and previously described [13]. Data is presented in the Table 1. The estimation of the fluorescence quantum yield value from these data can be carried out in 2 ways. First: using the fluorescence lifetimes values in the study of Lin et al. [15] it was shown that it is proportional to the fluorescence quantum yield. In this case, the expected difference in the fluorescence quantum yield of the two polynyas was 25%. Second: using measurements of the relative quantum yield of photochemical quenching. Assuming the contribution of NPQ in PIP and ASP is the same, we expect to observe the difference in the magnitude of fluorescence quantum yield of the order of the difference in the quantum yield of photochemical quenching of these regions-that is about 27%.

Table 1. Results of in situ data from paper [13].

	Concentration of chlorophyll-a, mg m^{-3}	Quantum yield of photochemistry, a.e.	Fluorescence lifetime, ns
ASP	3.05 ± 1.77	0.35 ± 0.06	1.25 ± 0.15
PIP	2.25 ± 0.54	0.48 ± 0.03	1.0 ± 0.3

Fluorescence quantum yield estimation algorithm proposed by Huot [16] was taken to compare with in situ data. To compare mean values of fluorescence quantum yield values of two polynyas the algorithm should meet the following requirements:

(1) Investigated regions have different mean chlorophyll concentrations, moreover chlorophyll concentration value is varying in wide range in each polynya. Fluorescence quantum yield is a parameter that characterizes physiological state of phytoplankton and it should not be correlated with chlorophyll concentration. Thus it is necessary for calculated quantum yield value to be independent on chlorophyll concentration.
(2) Fluorescence quantum yield value is calculated using mainly reflected signal at 678 nm. In the investigated region there is a lot of drifting ice that increases scattered signal in microwave spectral range. Microwave range is used in signal atmosphere correction that leads to corrected signal increase for regions of high ice density. Atmosphere correction mainly manifests itself in red spectral region and thus in fluorescence estimation. Thus the algorithm should be applied to ice-free regions only.

Let us study the distributions of the quantum yield of fluorescence from the values, calculated using the Huot model [16] across all polynya. The results are presented in Fig. 1. We analyzed the data of the 8-day averaging at the flowering peak in January–February 2012. The data include the dates of the expedition [13] and the two preceding weeks. According to satellite data, the blooms peak in the ASP fell on the first week of February (February 2–9). It is expected that in these terms the manifestation of the iron limitation and the difference in the fluorescence quantum yields will be most pronounced.

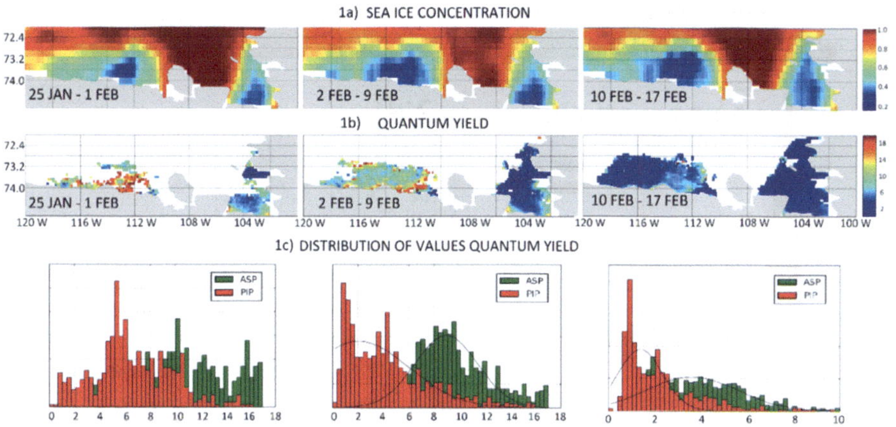

Fig. 1. 1(a) Sea ice concentrations obtained from AVHRR data. 1(b) Map of the fluorescence quantum yield obtained from MODIS data. 1(c) The distributions of the quantum yield of fluorescence from the values and their Gaussian fitting obtained from MODIS data.

Obtained results (see Fig. 1c) demonstrate notable differences in the quantum yield fluorescence of two polynyas for the period from February 2–9, when the blooms peak in the ASP was observed. Supposedly such a noise character of the distributions is related with that points with the concentration of ice (sea ice) more than 50% per cent make a significant contribution to the statistics of values.

We restrict the consideration to regions with ice concentrations less than 30%. Since it is in these areas the highest concentration of chlorophyll is observed, this analysis allows one to analyze the physiological state of the phytoplankton of each polynya.

Figure 2 shows the distributions of the quantum yield of fluorescence from the values and the approximation by the Gaussian distribution. The resulting average values presumably correspond to the average values of the quantum yield of fluorescence in each polynya. Figure 2a corresponds to the dates January 25–February 1, Fig. 2b: February 2–9, Fig. 2c: February 10–17. With such data selection, we can observe the dynamics of the manifestation of the effect of iron limitation. So for the period January 25–February 1, the effect of the difference in the quantum yield of fluorescence is already expressed, although the greatest difference was observed in the following week, it correlates with the data of the satellite monitoring of chlorophyll concentration in the ASP polynyas. In the period corresponding to the expedition (2–9 February), there was also a manifestation of iron limitation, the difference in the

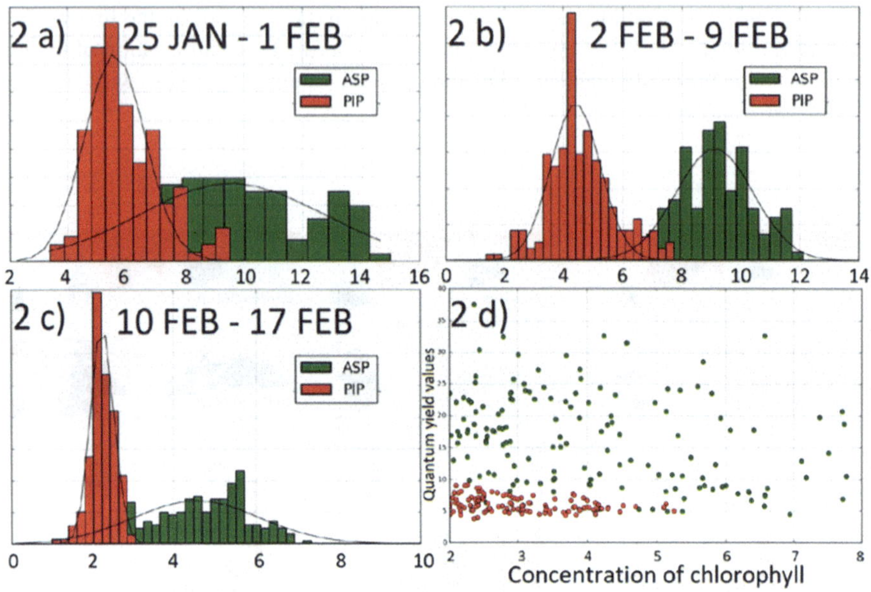

Fig. 2. The distributions of the quantum yield of fluorescence from the values and their Gaussian fitting obtained from MODIS data: (a) during 25 January–1 February (b) during 2–9 February (c) during 10–17 February. (d) Dependence of values of the fluorescence quantum yield from chlorophyll-a concentrations obtained from MODIS data.

average values of the fluorescence quantum yield was about 40%, which correlates with the data "in situ".

Figure 2d illustrates the dependence of the calculated values of the quantum yield fluorescence on the concentration of chlorophyll. For the analysis of the result's dependence of the operation of the algorithm for calculating fluorescence quantum yield on chlorophyll concentrations, the used algorithm demonstrates independence of the results of the work from the estimated values of chlorophyll concentration, what is necessary to compare regions with different chlorophyll concentration values.

4 Conclusions

The purpose of our investigation was correction of procedure to estimation of the fluorescence quantum yield from satellite data in order to observe the manifestation of the effect of iron limitation in the ASP clearing of the Amundsen sea (compared to the PIP clearing of this sea). We have put forward the requirements for the algorithm: independence from the chlorophyll concentration and errors in the microwave range associated with problems of atmospheric correction and the presence of ice cover. Polar regions are areas where satellite analysis is difficult due to high cloudiness and a large amount of drifting ice. We have developed a method for estimating the value of the quantum yield of fluorescence, which characterizes the physiology of the phytoplankton of this region on the basis of a comparison of the data of MODIS and AVHRR satellite systems. The use of MODIS microwave data is difficult due to the small amount of data for this area. The 2012 data, corresponding to the expedition period and 2 weeks preceding it, was analyzed by the developed method. We observed a difference in fluorescence quantum yield in all data of the indicated period, it was the most expressed during the period of the most intensive flowering in the ASP polynya. The difference in average values of fluorescence quantum yield obtained from data satellite sounding during the period corresponding to the expedition was 40%, which exceeds the difference in 27% in the quantum yield of photochemistry, measured "in situ". Firstly, this is due to the sample (irregular distribution of the stations in the region), and secondly due to the fact that the dependence of fluorescence quantum yield on a quantum yield of photochemistry is not strictly proportional, i.e. with the effect of non-photochemical quenching.

Acknowledgements. This research was supported by the Russian Foundation for Basic Research (project No. 16-05-01110 A).

References

1. Sarmiento, J.L., Toggweiler, J.R.: A new model for the role of the oceans in determining atmospheric PCO2. Nature **308**(5960), 621–624 (1984)
2. Sarmiento, J.L., et al.: Response of ocean ecosystems to climate warming. Glob. Biogeochem. Cycles **18**(3) (2004)
3. Sigman, D.M., Boyle, E.A.: Glacial/interglacial variations in atmospheric carbon dioxide. Nature **407**(6806), 859–869 (2000)
4. Rignot, E.: Changes in West Antarctic ice stream dynamics observed with ALOS PALSAR data. Geophys. Res. Lett. **35**(12) (2008)
5. Arrigo, K.R., Lowry, K.E., van Dijken, G.L.: Annual changes in sea ice and phytoplankton in polynyas of the Amundsen Sea, Antarctica. Deep Sea Res. Part II Topical Stud. Oceanogr. **71**, 5–15 (2012)
6. Alderkamp, A.-C., et al.: Fe availability drives phytoplankton photosynthesis rates during spring bloom in the Amundsen Sea Polynya, Antarctica. Elem. Sci. Anth. **3** (2015)
7. Gerringa, L.J.A., et al.: Iron from melting glaciers fuels the phytoplankton blooms in Amundsen Sea (Southern Ocean): iron biogeochemistry. Deep Sea Res. Part II Topical Stud. Oceanogr. **71**, 16–31 (2012)
8. Krause, G.H., Weis, E.: Chlorophyll fluorescence and photosynthesis: the basics. Ann. Rev Plant Biol. **42**(1), 313–349 (1991)
9. Falkowski, P.G., Greene, R.M., Geider, R.J.: Physiological limitations on phytoplankton productivity in the ocean. Oceanography **5**(2), 84–91 (1992)
10. Greene, R.M., et al.: Iron-induced changes in light harvesting and photochemical energy conversion processes in eukaryotic marine algae. Plant Physiol. **100**(2), 565–575 (1992)
11. Vassiliev, I.R., et al.: Effects of iron limitation on photosystem II composition and light utilization in Dunaliella tertiolecta. Plant Physiol. **109**(3), 963–972 (1995)
12. Greene, R.M., Geider, R.J., Falkowski, P.G.: Effect of iron limitation on photosynthesis in a marine diatom. Limnol. Oceanogr. **36**(8), 1772–1782 (1991)
13. Park, J., et al.: Light availability rather than Fe controls the magnitude of massive phytoplankton bloom in the Amundsen Sea polynyas, Antarctica. Limnol. Oceanogr. (2017)

14. https://www.ncei.noaa.gov
15. Lin, H., et al.: The fate of photons absorbed by phytoplankton in the global ocean. Science **351**(6270), 264–267 (2016)
16. Huot, Y., Brown, C.A., Cullen, J.J.: New algorithms for MODIS sun-induced chlorophyll fluorescence and a comparison with present data products. Limnol. Oceanogr. Meth. **3**(2), 108–130 (2005)

The Exact Mathematical Models of Nonlinear Surface Waves

Anatoly Kistovich(✉)

All-Russian Scientific Research Institute of Physico-Technical and Radio-Technical Measurements, Moscow, Russia
kavmendeleevo@mail.ru

Abstract. The problem of exact mathematical models of potential surface waves propagation is considered. The existing models converting the original system of hydrodynamic equations and boundary conditions into some new form are analyzed. A new approach to the above mentioned problem is presented resulting in the single equation for the waveform. The relations permitting to reconstruct all physical fields on the base of this waveform are formulated. It is shown how on the base of the equation received to construct all well-known results of approximate theories of stationary surface waves.

Keywords: Potential waves · Waveform · Stream function

1 Introduction

For a long time the potential surface wave's description is one of the classic hydrodynamic problems. The well-known results by Boussinesq, Stokes, Lord Rayleigh and many other researchers that made up a foundation of the surface waves theory were produced as solutions of approximate equations for the waveform. The majority of applied methods are based on using of a priory assumptions related the dependence of physical fields on the depth. At the same time, the heightened interest in producing either exact solutions or exact mathematical models simplifying the formulation of original problem is remained until now.

The classic system of governing equations and boundary conditions of the problem under consideration has the form

$$\mathbf{v}'_t - (\mathbf{v}\nabla)\mathbf{v} = -\frac{1}{\rho}\nabla p + \mathbf{g}\,, \quad \nabla\cdot\mathbf{v} = \mathbf{0}\,, \quad \nabla\times\mathbf{v} = \mathbf{0}$$

$$\mathrm{w} - \mathrm{u}\,\eta'_x - \mathrm{v}\,\eta'_y\big|_{z=\eta} = 0\,, \quad \mathrm{w} - \mathrm{u}\,\zeta'_x - \mathrm{v}\,\zeta'_y\big|_{z=\zeta} = \zeta'_t \tag{1}$$

$$p - \alpha K\big|_{z=\zeta} = p_a\,, \quad K = -\nabla\cdot(\nabla\zeta/\sqrt{1+(\nabla\zeta)^2})\,.$$

Here $\mathbf{v}(\mathbf{r}\,,t) = \mathrm{u}\mathbf{e_x} + \mathrm{v}\mathbf{e_y} + \mathrm{w}\mathbf{e_z}$ is the velocity field, $p(\mathbf{r}\,,t)$ and p_a are the pressures in liquid and air correspondingly, ρ is the liquid density, α is the

V. Karev et al. (Eds.): PMMEEP 2017, SPRINGERGEOL, pp. 305–316, 2018.
https://doi.org/10.1007/978-3-319-77788-7_32

surface-tension coefficient, $\eta(x, y)$ is a shape of rigid bottom, and $\zeta(x, y, t)$ is the waveform of free surface.

The first successful attempt of exact transformation of system (1) into some new form is Nekrasov's Eq. [1] for the steady periodic waves

$$\omega(\theta) = \frac{1}{6\pi} \int_0^{2\pi} \frac{\mu \sin \omega(p)}{1 + \mu \int_0^p \sin \omega(q)\, dq} \ln \left| \frac{\sin \dfrac{p+\theta}{2}}{\sin \dfrac{p-\theta}{2}} \right| dp\,. \tag{2}$$

In this equation the unknown function desired is the slope angle $\omega(\theta)$ of free surface to horizon, $R(\theta)$ is the ratio of stream velocity at infinity to local velocity of liquid particle. For the waves of the above mentioned class, the announced equation is exact one, but nobody had received its exact solution. The approximate solutions give well-known results for infinitesimal and lined waves. The results followed from (1) were confirmed later in papers [2,3].

The representation of wave characteristics by means of non-local relations was developed in alternative to standard methods applied for a description of surface waves. In paper [4] the original problem (when $\alpha = 0$) was transformed into the system of equations for the deviation $\Delta h = h - h_0$ of free surface and the potential φ of velocity field

$$\begin{gathered} \Delta h = \left\{ \varphi'_t + \frac{\partial(\xi,\zeta)}{\partial(x,z)} \left[\varphi'_\xi \frac{\partial(x,z)}{\partial(\zeta,t)} - \varphi'_\zeta \frac{\partial(x,z)}{\partial(\xi,t)} + \frac{1}{2}\left({\varphi'_\xi}^2 + {\varphi'_\zeta}^2 \right) \right] \right\} \Bigg|_{\zeta=1} \\ \varphi = -\frac{1}{2\pi} \int_{-\infty}^{+\infty} \frac{\partial(x,z)}{\partial(\xi,t)} (\xi - \xi', t) \ln(\cos(\pi\zeta) + \cosh(\pi\xi'))\, d\xi' \end{gathered} \tag{3}$$

where $h(x, t)$ is the distance from bottom to surface and the transformation of the region $-\infty < x < +\infty$, $0 \le z \le h(x, t)$ into the band $-\infty < \xi < +\infty$, $0 \le \zeta \le 1$ is defined by the relations

$$z(\xi,\zeta,t) = \frac{\sin(\pi\zeta)}{2} \int_{-\infty}^{+\infty} \frac{h(\xi - \xi', t)\, d\xi'}{\cos(\pi\zeta) + \cosh(\pi\xi')}, \quad x(\xi,\zeta,t) = \int_0^{\xi} \frac{\partial z}{\partial \zeta}(\xi', \zeta, t)\, d\xi'\,.$$

The system produced (3) is so complex and bulky that its explicit form needs for several sites. The practical application of (3) to non-linear problems is more difficult than original system (1). The numerical solution of this system also is unobtainable.

Other approach to description of non-stationary waves was realized in paper [5] where the original problem (1) (when $\alpha \neq 0$) was transformed into the system of equations for the surface perturbation ζ and for the quantity $q(x, y, t) = \varphi|_{z=\zeta}$

$$q'_t + \frac{1}{2}(\nabla q)^2 + g\zeta - \frac{\left(\zeta'_t + \zeta'_x q'_x + \zeta'_y q'_y\right)^2}{2\left(1 + (\nabla\zeta)^2\right)} = \frac{\alpha}{\rho}\nabla \cdot \frac{\nabla\zeta}{\sqrt{1 + (\nabla\zeta)^2}}$$

$$\int\limits_{-\infty}^{+\infty}\int\limits_{-\infty}^{+\infty} dx\, dy\, \exp(i(k_x x + k_y y))\times \tag{4}$$

$$\times\left[i\zeta'_t \cosh(|\mathbf{k}|(\zeta + \mathbf{h})) + \frac{\sinh(|\mathbf{k}|(\zeta + \mathbf{h}))}{|\mathbf{k}|}\mathbf{k} \cdot \nabla\mathbf{q}\right] = \mathbf{0}\,.$$

The solution of (4) also is a formidable task. The first equation in (4) is defined in configuration space whereas the second is defined in the wave-number space. This fact makes the analysis of (4) as the one of difficult problems. Approximate solutions of (4) coincide with well-known results of wave theory.

In paper [6] for the case of stationary waves ($\alpha = 0$, $\eta = -h$) the non-linear integral equation was received

$$\zeta(\xi) = -\frac{1}{\pi}\int\limits_{0}^{\infty} \frac{\sinh(k(\zeta(\xi) + h))}{\sinh(kh)}\left[\int\limits_{-\infty}^{+\infty} \cos(k(\xi - \eta))F(\eta)\right] dk = 0\,,\ \xi = x - ct \tag{5}$$

in which a single unknown function is the free surface deviation ζ and

$$F(\xi) = \sum_{n=1}^{\infty}(-\zeta)^n F_n\,, \quad F_0 = 0\,, \quad F_1 = -\sqrt{\frac{1 - 2g\zeta/c^2}{1 + (\zeta')^2}}$$

$$F_n = \frac{1}{n}\frac{F_{n-1}\zeta'' + 2F'_{n-1}\zeta' - F''_{n-2}/(n-1)}{1 + (\zeta')^2}\,, \quad n \geq 2\,.$$

The approximate solutions of (5) coincide to the known results of linear and weakly non-linear theories of waves including soliton solutions. For the case of non-stationary waves ($\alpha = 0$) in paper [7] the system was produced

$$\frac{\partial \Phi_0}{\partial t}\left(1 + {\zeta'_x}^2\right) + \frac{1}{2}\left(\frac{\partial \Phi_0}{\partial x}\right)^2 - \frac{\partial \Phi_0}{\partial x}\zeta'_x\zeta'_t - \frac{1}{2}{\zeta'_t}^2 + g\zeta\left(1 + {\zeta'_x}^2\right) = 0$$

$$\Phi_0(x,t) = \frac{1}{\pi}\int\limits_{0}^{\infty} \frac{\cosh(k(\zeta(x,t) + h))}{\cosh(kh)} \int\limits_{-\infty}^{+\infty} F(x',t)\cos(k(x - x'))\, dx\, dk$$

$$\Phi_n = \frac{1}{n}\frac{\Phi_{n-1}\zeta''_{xx} + 2{\Phi_{n-1}}'_x\zeta'_x - {\Phi_{n-2}}''_{xx}/(n-1)}{1 + {\zeta'_x}^2}\,, \quad n \geq 2 \tag{6}$$

$$\Phi_1 = \frac{{\Phi_0}'_x\zeta'_x + \zeta'_t}{1 + {\zeta'_x}^2}\,, \quad F(x,t) = \sum_{n=0}^{\infty}(-\zeta)^n\Phi_n\,.$$

The approximate solutions of (6) give well-known results for the non-stationary waves including the soliton of envelope. In the limit of the stationary waves the relations (6) transform into (5).

As it was shown in paper [8] the a priory assumption of exponential decreasing of all physical fields with depth leads to the absence of exact solutions in the class of real functions. However, at the same time there are two complex-valued exact solutions of (1) (when $\alpha = 0$) for the deviation of the free surface and stream function

$$\zeta_{\pm}(\xi, \zeta_0) = -a_{\pm} - \frac{1}{k} W(-k\zeta_0 \exp(\pm ik\xi))$$
$$\psi_{\pm}(\xi, z) = c\zeta_0 \exp(kz) \exp(\pm ik\xi)\,. \tag{7}$$

Here $W(x)$ is the Lambert function, which is the solution of the functional equation $W(x)\exp(W(x)) = x$, $a_{\pm}$ and ζ_0 are free parameters, ξ is the same as in (5).

By means of exact solutions (7) the approximate solutions for infinitesimal and inclined waves of finite amplitude may be produced.

Unsatisfactory features characterize all papers reviewed above. In particular, the Nekrasov's Eq. (2) describes only spatially periodic waves, which are symmetric with respect to its crests and cavities. The majority of the above results of the exact mathematical modeling are based on harmonic analysis application to Laplace equation for stream function. As a result, the systems (4–6) include integral transformations by means of harmonic functions, which are not suitable and adequate for arbitrary waveform observed in nature. An a priory assumption of physical fields dependence on vertical coordinate leads to the complex-valued solutions (7) which can't be realized in laboratory or natural experiments.

The goal of the presented work is to transform the system (1) as far as possible to the single equation for the free surface deviation from the undisturbed state and to get relations that permit to calculate the velocity and the pressure fields by means of the known waveform. Moreover, to avoid the above mentioned unsatisfactory features of (2–7) the expected results should be produced under refuse of harmonic analysis methods and any a priory assumptions of the dependence of the physical fields on vertical coordinate.

2 The Equation for Waveform

Thus 2-D problem (1) for the liquid of constant depth h without surface tension ($\alpha = 0$) is considered in the paper [9]. Initially the undisturbed liquid is moving along horizontal axis x with the speed U. It is presumed that the steady surface wave propagated in x-direction is characterized by velocity c so the variable $\xi = x - ct$ may be introduced. The condition of impressibility $\nabla \cdot \mathbf{v} = \mathbf{0}$ permits to introduce stream function ψ so $v_x = \psi'_x$, $v_z = -\psi'_x$. Also dimensionless variables are used

$$\tilde{x} = x/L,\ \tilde{z} = z/h,\ \tilde{t} = ct/L,\ \tilde{\xi} = \xi/L$$
$$\eta = \zeta/a,\ \tilde{p} = p/\rho c^2,\ \tilde{p}_a = p_a/\rho c^2,\ \tilde{\psi} = \psi/ac. \tag{8}$$

Substitution of relations (8) into the system (1) and integration of Euler′s equation by means of the relation

$$p = \beta\lambda\psi'_z - \frac{\lambda^2}{2}\left(\mu^2\psi'^2_\xi + \psi'^2_z\right) - \sigma z + p_a \tag{9}$$

transform the original hydrodynamic system to the new one

$$\mu^2\psi''_{\xi\xi} + \psi''_{zz} = 0 \quad \beta\psi'_z - \frac{\lambda}{2}\left(\mu^2\psi'^2_\xi + \psi'^2_z\right)\Big|_{z=\lambda\eta} = \sigma\eta$$
$$\psi'_\xi + \lambda\psi'_z\eta'_\xi\Big|_{z=\lambda\eta} = \beta\eta'_\xi,\quad \psi'_\xi\Big|_{z=-1} = 0. \tag{10}$$

Here $\beta = 1 - U/c$, $\lambda = a/h$, $\mu = h/L$, $\sigma = gh/c^2$, a is the maximum deviation of the free surface from its undisturbed level $z = 0$ and L is some characteristic horizontal scale. Symbol tilde over dimensionless variables in (9, 10) is omitted for brevity. The system received (10) is the governing system for the problem under consideration.

2.1 The Transformation of Governing System

At the first step the integration of kinematic boundary conditions is done. Because the condition on the free surface may be presented in the from

$$\psi'_\xi + \lambda\psi'_z\eta'_\xi\Big|_{z=\lambda\eta} = \frac{\partial}{\partial\xi}\left(\psi\Big|_{z=\lambda\eta}\right) = \beta\eta'_\xi$$

then after integration it follows

$$\psi\Big|_{z=\lambda\eta} = \beta\eta \tag{11}$$

and the constant of integration equals zero for the reason of absence of any perturbation when $\psi = 0$ in the whole space.

Analogously, from the kinematic condition on the bottom it follows

$$\psi'_\xi\Big|_{z=-1} = \frac{\partial}{\partial\xi}\left(\psi\Big|_{z=-1}\right) = 0$$

so the relation is valid

$$\psi\Big|_{z=-1} = 0 \tag{12}$$

and the constant of integration also equals zero for simplicity and coordination with (11).

At the second step it is necessary to integrate the Laplace equation of the system (9) without some method based on the principle of variables splitting. It is possible for the reason that the Laplace equation permits factorization

$$\mu^2 \psi_\xi'^2 + \psi_z'^2 = (\partial_z - i\mu\partial_\xi)(\partial_z + i\mu\partial_\xi)\psi = 0$$

where i is imaginary unit, so the solution may be presented in the form

$$\psi = F(\xi - i\mu(1+z)) + G(\xi + i\mu(1+z)) \tag{13}$$

where F and G are some arbitrary function of their own arguments.

Substitution of the relation (13) into relations (11, 12) and dynamic boundary condition of the system (10) transforms the governing system into new form

$$\begin{aligned} 2\lambda\mu^2 F'G' + i\beta\mu(F' - G')\Big|_{z=\lambda\eta} + \sigma\eta = 0 \\ F + G\Big|_{z=\lambda\eta} = \beta\eta\,,\ F(\xi) + G(\xi) = 0 \end{aligned} \tag{14}$$

where superscript means differentiation with respect to function argument.

Thereby, the governing system (10) is transformed into new system (14) which includes only boundary conditions.

2.2 The Deduction of Waveform Equation

Integration of the two first relations of the system (14) leads to results

$$F(\xi - i\mu(1+\lambda\eta)) = \frac{i}{2\lambda\mu}\left(\beta(\xi - i\lambda\mu\eta) - \int_{-\infty}^{\xi} R(\tau)\, d\tau\right) + \Phi_0 \tag{15}$$

where $\xi - i\mu(1+\lambda\eta)$ is the argument of function F, and

$$G(\xi + i\mu(1+\lambda\eta)) = -\frac{i}{2\lambda\mu}\left(\beta(\xi + i\lambda\mu\eta) - \int_{-\infty}^{\xi} R(\tau)\, d\tau\right) - \Phi_0 \tag{16}$$

where $\xi + i\mu(1+\lambda\eta)$ is the argument of function G.

Symbol R in the relations (15, 16) denotes the function

$$R(\xi) = \sqrt{(\beta^2 - 2\sigma\lambda\eta(\xi))(1 + \lambda^2\mu^2\eta'(\xi)^2)}\,. \tag{17}$$

After integration done the system (14) is reduced to the single equation

$$F(\xi) + G(\xi) = 0 \tag{18}$$

but the problem is arisen: how to use the relations (15, 16) with functions F and G of arguments $\xi - i\mu(1+\lambda\eta)$ and $\xi + i\mu(1+\lambda\eta)$ correspondingly in the Eq. (18) where arguments are simple ξ?

For solving this problem the transformations of ξ in (15, 16) are produced. In Eq. (15) the transformation $\xi \to \xi + f(\xi)$ is done and the condition

$$\xi + f(\xi) - i\mu(1 + \lambda\eta(\xi + f(\xi))) = \xi \sim f(\xi) = i\mu(1 + \lambda\eta(\xi + f(\xi))) \quad (19)$$

should be valid.

Analogously, in Eq. (16) the transformation $\xi \to \xi + g(\xi)$ is done and the condition

$$\xi + g(\xi) + i\mu(1 + \lambda\eta(\xi + g(\xi))) = \xi \sim g(\xi) = -i\mu(1 + \lambda\eta(\xi + g(\xi))) \quad (20)$$

should be valid.

Under action of (19, 20) the relations (15, 16) transform into

$$F(\xi) = \frac{i}{2\lambda\mu}\left(\beta\xi - \int\limits_{-\infty}^{\xi+f(\xi)} R(\tau)\, d\tau\right) + \Phi_0 - \frac{\beta}{2\lambda} \quad (21)$$

and

$$G(\xi) = -\frac{i}{2\lambda\mu}\left(\beta\xi - \int\limits_{-\infty}^{\xi+g(\xi)} R(\tau)\, d\tau\right) - \Phi_0 - \frac{\beta}{2\lambda} \quad (22)$$

Now the substitution of the relations (21, 22) into the last boundary condition (18) forms the equation

$$2i\beta\mu + \int\limits_{\xi+f(\xi)}^{\xi+g(\xi)} R(\tau)\, d\tau = 0\,. \quad (23)$$

Differentiation of the relation (23) with respect to ξ transforms it into equation

$$(\beta^2 - 2\sigma\lambda\eta(\xi + f(\xi)))(1 + 2f'(\xi)) = (\beta^2 - 2\sigma\lambda\eta(\xi + g(\xi)))(1 + 2g'(\xi))) \quad (24)$$

in which the unknown functions $f(\xi)$ and $g(\xi)$ present.

To eliminate these functions the formal solutions of the relations (19, 20) are used

$$f(\xi) = \xi_+ - \xi\,, \quad g(\xi) = \xi_- - \xi \quad (25)$$

where

$$\xi_\pm = \xi \pm i\mu(1 + \lambda\eta(\xi \pm i\mu(1 + \lambda\eta(\xi \pm i\mu(1 + \lambda\eta(\xi\pm))))))\,. \quad (26)$$

It is necessary to emphasize that in (26) the infinite enclosing has place.

Finely the substitutions of (25, 26) into (24) form the equation

$$\begin{aligned} &i\mu\beta^2\big(\eta'_\xi(\xi_+) + \eta'_\xi(\xi_-)\big) - \sigma\big(\eta(\xi_+) - \eta(\xi_-)\big) - \\ &-2i\mu\lambda\sigma\big(\eta(\xi_+)\eta'_\xi(\xi_+) + \eta(\xi_-)\eta'_\xi(\xi_-)\big) = 0 \end{aligned} \quad (27)$$

in which the single unknown function namely waveform η is presented.

The equation received (27) is constructed on the base of the system (1) without any approximation, hence it is the exact equation for the waveform.

2.3 Reconstruction of Physical Fields

The solution of (27) permits to reconstruct stream function by means of the known waveform. So accordingly (13) and with the help of (21, 22) one can to deduce

$$\psi = \frac{\beta z}{\lambda} - \frac{i}{2\lambda\mu} \int\limits_{A(\xi,z)}^{B(\xi,z)} R(\tau)\, d\tau \tag{28}$$

where

$$\begin{aligned} A(\xi, z) &= \xi + i\mu(1+z) + g(\xi + i\mu(1+z)) \\ B(\xi, z) &= \xi - i\mu(1+z) + f(\xi - i\mu(1+z))\,. \end{aligned} \tag{29}$$

From the relations (21, 22, 28) it follows

$$v_x = \psi'_z = \frac{\beta}{\lambda} - \frac{1}{2\lambda}(\Phi_f + \Phi_g), \quad v_z = -\psi'_\xi = \frac{i}{2\lambda\mu}(\Phi_f - \Phi_g)\,. \tag{30}$$

The functions Φ_f and Φ_g are defined by relations

$$\begin{aligned} \Phi_f &= (1 + i\lambda\mu\eta'(\xi - i\mu(1+z)))R(\xi - i\mu(z + \lambda\eta(\xi - i\mu(1+z)))) \\ \Phi_g &= (1 - i\lambda\mu\eta'(\xi + i\mu(1+z)))R(\xi + i\mu(z + \lambda\eta(\xi + i\mu(1+z))))\,. \end{aligned} \tag{31}$$

Here superscripts in the relations (31) mean differentiation with respect to corresponding argument of function. The results received permits to calculate the pressure by means of the relation (9).

2.4 Alternative Equation for Waveform

One can to construct the solutions of the relations (19, 20) with respect to functions $f(\xi)$ and $g(\xi)$ correspondingly by means of formal rows on parameter μ. Substitution of the above mentioned solutions in (23) with consequent integration also by means of formal rows leads to the alternate equation for the waveform

$$\sum_{n=0}^{\infty}(-1)^n \frac{\mu^{2n}}{(2n+1)!}\frac{d^{2n}}{d\xi^{2n}}((1 + \lambda\eta(\xi))^{2n+1}R(\xi)) - \beta = 0 \tag{32}$$

where $R(\xi)$ was defined in (17).

In this equation also as in (27) the single unknown function namely waveform is presented but in contrast to (27) without infinite enclosing. From the other hand the degree of the differential Eq. (32) equals infinity that imply own characteristic problems.

3 Validation of the Waveform Equation

The best method of the inspection of results received is to get an exact solution of (27) or (32) then to reconstruct the required physical fields and to substitute the results into the system (1). Unfortunately, at the present time no one exact solution of above mentioned equations is known. For this reason, the validation of Eqs. (27, 32) will be done for classical problems of the approximate wave theory.

3.1 Infinitesimal (Lined) Waves

For infinitesimal (lined) waves the ratio $\lambda\mu \ll 1$ is valid. As a result it follows from (27)

$$i\mu\beta^2(\eta'_\xi(\xi+i\mu)+\eta'_\xi(\xi-i\mu))-\sigma(\eta(\xi+i\mu)-\eta(\xi-i\mu))=0\,. \tag{33}$$

The usage of the formal rows

$$\eta(\xi \pm i\mu)=\eta(\xi)\pm i\mu\eta'(\xi)-\frac{\mu^2}{2}\eta''(\xi)\mp i\frac{\mu^3}{3!}\eta'''(\xi)+\dots$$

permits to transform (33) into equation

$$\begin{aligned}&\beta^2\Big(\eta(\xi)-\frac{mu^2}{2!}\eta'''(\xi)+\frac{mu^4}{4!}\eta^{(4)}(\xi)-\dots\Big)-\\&-\sigma\Big(\eta'(\xi)-\frac{mu^2}{3!}\eta'''(\xi)+\frac{mu^4}{5!}\eta^{(5)}(\xi)-\dots\Big)=0\end{aligned} \tag{34}$$

with solution $\eta=\cos(\xi)$ and dispersion equation

$$\sigma=\beta^2\mu\,\cot(\mu)\sim(c-U)^2=\frac{g}{k}\tan(kh) \tag{35}$$

where $k=1/L$ is a wave-number, $kh=\mu$.

Analogously, from (32) the equation

$$\eta(\xi)(1-\sigma/\beta^2)-\frac{\mu^2}{3}\eta''(\xi)-\frac{\mu^4}{45}\eta^{(4)}(\xi)-\frac{2\mu^6}{945}\eta^{(6)}(\xi)-\dots=0$$

follows from which for solution $\eta=\cos(\xi)$ one can get dispersion Eq. (35).

For the case under consideration, the relations have place

$$\begin{aligned}&f(\xi-i\mu(1+z))\approx i\mu(1+\lambda\,\cos(\xi-i\mu z))\\&g(\xi+i\mu(1+z))\approx -i\mu(1+\lambda\,\cos(\xi+i\mu z))\\&A(\xi,z)\approx\xi+i\mu z-i\lambda\mu\,\cos(\xi+i\mu z)\\&B(\xi,z)\approx\xi-i\mu z+i\lambda\mu\,\cos(\xi-i\mu z)\end{aligned} \tag{36}$$

Substitution of the (36) and $R(\xi) \approx \beta - \sigma\lambda\cos(\xi)/\beta$ into (28) leads to expression for the stream function

$$\psi(\xi, z) \approx \cos(\xi)\frac{\sinh(\mu(1+z))}{\sinh(\mu)}\,. \tag{37}$$

This is well-known result of the approximate theory of infinitesimal surface waves. Also in the limit $\mu \to \infty$ from the relations (35, 37) follow standard results for infinitely deep water.

3.2 Waves of Small but Finite Amplitude

For this case the solution of (27, 32) was constructed but tedious and bulky form for water of finite depth does not permit to display the result here. In the case of infinitely deep water ($\mu \to \infty$) the waveform is

$$\eta = \cos(\xi) + \frac{ka}{2}\cos(2\xi) + \frac{3k^2a^2}{8}\cos(3\xi) + \frac{k^3a^3}{3}\cos(4\xi) + \ldots\,. \tag{38}$$

This wave propagates with the speed

$$c = U + \sqrt{g/k}(1 + k^2a^2/2) \tag{39}$$

where k is defined in the relation (35).

This result, so called Stokes wave, coincides with results of the paper [10].

3.3 Localized Waves on Shallow Water

The term "localized wave" suggests the validity of conditions

$$\eta^{(n)}(\xi)\big|_{\xi=\pm\infty} = 0\,, \quad n \in N\,. \tag{40}$$

In the case of shallow water the condition $\mu \ll 1$ is valid and the expansion of (27) transforms it into equation

$$(\beta^2 - \sigma)\eta + \frac{\mu^2}{3!}(\sigma - 3\beta^2)\eta'' + \frac{\mu^4}{5!}(5\beta^2 - \sigma)\eta^{(4)} + \frac{\mu^6}{7!}(\sigma - 5\beta^2)\eta^{(6)} + \cdots -$$
$$-3\lambda\sigma\eta^2 + \lambda\mu^2\Big((3\sigma - 2\beta^2)(\eta\eta'' + \eta'^2) + \lambda\eta(5\sigma - 2\beta^2)(\eta\eta'' + 2\eta'^2) +$$
$$+\tfrac{7}{2}\lambda^2\eta^2(\eta\eta'' + 3\eta'^2)\Big) + \cdots = 0\,.$$

When $\lambda \sim \mu^2$, $\sigma = \beta^2(1 - s\mu^2)$ (the standard assumptions for soliton solutions) the above equation to the degree of the second order with respect to μ inclusively transforms into equation

$$\frac{1}{3}\eta'' - s\eta + \frac{3}{2}\eta^2 = 0 \tag{41}$$

which solution is well-known Rayleigh's soliton [11].

3.4 The Speed of Steady Waves

As it follows from (26) the relation is valid

$$a\frac{\partial\eta(\xi_\pm)}{\partial a}+h\frac{\partial\eta(\xi_\pm)}{\partial h}+L\frac{\partial\eta(\xi_\pm)}{\partial L}=0 \tag{42}$$

which can be proved by the direct substitution of (26) into (42).

The application of (42) to the Eq. (27) for waveform permits to get some relation for relative speed $\tilde{c}=c-U$

$$a\frac{\partial\tilde{c}}{\partial a}+h\frac{\partial\tilde{c}}{\partial h}+L\frac{\partial\tilde{c}}{\partial L}-\frac{\tilde{c}}{2}=0 \tag{43}$$

or in equivalent formulation

$$\tilde{c}=ga\,f_1(\lambda,\mu)=gh\,f_2(\lambda,\mu)=gL\,f_3(\lambda,\mu)\,. \tag{44}$$

Here f_1, f_2, f_2 are arbitrary functions of λ and μ parameters. Validation of the relations (43, 44) shows that the speeds of all known wave solutions of approximate theories satisfy the above equations.

4 The Exact Solution for Complex-Valued Waves

If in the case of infinitely deep water ($\mu\to\infty$) the a priory assumption about exponential decreasing of physical fields with depth is done, i.e.

$$\psi(\xi,z)=\Psi(\xi)\exp(\mu z)\quad\Rightarrow\quad\psi'_z(\xi,z)=\mu\psi(\xi,z) \tag{45}$$

then substitution of (28) into the second relation of (45) leads to equation

$$\psi'_z(\xi,z)=\frac{\beta}{\lambda}+\frac{i}{2\lambda\mu}\left(R(A(\xi,z))\frac{\partial A(\xi,z)}{\partial z}-R(B(\xi,z))\frac{\partial B(\xi,z)}{\partial z}\right). \tag{46}$$

On the free surface of water the relations $z=\eta(\xi)$ and $\psi(\xi,\eta(\xi))=\beta\eta(\xi)$ have place and the relations are valid

$$A(\xi,\eta(\xi))=B(\xi,\eta(\xi))=\zeta$$

$$\left.\frac{\partial A(\xi,z)}{\partial z}\right|_{z=\eta(\xi))}-\frac{i\mu\eta'(\xi)}{1+i\lambda\mu\eta'(\xi)},\quad\left.\frac{\partial B(\xi,z)}{\partial z}\right|_{z=\eta(\xi))}=-\frac{i\mu\eta'(\xi)}{1-i\lambda\mu\eta'(\xi)}\,. \tag{47}$$

Substitution of (47) into (46) forms equation

$$\beta(1-\lambda\mu\eta(\xi))=\sqrt{\frac{\beta^2-2\sigma\lambda\mu\eta(\xi)}{1+\lambda^2\mu^2\eta'^2(\xi)}}$$

from which the equations for waves on infinitely deep water follow

$$ka\eta'(\xi)=\pm i\frac{ka\eta(\xi)}{1-ka\eta(\xi)} \tag{48}$$

where k is defined in (35).

The solutions of (48) are presented in the relations (7) as complex-valued exact solutions of the system (1).

5 Conclusions

Presented here the new approach to formation of exact mathematical models of surface waves propagation transforms the original boundary problem into the single functional-differential equation for the waveform and rules for reconstruction of physical fields by means of the above mentioned waveform. The approximate solutions of the equation received coincide to well-known results of approximate theories of surface waves.

References

1. Nekrasov, A.I.: The exact theory of steady waves on the surface of heavy liquid, collection of works by A.I. Nekrasov. Academy of Science, Moscow, USSR, pp. 358–439 (1961). (In Russian)
2. Levi-Civita, T.: Determinazione rigorosa delle onde irrotazionale periodiche in aqua profonda. Atti. Accad. Licei. **33**(5), 141–150 (1924)
3. Struik, D.J.: Dtermination rigoureuse des ondes irrotationelles permanentes dans un canal profondeur finie. Math. Ann. **95**, 595–634 (1926)
4. Byatt-Smith, J.G.B.: An integral equation for unsteady surface waves and a comment on the Boussinesq equation. J. Fluid Mech. **49**, 625–633 (1971)
5. Ablowitz, M.J., Fokas, A.S., Musslimani, Z.H.: On an new non-local theory of water waves. J. Fluid Mech. **562**, 313–343 (2006)
6. Kistovich, A.V., Chashechkin, Y.D.: An integral description of the propagation of steady-stable perturbations of the heavy liquid surface. Izv. Atmos. Ocean. Phys. **45**, 654–659 (2009). https://doi.org/10.1134/S0001433809050120
7. Kistovich, A.V., Chashechkin, Y.D.: Analytical models of stationary non-linear gravitational waves. Water Resour. **43**(1), 86–94 (2016). https://doi.org/10.1134/S0097807816120083
8. Kistovich, A.V.: The exact complex-valued solution for steady surface waves. Procedia IUTAM **8**, 161–165 (2013). https://doi.org/10.1016/j.piutam.2013.04.020
9. Kistovich, A.V.: A new approach to description of potential surface waves. Process. in Geomedia **3**(3), 34–40 (2015). IPMech. RAS, Moscow (In Russian)
10. Stokes, G.G.: On the theory of oscillatory waves. Trans. Cam. Phil. Soc. **8**, 441–455 (1847)
11. Lord Rayleigh: On waves. Phil. Mag. and J. Sci. 5, **1**(4), 257–279 (1876)

Numerical Analysis and Prediction of the Consequences of Natural and Technological Impacts in Coastal Areas of the Azov Sea

T. Ya. Shul'ga[1(✉)], S. M. Khartiev[2], and A. R. Ioshpa[2]

[1] FSBSI Marine Hydrophysical Institute, Kapitanskaya. 2, 299011 Sevastopol, Russia
shulgaty@mail.ru

[2] Southern Federal University, 105/42 Bolshaya Sadovaya Street, Rostov-on-Don 344006, Russia
aioshpa@yandex.ru

Abstract. In this work, the waves and currents generated by prognostic wind in the Sea of Azov are investigated using a three-dimensional nonlinear sigma-coordinate Princeton Ocean Model. The mathematical model was also used for studying the transformation of passive admixture in the Sea of Azov, caused by the spatiotemporal variations in the fields of wind and atmospheric pressure, obtained from the prediction SKIRON model. Comparison of the results of numerical calculations and the data of field observations, obtained during the action of the wind on a number of hydrological stations was carried out. The growth of storm surges, velocities of currents and the characteristics of the pollution region at different levels of intensity of prognostic wind and stationary currents were found. The obtained results are presented in the table of the sea level changes caused by the onshore and offshore winds and the current velocities for different characteristics of constant and variable wind. The results of a comprehensive study allow reliably estimate modern eco-logical condition of offshore zones, develop predictive models of catastrophic water events and make science-based solutions to minimize the possible damage.

Keywords: Sea of Azov · Storm · Surge phenomena processes
Surface currents · Evolution of passive admixture
Three-dimensional hydrodynamic model

1 Introduction

Currently steady growth of interest in the mathematical modeling of wave motions of various natural stratified environments is observed. This is due to problems of geophysics, oceanography, atmospheric physics protection and study of the environment, the operation of complex hydraulic structures, including offshore oil complexes and other. Industrial activity in the offshore including those related to mining is important task wave dynamics, and obtained characteristics are used to assess the environmental impact on marine technology design and development of effective methods of

V. Karev et al. (Eds.): PMMEEP 2017, SPRINGERGEOL, pp. 317–326, 2018.
https://doi.org/10.1007/978-3-319-77788-7_33

forecasting of extreme hydrological events. Numerical study of hydrodynamic processes in the Azov Sea, arising from the different types of atmospheric circulation will be performed using the three-dimensional nonlinear sigma coordinate model with high spatial resolution adapted to the peculiarities of this basin. New modules and routines were included in the model in order to obtain assessment of the response patterns of surface and bottom currents at strengthening wind stress. Based on the analysis of results of a series of numerical experiments are performed to identify extreme synoptic perturbation level and flow velocity of the Azov Sea.

2 The Three Dimensional Primitive Equations. Boundary and Initial Conditions

2.1 Model Configuration

We investigate the waves and currents generated by prognostic wind [1, 2] in the Sea of Azov. We used a three-dimensional nonlinear Princeton Ocean (POM) model to do it [3]. We shall consider a rectangular coordinate system in which the x-axis is directed to the east, the y-axis is directed to the north, and the z-axis is directed vertically upwards. The mathematical model is based on the equations of motion and continuity using the hydrostatic approximation [3, 4].

$$\frac{du}{dt} - fv + \frac{1}{\rho}\frac{\partial p}{\partial x} = 2\frac{\partial}{\partial x}\left(A_M \frac{\partial u}{\partial x}\right) + \frac{\partial}{\partial y}\left[A_M\left(\frac{\partial u}{\partial x} + \frac{\partial v}{\partial y}\right)\right] + \frac{\partial}{\partial z}K_M\frac{\partial u}{\partial z}, \quad (1)$$

$$\frac{dv}{dt} + fu + \frac{1}{\rho}\frac{\partial p}{\partial y} = 2\frac{\partial}{\partial y}\left(A_M \frac{\partial v}{\partial y}\right) + \frac{\partial}{\partial x}\left[A_M\left(\frac{\partial u}{\partial x} + \frac{\partial v}{\partial y}\right)\right] + \frac{\partial}{\partial z}K_M\frac{\partial v}{\partial z}, \quad (2)$$

$$\frac{\partial p}{\partial z} + g\rho = 0, \quad (3)$$

$$\frac{\partial u}{\partial x} + \frac{\partial v}{\partial y} + \frac{\partial w}{\partial z} = 0. \quad (4)$$

We denote that u, v, and w are velocity projections on the axes x, y, and z, respectively; t is the time; $p|_z = p(x, y, z, t)$ and $p|_\zeta = p_{atm}$ is the standard atmospheric pressure; ρ is the density; g is the acceleration due to gravity; and f is the Coriolis parameter. The coefficient of the horizontal viscosity A_M is calculated using the Smagorinskii model of subgrid viscosity [5] depending on the horizontal velocity gradients. The relations for the calculation of the vertical viscosity coefficients K_M and the turbulent diffusion K_H according to the semi-empirical model of Mellor-Yamada (level 2.5) are written as follows [6]:

$$K_M = qlS_M, \; K_H = qlS_H, \quad (5)$$

where S_M and S_H in a neutrally stratified flow are equal to 0.30 and 0.49, respectively. This parameterization is based on the solution of two additional equations in partial derivatives to determine the turbulent kinetic energy ($q^2/2$) and the turbulence macroscale (l).

The boundary conditions at the free surface ($z = \zeta(x, y, t)$) for the equations of motion are written as follows:

$$w|_{z=\zeta} = \frac{\partial \zeta}{\partial t} + u\frac{\partial \zeta}{\partial x} + v\frac{\partial \zeta}{\partial y}, \ K_M\left(\frac{\partial u}{\partial z}, \frac{\partial v}{\partial z}\right)\bigg|_{z=\zeta} = \left(\tau_{0x}, \tau_{0y}\right). \tag{6}$$

We note that $(\tau_{0x}, \tau_{0y}) = \rho_a C_a|\mathbf{U}_W|(U_W, V_W)$ are projections of the tangential wind stress [3]; $\mathbf{U}_W$ is the wind speed at 10 m above the sea water surface, U_W and V_W are two components of the wind speed vector; ρ_a is the density of air at the standard atmospheric conditions, and C_a is an empirical coefficient of the surface friction [7], which varies depending on the wind velocity:

$$10^3 C_a = \begin{cases} 2,5, & & |\mathbf{U}_W| & > & 22 & \mathrm{m \cdot s^{-1}}, \\ 0,49 + 0,065|\mathbf{W}|, & 8 \leq & |\mathbf{U}_W| & \leq & 22 & \mathrm{m \cdot s^{-1}}, \\ 1,2, & 4 \leq & |\mathbf{U}_W| & \leq & 8 & \mathrm{m \cdot s^{-1}}, \\ 1,1, & 1 \leq & |\mathbf{U}_W| & \leq & 4 & \mathrm{m \cdot s^{-1}}. \end{cases} \tag{7}$$

The normal component of the velocity at the bottom is zero ($z = -H(x, y)$); the bottom tangential stresses are related to the velocity by the quadratic equation [3]

$$\left(w + u\frac{\partial H}{\partial x} + v\frac{\partial H}{\partial y}\right)\bigg|_{z=-H} = 0, \ K_M\left(\frac{\partial u}{\partial z}, \frac{\partial v}{\partial z}\right)\bigg|_{z=-H} = (\tau_{1x}, \tau_{1y}), \tag{8}$$

where $(\tau_{1x}, \tau_{1y}) = c_b|\mathbf{U}_b|(u_b, v_b)$, u_b and v_b are the horizontal flow velocities at the grid point closest to bottom and c_b is the bottom drag coefficient determined as the maximum between a value calculated according to the logarithmic law of the wall and a value equal to 0.0025: $c_b = \max\{k^2(\ln(H + z_b)/z_0)^{-2}; 0{,}0025\}$, where z_b is the vertical step in the bottom layer; and $z_0 = 0.03$ mm is the roughness parameter that characterizes the hydrodynamic properties of the underlying bottom surface. The non-slip conditions are specified at the lateral boundaries. As initial (at $t = 0$), conditions are assumed for the absence of fluid motion and the horizontal nature of the free surface before the onset of atmospheric disturbances.

Let use the equation of transport and diffusion to calculate the spreading of the admixture with the concentration $C(x, y, z, t)$ [3]:

$$\frac{dC}{dt} = \frac{\partial}{\partial x}\left(A_H\frac{\partial C}{\partial x}\right) + \frac{\partial}{\partial y}\left(A_H\frac{\partial C}{\partial y}\right) + \frac{\partial}{\partial z}\left(K_H\frac{\partial C}{\partial z}\right). \tag{9}$$

Here, $A_H = 10\ \mathrm{m^2/s}$ [3, 5] is the coefficient of horizontal turbulent diffusion and K_H is the vertical diffusion; K_H is found from Eqs. (5), (1). The conditions of zero admixture fluxes through the free surface, lateral walls (S), and bottom of the basin are added to the dynamic boundary conditions at the free surface and bottom layer [3]:

$$\left(K_H \frac{\partial C}{\partial \mathbf{n}}\right)\bigg|_{z=\zeta} = 0, \left(A_H \frac{\partial C}{\partial \mathbf{n}}\right)\bigg|_{S} = 0, \left(K_H \frac{\partial C}{\partial \mathbf{n}}\right)\bigg|_{z=-H} = 0. \tag{10}$$

The initial pollution region for all the atmospheric perturbations considered below is located in the surface layer:

$$C_0(x, y, z) = \begin{cases} 1, & r \le R, \ 0 \ge z \ge -z_1, \\ 0, & r > R, \ z < 0; \ r \le R, \ \ z < -z_1, \end{cases} \tag{11}$$

where z_1 is the thickness of the pollution region, $r = ((x - x_0)^2 + (y - y_0)^2)^{1/2}$ is the distance from the center of this region (x_0, y_0) to the point at which the concentration is calculated and R is its radius. We select the time of the dispelling of the admixture (t_d) and the coefficient of the maximum square of its spreading at different levels (K_{max}) as examples of the parameters characterizing the evolution of the passive admixture. Then, $K_{max} = S_{max}/S_0$, where S_0 is the square of the initial region of pollution in the surface layer, and S_{max} is the maximum square of the pollution at the considered depth during the transformation of the admixture. The condition of complete pollution dispelling is for a concentration that does not exceed 2.5×10^{-2} over the entire basin of the sea ($C_d = 2.5 \times 10^{-2}$).

The transition from the z-coordinate to the sigma-coordinate in Eqs. (1)–(4), (6), (8)–(10) is made to perform the numerical realization [3]. In this case, the solution algorithm is based on the application of two-layer differential schemes. The transport operators are approximated [8] using a *TVD* scheme (a linear combination of the scheme of directed differences and the Lax-Wendroff scheme); the spatial digitization of the equations is performed using a *C* grid. Uniform steps over the horizontal coordinates Δx and Δy and the σ-coordinate are used. The resolutions of the model by latitude and longitude are $(1/59)^\circ \times (1/84)^\circ$, at which the linear sizes of the cell are $\Delta x = \Delta y = 1.4$ km; the number of the horizontal grid nodes is 276×176. The number of σ-sigma levels by the vertical is 11. The equations were integrated with a step of

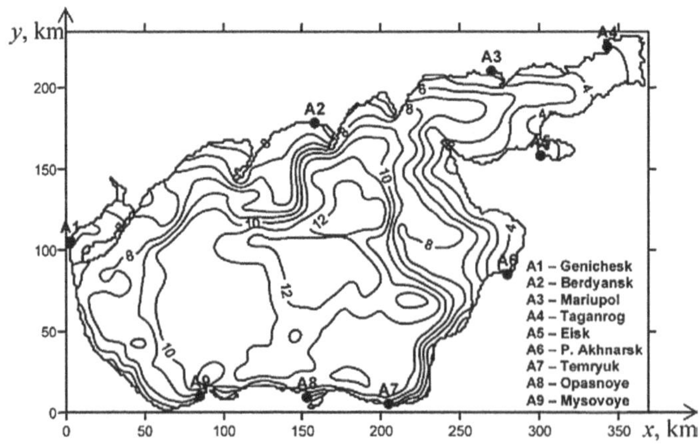

Fig. 1. The bottom relief (m) of the Sea of Azov, the position of coastal stations.

$\Delta t = 18$ s to determine the average two-dimensional velocity components and the sea level and with a step of $\Delta t_A = 10\Delta t = 3$ min to calculate the deviations from the found mean values and vertical velocity component. The choice of the integration steps over the temporal and spatial coordinates was performed according to the stability criterion for barotropic waves [9]. The bottom topography is interpolated to the model grid using the depths given in the navigation charts. The deviations of the Azov Sea level were analyzed at nine stations located near large towns (see Fig. 1).

2.2 Wind Stress and Atmospheric Pressure Conditions

Stationary motions in the Sea of Azov are generated by the western wind field, whose velocity at the sea surface $\left(|\mathbf{U}|_1^{st} = 5 \text{ and } |\mathbf{U}|_2^{st} = 10\ \text{m/s}\right)$ does not depend on x and y; in the first three hours ($0 < t < 3$ h), it increases with time according to a linear law $\left(|\mathbf{U}|_{1,2}^{st} = 0 \text{ at } t = 0\right)$ and reaches the maximum value; later ($t \leq 3$ h) it does not change. The moment when the currents reach the stable regime ($t = t_0$) is determined by the fact that no notable deviations of the level and current velocities (the variations do not exceed 3%) occur between two neighboring time moments ($T > t_0$ and $T + \Delta t$). From this we find the time when the fluid motion reaches the stable regime $(t = t_{01},\ |\mathbf{U}|_1^{st};\ t = t_{02},\ |\mathbf{U}|_2^{st})$. At the moment $t = t_0$) when the fluid motion stably, a temporally and spatially inhomogeneous wind field and atmospheric pressure obtained from the reanalysis data (SKIRON) is added to the stationary wind and the standard atmospheric pressure. At $t > t_0$, the wind $\mathbf{U}_{1,2}^{st}$ maintains the stationary motion, and $\mathbf{U}_{SKIRON}$ adds a non-stationary component to the motion.

We study the joint influence of the stationary ($\mathbf{U}_{1,2}^{st}$ and variable $\mathbf{U}_{SKIRON}$) wind on the maximum velocities of the currents and the extreme values of the offshore and onshore transport using a series of numerical experiments. These results are compared with the data obtained from the calculation of the winds and currents caused only by constant wind forcing or only by wind determined from the reanalysis.

The investigation of the stationary currents is performed for the constant western wind, whose velocity is 5 and 10 m/s. The surface wind field from the SKIRON model [15] in the period from September 8 to 18, 2007, is used as the variable wind in time and space. Its forcing in the existence of the stationary currents in the Sea of Azov occurs during 10 days starting from September 8 at 00:00 h.

Table 1 presents the maximum wind velocities and wind directions ($\mathbf{U}_{SKIRON}$) over the Sea of Azov from September 11 to 18, 2007, as functions of time. The wind velocity changes monotonously between two neighboring times. The deviations of the wind velocity vector from the x-axis (the x-axis is directed to the east at an angle of 50° to the latitude) are given in degrees. It is seen from this that, during the studied period (192 h), the maximum and minimum velocities were 12.7 and 1.6 m/s, respectively. The dominating directions of the wind are northeastern and northwestern. We note that the results of the long-term observations of the atmospheric perturbations in the region of the Sea of Azov [1] in this period (September) agree well with the data of the SKIRON model given in this table.

Table 1. The maximum wind velocity (m/s) as a function of time (h) and its direction (deg.) obtained from the SKIRON model data from September 11, 2007 at 0 h to 18, 2007 at 24 h

Time	Wind velocity	Wind direction	Time	Wind velocity	Wind direction
2	6.8	107	76	9.4	354
14	2.6	344	92	12.7	350
28	5.8	9	104	9.6	100
30	3.0	10	106	7.5	210
32	5.2	107	108	5.7	200
44	5.8	344	112	3.5	110
48	8.1	100	124	5.2	344
50	4.2	213	130	3.3	354
54	5.8	195	132	1.6	347
56	7.9	192	140	2.1	10
58	9.6	108	152	4.9	106
62	11.6	350	192	5.8	200

3 Analysis of Numerical Experiments

Numerical experiments were realized for two velocities of the stationary western wind to study the influence of the induced currents on the sea level fluctuations and the variations in the velocity field of the nonstationary currents caused by the SKIRON wind. Table 2 gives the maximum sea level deviations caused by the stationary wind ($U^{st}_{1,2}$), only by the wind based on the reanalysis data ($\mathbf{U}_{SKIRON}$), and by their joint forcing ($U^{st}_{1,2} + U_{SKIRON}$) at the coastal stations of the Sea of Azov. The sea level variations caused by onshore winds are given in the upper part of the table, and the offshore values are given in the lower part of the table. It follows from the analysis of these results that the maximum onshore sea level changes generated by the system of stationary and variable wind were recorded at the following stations: 20.7 cm (U^{st}_1) and 62.4 cm (U^{st}_2) in Taganrog, 57.1 cm ($\mathbf{U}_{SKIRON}$) and 80.4 cm ($U^{st}_1 + U_{SKIRON}$) in Primosko-Akhtarsk, and 102.2 cm ($U^{st}_2 + U_{SKIRON}$) in Eisk. It is seen from here that the maximum sea level change caused by onshore winds in the case of ($U^{st}_2 + U_{SKIRON}$) (102.2 cm) is 1.27 times greater than in the case of ($U^{st}_1 + U_{SKIRON}$) (80.4 cm). The minimum sea level changes caused by onshore winds appear in Mysovoye (7.5 cm for U^{st}_1, 13.9 cm for U^{st}_2) and in Opasnoye (9.4 cm for $\mathbf{U}_{SKIRON}$, 16.1 cm for $U^{st}_1 + U_{SKIRON}$, and 24.8 cm for ($U^{st}_2 + U_{SKIRON}$).

The maximum sea level changes become greater under the joint influence of variable and constant offshore winds than in the stable regime and in the case of zero stationary currents. Among all the types of winds, the maximum sea level changes under offshore winds occur in Genichesk: 12.2 cm (U^{st}_1), 51.7 cm ($\mathbf{U}_{SKIRON}$), 76.5 cm ($U^{st}_1 + U_{SKIRON}$), and 87.0 cm ($U^{st}_2 + U_{SKIRON}$). The minimum sea level changes under offshore winds occur in Opasnoye (3.3 cm (U^{st}_1), 11.1 cm (U^{st}_2) and in Temryuk (8.7 cm ($\mathbf{U}_{SKIRON}$), 15.2 cm ($U^{st}_1 + U_{SKIRON}$), and 23.0 cm ($U^{st}_2 + U_{SKIRON}$).

Let us compare the modeling results and the data of the field measurements presented in the tables of the hourly sea level values of the State Meteorological Service of

Table 2. Maximum sea level displacements caused by onshore and offshore winds (cm) at coastal stations of the Sea of Azov under a stationary regime and caused by a prognostic wind in the presence of stationary currents

Coastal stations	U_1^{st}	U_2^{st}	U_{SKIRON}	$U_1^{st} + U_{SKIRON}$	$U_2^{st} + U_{SKIRON}$
Genichesk	–	–	25,4	32,2	62,3
Berdyansk	–	–	9,6	16,9	44,3
Mariupol	9,8	37,3	29,3	46,4	80,4
Taganrog	20,7	62,4	50,6	63,1	89,5
Eisk	13,8	52,2	38,1	76,0	102,2
P. Akhnarsk	8,1	43,2	57,1	80,4	91,1
Temryuk	10,2	26,9	24,5	29,7	49,9
Opasnoye	–	–	9,4	16,1	24,8
Mysovoye	7,5	13,9	12,1	19,6	34,2
Genichesk	12,2	51,7	42,5	76,5	87,0
Berdyansk	4,0	17,6	17,3	30,9	62,1
Mariupol	–	–	18,2	26,0	39,7
Taganrog	–	–	29,0	42,4	72,9
Eisk	–	–	18,9	41,1	45,3
P. Akhnarsk	–	–	14,1	23,8	35,5
Temryuk	–	–	8,7	15,2	23,0
Opasnoye	3,3	11,1	10,6	20,5	34,2
Mysovoye	–	–	22,3	39,4	63,9

Ukraine in the period from September 8 to 18, 2007. Let us estimate numerically the obtained extreme values of the sea level changes under onshore winds caused by the $\mathbf{W}_{SKIRON}$ wind with the hourly data from these tables. The simulated maximum of the onshore sea level change in Genichesk is 25.4 cm, which is 4.7 cm (16%) smaller than from the data of the observations. It follows from this that the amplitudes of the sea level fluctuations obtained from the field data and from the numerical calculations agree quite well. The indicated differences are likely to be caused by the errors of the measurements and mathematical modeling.

Variations in the profile of the sea surface caused by wind forcing are shown in Fig. 2. It is seen from the figure that, in the stable regime (Fig. 2*a*), the sea level decreases along the western coast and increases near the eastern coast. The node line (dashed line) crosses the central part of the sea normal to the wind velocity. The regions in which the maximum and minimum sea level deviations appear three change during the wind forcing (Fig. 2*b*–*d*).

The objective of the following numerical experiments is to estimate the influence of the wind fields and generated currents on the spreading of the passive admixture (9) transported to the central region of the sea. The initial position of the center of the region where the admixture was released is located at a point with the coordinates $x_0 = 180$ km, $y_0 = 120$ km; the depth of the sea at this point is 12 m. The region where the admixture was released is a cylinder with radius $R = 9$ km and depth h_1 $(0 > z \geq h_1)$,

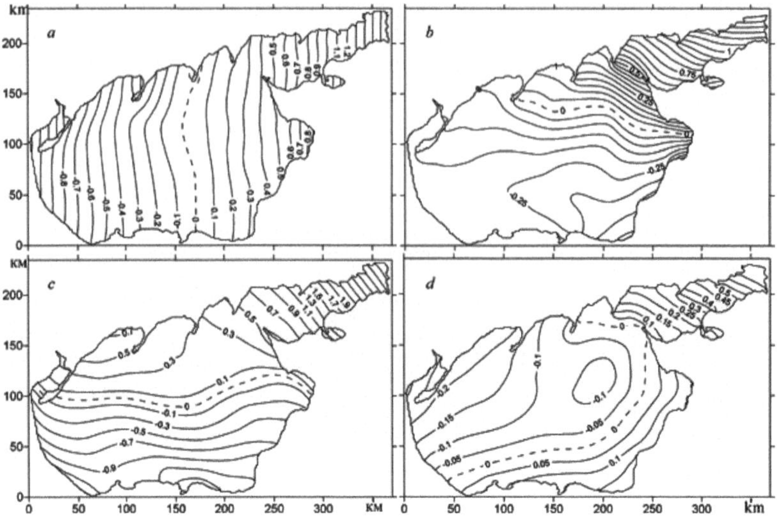

Fig. 2. Sea level fields of the Sea of Azov at different time moments stationary regime $t = 48$ h (*a*), nonstationary regime: $t = 68$ h (*b*); $t = 90$ h (*c*); $t = 140$ h (*d*).

where h_1 (1 m) is the vertical step in the surface layer (11). The initial concentration is constant in this region and equal to unity ($C_0(x, y, z, 0) = 1$).

The time of the release of the admixture in different experiments is not the same and depends on the characteristics of the wind. In the case of a nonstationary wind, the moments of the admixture release ($t = t_{01}$, U_1^{st}; $t = t_{02}$, $\mathbf{U}_2^{st}$) and the stabilization of the fluid motion ($t = t_{11}$, U_1^{st}; $t = t_{12}$, U_2^{st}) coincide: $t_{01} = t_{11} = 38$ and $t_{02} = t_{12} = 43$ h. If only the wind $\mathbf{U}_{SKIRON}$ forms the forcing, the time moment of release is September 11, 2007, at 00:00 ($t_{03} = 72$ h). If we consider the joint forcing of the stationary and non-stationary wind, the time of release is determined as follows: $t_{04} = t_{01} + t_{03}$ ($U_1^{st} + U_{SKIRON}$) and $t_{05} = t_{02} + t_{03}$ ($U_2^{st} + U_{SKIRON}$). For the convenience of the analysis of the results, we assume that, in all the cases, the time of the release of the admixture is zero ($t_0 = 0$).

Table 3 presents the coefficients of the maximum spreading of the passive admixture (K_{max}), the time when it occurred (t_{max}, h), and the time of the complete dispelling of the admixture (t_d, h) at three depths for constant wind ($\mathbf{U}_{1,2}^{st}$) and three variables ($\mathbf{U}_{SKIRON}$ and $U_{1,2}^{st} + U_{SKIRON}$). One can see from the data analysis that the maximum square of the pollution at each depth depends on the wind velocity that leads to the stable motion.

In the case of greater velocity of the constant wind ($\mathbf{U}_2^{st} > \mathbf{U}_1^{st}$), the velocities of the currents increase, the square of the admixture region spreading increases (K_{max}), and time of its complete dispelling also increases (t_d). The joint forcing of the stationary and $\mathbf{U}_{SKIRON}$ wind causes an increase in the pollution region.

In this case, the maximum pollution square occurs if the velocity of the forcing wind is maximal ($U_2^{st} + U_{SKIRON}$). In this case, K_{max} at the free surface is 1.32 40 h

Table 3. Parameters (K_{max}; t_{max}, h; t_d, h) of the spreading evolution of the admixture at different depths of the Sea of Azov

Depth, m	K_{max}, t_{max}, t_d	U_1^{st}	U_2^{st}	U_{SKIRON}	$U_1^{st} + U_{SKIRON}$	$U_2^{st} + U_{SKIRON}$
$z = z_1$	K_{max}	1,14	1,18	1,25	1,30	1,32
	t_{max}	5,7	4,9	31	40	40
	t_d	17,3	18,2	57,1	84,5	86,5
$z = -H/2$	K_{max}	1,16	1,18	1,27	1,33	1,35
	t_{max}	14,3	14,7	34	42	42
	t_d	36,7	37,6	104	106	110
$z = -H+h_b$	K_{max}	1,16	1,19	1,33	1,37	1,38
	t_{max}	26,9	25,5	55	58	59
	t_d	53,4	55,2	108	110	115

after the release of the admixture, and the time of its complete dispelling (U_1^{st}) is 86.5 h. The maximum square of the pollution region at the depth $z = -H/2$ is gained 42 h after the release of the admixture ($K_{max} = 1.35$), and the complete dispelling of the pollution occurs in 110 h. In the bottom layer ($z = H + h_b$), the coefficient of the maximum spreading of the admixture is 1.38 ($t_{max} = 59$ h). We note that, in the case considered here ($U_2^{st} + U_{SKIRON}$), the concentration of the admixture in the entire basin 115 h after its release does not exceed 2.5% of its initial value ($C_d = 2.5 \times 10^{-2}$).

4 Conclusion

In this work, we present the results of investigations of the phenomena caused by onshore–offshore winds and the evolution of a passive admixture by the current system generated by the constant and variable winds in the Sea of Azov. The reliability of these results is confirmed by the comparison of the simulated values of the extreme sea level changes caused by onshore and offshore winds with the field data obtained during wind forcing by the surface wind obtained from the SKIRON model at coastal hydrometeorological stations. The obtained results are presented in the table of the sea level changes caused by the onshore and offshore winds and the current velocities for different characteristics of constant and variable wind. We also performed the analysis of the influence of the wind velocity and the generated currents on the characteristics of the transformation of the passive admixture.

The analysis of the modeling results and the dynamic processes in the Sea of Azov allowed us to reach the following conclusions:

(1) It is found from the analysis of the stationary motions that, under constant wind forcing with a two-fold increase in the velocity (5 and 10 m), the maximum deviations of the sea level increase by a factor of 3.45 (0.2 and 0.69 cm), the minimum deviations increase by a factor of 3.9 (0.1 and 0.39 cm), and the maximum velocities of the stable currents increase by 12 times (0.16 and 1.17 m/s);

(2) An increase in the maximum wind velocity leads to an increase in the volume of the pollution region; the minimum pollution square appears in the absence of wind.
(3) The time needed for the pollution region to reach the maximum volume decreases when the wind velocity increases

Acknowledgments. Work is performed under a grant VnGr «Development of methodical bases and guidelines for integrated coastal zone management of the Azov Sea in conditions of growth of dangerous ex-ogenic processes, recreational load, climatic variability».

References

1. Kallos, G., Nickovic, S., et al.: The regional weather forecasting system SKIRON and its capability for forecasting dust uptake and transport. In: Proceedings of the WMO Conference on Dust Storms, p. 9, Damascus (1997)
2. http://forecast.uoa.gr. Accessed 18 Sept 2017
3. Blumberg, A.F., Mellor, G.L.: A description of three dimensional coastal ocean circulation model. In: Heaps, N. (eds.) Three Dimensional Coastal Ocean Circulation Models Coastal Estuarine Science, pp. 1–16. American Geophysical Union, Washington, D.C. (1987)
4. Cherkesov, L.V., Ivanov, V.A., Khartiev, S.M.: Introduction into Hydrodynamics and Wave Theory, 264 p. Gidrometeoizdat, St. Petersburg (1992)
5. Smagorinsky, J.: General circulation experiments with primitive equations, I. The basic experiment. Mon. Weather Rev. **91**, 99–164 (1963)
6. Mellor, G.L., Yamada, T.: Development of a turbulence closure model for geophysical fluid problems. Rev. Geophys. Space Phys. **20**(4), 851–875 (1982)
7. Wannawong, W., Wongwises, U., Vongvisessomjai, S.: Mathematical modeling of storm surge in three dimensional primitive equations. Int. J. Math. Comput. Phys. Electr. Comput. Eng. **5**(6), 797–806 (2011)
8. Pietrzak, J.: The use of TVD limiters for forward-in-time upstream-biased advection schemes in ocean modeling. Mon. Weather Rev. **126**, 812–830 (1998)
9. Courant, R., Friedrichs, K.O., Lewy, H.: On the partial difference equations of mathematical physics. IBM J. **3**, 215–234 (1967)

The Problem of Forecasting of Vertical Temperature Distribution in Inland Hydrophysical Objects with Experimental Data

D. Gladskikh[1,2(✉)], D. Sergeev[1], G. Baydakov[1], I. Soustova[1], and Yu. Troitskaya[1]

[1] Division of Geophysical Research, Institute of Applied Physics of the Russian Academy of Sciences, Nizhny Novgorod, Russia
daria.gladskikh@gmail.com

[2] Institute of Radio Electronics and Information Technologies, Department of Information Systems and Technology, Nizhny Novgorod State Technical University n.a. R.E. Alekseev, Nizhny Novgorod, Russia

Abstract. The problem of computation of vertical temperature distribution in inland water objects is raised in the research, and the modified version of the program complex LAKE is proposed for its solving. Also the results of validation of the modified version in accordance with the parameters of lake part of Gorky water reservoir are provided. The modification caused changing the procedure of input data assignment, ensured the possibility of using the results from field measurements as starting profiles, led to the greater consideration of wind influence on mixing and allowed to reproduce general forms of distribution more accurately. The data from the experimental sites of Gorky water reservoir in combination with the files of global reanalysis and the data of hydrometeorological station were used for validation. Measurements of the vertical temperature profiles were carried out with freely sinking CTD-probe, which recorded the temperature values several times per second. Time and temperature dependences in control points, comparison of the forms of the profiles and standard deviation were analyzed to assess the results of numerical modeling. Numerical simulation with the modified model gave results which were in good agreement with the real seasonal variations of thermal regime in lake part of Gorky water reservoir. The best quantitative and qualitative agreement was obtained for the thermal regimes with strong mixing: the results of the forecasting reproduced the forms of distribution and the values of temperature extremely accurately. The results for the ones with strong stratification demonstrated less accuracy in the values of temperature in control points, particularly near the surface, but all specificities of the forms of vertical temperature distribution were correctly reproduced.

Keywords: Vertical temperature distribution · Temperature profile Numerical modeling · Inland water objects

V. Karev et al. (Eds.): PMMEEP 2017, SPRINGERGEOL, pp. 327–335, 2018.
https://doi.org/10.1007/978-3-319-77788-7_34

1 Introduction

One of the main parameters of natural hydrophysical objects is vertical temperature distribution. It significantly affects ecological conditions of lakes and water reservoirs and is one of the causes of eutrophication (enrichment of a water body with nutrients). Thermal regime is characterized by strong seasonal and short-term variability [1–3] by the reason of significant influence of wind-induced mixing on the temperature distribution in the small- and mid-sized water bodies. It should also be noted, that thermal regime should and could be predicted, and there is a growing number of requests for forecasts of vertical temperature distribution in inland water bodies [4] which cannot be obtained from global meteorological forecasts.

Different models are used for computation of the hydrological parameters, such as vertical temperature distribution. One-dimensional models have less requirements for computing resources than multi-dimensional ones, and demonstrate satisfactory agreement with experimental evidence, but these models require validation in accordance with the parameters of the given water body. One-dimensional model based on heat equation is submitted in the program complex LAKE, developed in [5]. Modification of this program complex and its validation in accordance with the parameters of mid-sized lowland water bodies with an example of lake part of Gorky water reservoir (56°42′N 43°19′E) are presented in this research for the forecasting of thermal regimes in this water object. The experimental data obtained from the observations during expeditions on the experimental sites of Gorky water reservoir, the data of hydrometeorological station in Gorodets and the files of global meteorological reanalysis NCEP/NCAR were used for validation.

2 Short Model Description

2.1 The Main Equation and Its Boundary Conditions

The model presented in [5] is based on the one-dimensional heat equation:

$$c\rho\frac{\partial T}{\partial t}=\frac{1}{h^2}\frac{\partial}{\partial \xi}\left(\lambda\frac{\partial T}{\partial \xi}\right)+c\rho\frac{dh}{dt}\frac{\xi}{h}\frac{\partial T}{\partial \xi}-c\rho\frac{1}{h}B_w\frac{\partial T}{\partial \xi}-\frac{1}{h}\frac{\partial S}{\partial \xi}. \quad (1)$$

The symbols are: c is heat capacity of the water, ρ – its density, λ - eddy diffusivity, T – temperature, $B_w = dh_0/dt$ – water balance at free surface of lake, S - solar radiation flux penetrating the water body. The beginning of the vertical downward z-coordinate is supposed to be coupled with water surface, so we can turn from z to the new independent variable $\xi = z/h$, where h = h(t) is a depth of water body, and t is a time.

The following generally accepted exponential relation is used for the calculation of the solar radiation flux penetrating the water body:

$$S(\xi) = S(0)\exp(-\alpha_e h\xi),$$

where S(0) is solar radiation flux on the surface of water, α_e- extinction coefficient. Temperature gradient (heat flux) is a boundary condition for the Eq. (1) on the air-water boundary and is calculated from the equation of heat balance:

$$S(1-\alpha)+E_a-E_s-H_s-LE_s=-\frac{\lambda}{h}\frac{\partial T}{\partial \xi}, \tag{2}$$

where S is total solar radiation flux, E_a – longwave radiation flux, E_s. – radiation of surface, H_s and LE_s are overt and covert heat fluxes respectively, α – albedo of a water surface.

2.2 Parameterization of Eddy Diffusivity

Eddy diffusivity λ is the main mechanism to describe vertical heat and mass exchange in water objects. There are different ways of its parameterization, we used parameterization based on equations of kinetic energy of turbulence and its dissipation rate, or "E – λ" -parameterization, developed in [6]. λ is defined as follows:

$$\lambda=c\rho k,$$

where k is the coefficient of turbulence evaluated as

$$k=C_e\frac{E^2}{\varepsilon}.$$

Here $E=1/2\left[\overline{(u')^2+(v')^2+(w')^2}\right]$ is kinetic energy of turbulence, the upper line is symbol of averaging, the values with primes are deviations from mean values, ε is dissipation rate of kinetic energy of turbulence, C_e is the dimensionless coefficient. This equation is used to estimate E:

$$\frac{\partial E}{\partial t}=\frac{\alpha_E}{h^2}\frac{\partial}{\partial \xi}k\frac{\partial E}{\partial \xi}+\frac{\xi}{h}\frac{dh}{dt}\frac{\partial E}{\partial \xi}+P-\varepsilon, \tag{3}$$

where α_E is the dimensionless constant, and P is total generation of turbulence energy through the shear of velocity and effect of density stratification.

Dissipation rate of turbulence energy is defined with the following equation:

$$\frac{\partial \varepsilon}{\partial t}=\frac{\alpha_\varepsilon}{h^2}\frac{\partial}{\partial \xi}k\frac{\partial \varepsilon}{\partial \xi}+\frac{\xi}{h}\frac{dh}{dt}\frac{\partial \varepsilon}{\partial \xi}+C_1\frac{\varepsilon}{E}(P-\varepsilon), \tag{4}$$

where α_ε is the dimensionless constant, and C_1 is the function of Reynolds number, which is defined as follows:

$$Re=\frac{(2E/3)^2}{\nu\varepsilon}.$$

Boundary condition for equation of kinetic energy of turbulence on the air-water boundary is as follows:

$$-\frac{k}{h}\frac{\partial E}{\partial \xi} = k_{we}\left(\frac{\tau}{\rho}\right)^{\frac{3}{2}}, \tag{5}$$

where k_{we} is coefficient of turbulence enhancement by wave breaking and τ is surface stress. Assumption about continuity of impulse flow is used for zonal and meridional components of current velocity on the air-water boundary, and these components are computed with the theory of similarity by Monin-Obukhov, whereas these components on the boundary water-soil are defined with Chezy formula [7].

3 Input Data and Parameters

The input data are the starting temperature profile, physical parameters and meteorological data. The input temperature profile was defined with temperature difference between the surface and the bottom and the thickness of mixed layer in previous version. Three points were accounted: temperature on the surface, on the lower boundary of mixed layer and on the water-soil boundary. This limited the possibility of using the results from field measurements. In the modified version the input profile is arbitrary defined as a two-dimensional array of points from the file with values of temperature on every step of the depth. This modification allows reproducing thermocline stratification typical for mid-sized lowland water bodies. The profiles obtained through the computation on every step become the input ones for the next steps of the forecasting.

The values of main physical parameters used in the model are either set through field measurement data (such as the coefficient of solar radiation extinction in water body is set through measurements with Secchi disk drawing on the theory developed in [8]) or adjusted on the basis of correspondence between numerical results and real data (for example, the coefficient of turbulence enhancement by wave breaking is taken equal to 10 for the best match between the forecast and the experimental evidence). These two parameters are recorded during all computation time.

The parameters of meteorological conditions include air temperature, atmospheric pressure, specific air humidity, precipitation, zonal and meridional wind components, downward solar (shortwave) radiation and downward longwave radiation. Meteorological data are read on the every step of computation – timestep for these data is 6 h. The parameters are collected with combined method: information is taken from the data of hydrometeorological station in Gorodets and the files of global meteorological reanalysis NCEP/NCAR [9] with spatial resolution 1°. Information about wind velocity is used to compute surface stress on air-water boundary. In previous version empirical parameterization [10] was applied, but it ignored features of underlying surface:

$$\tau = 1.273 * 10^{-3} * w_r,$$

where $w_r = \sqrt{u^2 + v^2}$, u and v – zonal and meridional wind components. We use updated formula based on multi-year measurements of wind conditions on the experimental sites of Gorky water reservoir (see [11]):

$$\tau = 1.274 * 10^{-3} * w_r + 3.4 * 10^{-4} * w_r^2 + 4.9 * 10^{-5} * w_r^3.$$

4 Experiment Description and Model Validation

Experimental data were used for validation of modified model. These data were obtained from the observations during expeditions on the experimental sites of Gorky water reservoir in the period 2014-2017. Measurement of the vertical temperature profiles were carried out with freely sinking CTD-probe, which recorded the temperature values 6 times per second. At the speed of lowering equal to 24 cm/sec this provides depth resolution about 4 cm. Several measurements were made during every expedition. The profile used as the starting one must be carried out at the same time with an hour from the first line in the file with meteorological conditions, and the

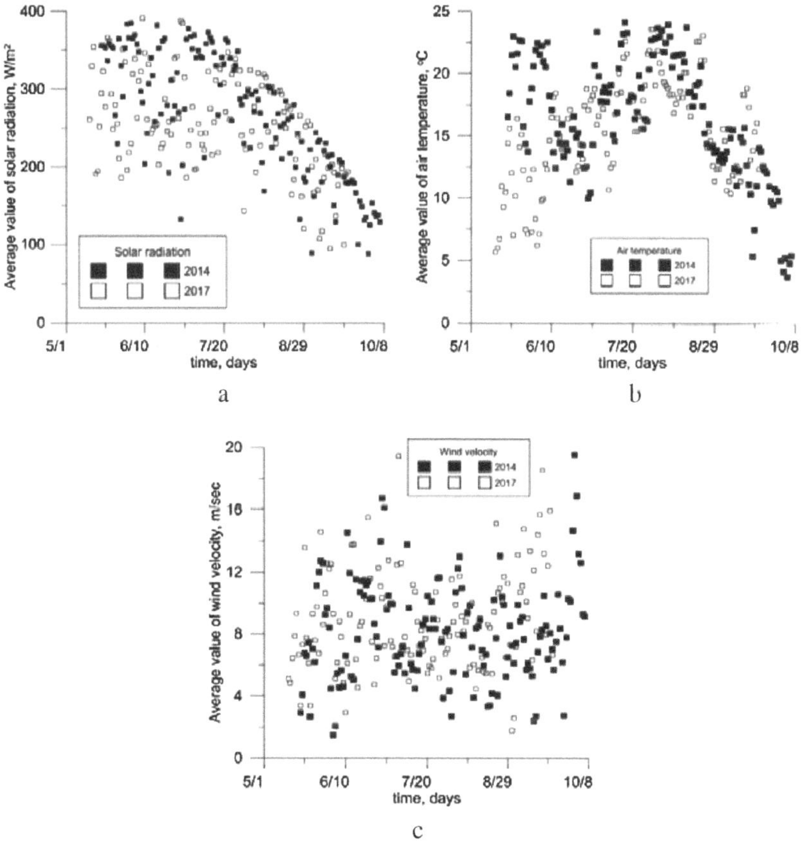

Fig. 1. Average daily values of (a) solar radiation, (b) air temperature, (c) wind velocity in 2014 and 2017.

profiles used for the validation of the model are obtained with averaging of the daily measurements for the purpose of comparison with average daily profiles produced with computation. 2014 and the current year were chosen from a large volume of data, because the seasons had a great difference in the thermal regimes due to the difference in seasonal meteorological conditions (see Fig. 1): 2014 was characterized by intensive solar radiation, a high temperature and a slightly weaker wind, whereas 2017 is marked with weak heat and strong wind-induced mixing. Thus, temperature profiles obtained in 2014 were characterized by significant temperature difference and strong stratification, and ones obtained in 2017 are homogeneous in depth.

5 Results

Time and temperature dependences in control points, comparison of the forms of the profile and standard deviation were reviewed to analyze numerical results.

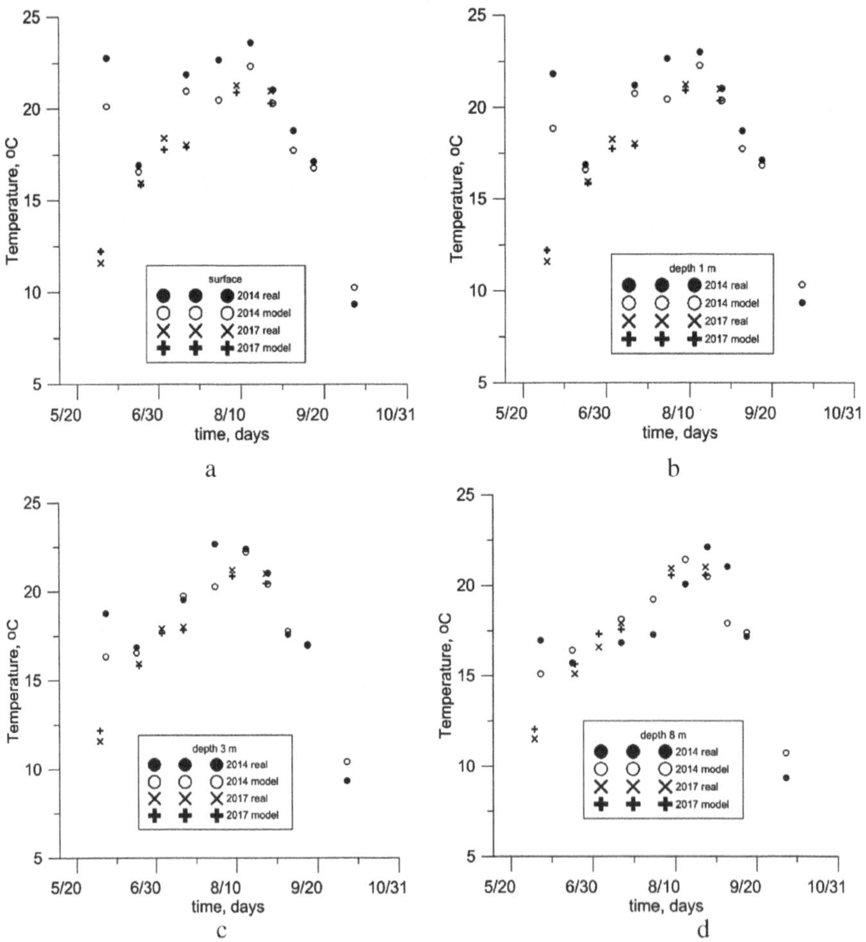

Fig. 2. Values of temperature in control points: (a) surface (b) depth 1 m (c) depth 3 m (d) depth 8 m in the real data and in the model results.

Time and temperature dependences in control points (surface, 1 m, 3 m and 8 m) are presented in Fig. 2. More of points were taken closer to the surface than to the bottom, because this layer is more exposed to variability and has a significant impact on eutrophication. So the difference between real and model data is less near the bottom, which confirms that variability and complexity of processes are greater near the surface due to wave-wind coupling.

Also the forms of profiles were compared (see Figs. 3 and 4). Numerical simulation reproduced general forms of distribution for all realizations. However, the best agreement was obtained for the regimes with strong mixing, when stratification is practically nonexistent.

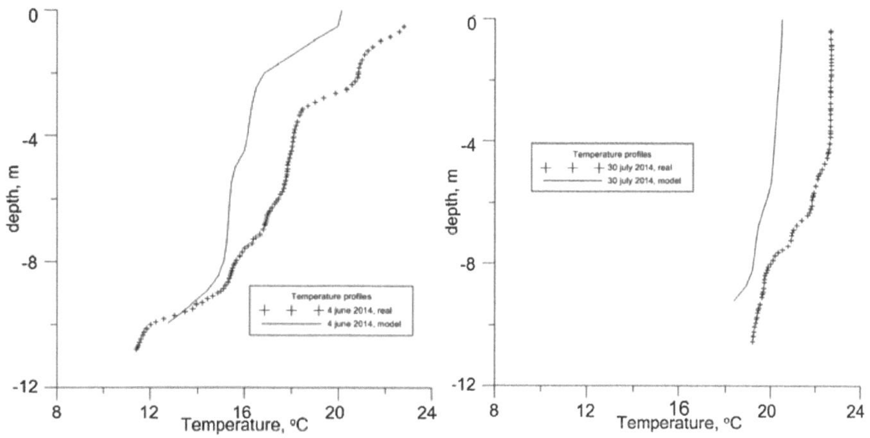

Fig. 3. Real and model temperature profiles with strong stratification, 2014.

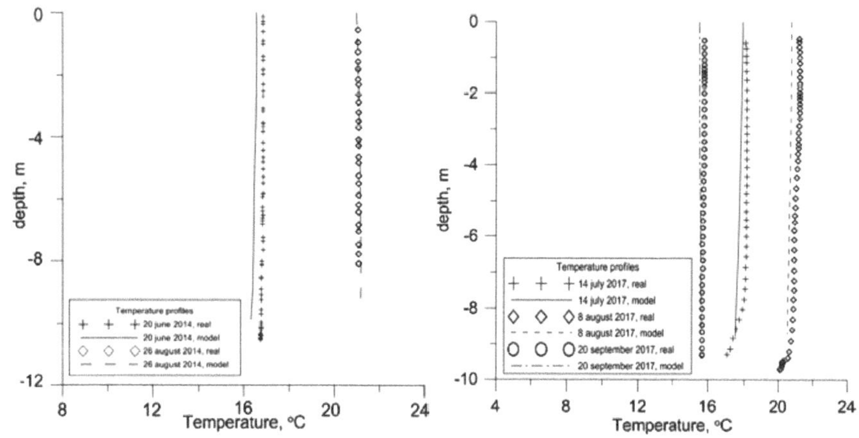

Fig. 4. Real and model temperature profiles with strong mixing, 2014 and 2017.

As for the integral characteristics of comparison between the real data and the model results, standard deviation does not exceed $0,8\,^{\circ}\mathrm{C}$ for all performed realizations for 2017 year, and $1,5\,^{\circ}\mathrm{C}$ for 2014 year, which is more accurate than the results in the previous version [5].

6 Conclusion

Modification of the program complex LAKE [5] and its validation are proposed for the forecasting of vertical temperature distribution in inland hydrophysical objects with an example of lake part of Gorky water reservoir. The modification caused changing the procedure of input data assignment, ensured the possibility of using the results from field measurements as starting profiles, led to the greater consideration of wind influence on mixing and allowed to reproduce general forms of distribution more accurately. Numerical simulation with the modified model gave results which were in good agreement with the real seasonal variations of thermal regime in lake part of Gorky water reservoir.

Regimes with strong mixing and weak heat are reproduced more accurately, than regimes with strong heat and stratification, so the model results for 2017 match the field measurement data better, than the model results for 2014. It can be explained by not quite adequate simulation of processes with weak wind and intensive solar radiation or by incorrect meteorological data, particularly data from reanalysis by reason of coarse grid. Also it may be due to physical parameters: they are recorded during all computation time, but it is planned to make them variable. The coefficient of solar radiation extinction which characterizes transparency of the water may slightly vary between 2 and 2,5 in Gorky water reservoir not only in different years but from month to month. The parameterization of the coefficient of turbulence enhancement by wave breaking is also foreseen, and the dependence between its value and the value of wind velocity has to be determined for this purpose. Also acquaintance with the data of total cloud cover is foreseen for the clarification of the information about solar radiation, which can provide more accurate forecast of vertical temperature distribution.

Acknowledgements. The work was partly supported by Russian Science Foundation project 15-17-20009 (numerical modeling) and RFBR projects 17-05-41117, 15-45-02580 (carrying out field measurements).

References

1. Stepanenko, V.M., Martynov, A., Jöhnk, K.D., Subin, Z.M., Perroud, M., Fang, X., Beyrich, F., Mironov, D., Goyette, S.: A one-dimensional model intercomparison study of thermal regime of a shallow, turbid midlatitude lake. Geosci. Model Dev. **6**, 1337–1352 (2013). https://doi.org/10.5194/gmd-6-1337-2013
2. Mironov, D.V., Golosov, S.D., Zilitinkevich, S.S., Kreiman, K.D., Terzhevik, A.Yu.: Seasonal changes of temperature and mixing conditions in a lake. In: Zilitinkevich, S.S. Modelling air-lake interaction, Physical Background, pp. 74–90. Springer, Heidelberg (1991)

3. Abbasi, A., Annor, F.O., Giesen, N.V.: Investigation of temperature dynamics in small and shallow reservoirs, case study: Lake Binaba, Upper East Region of Ghana. Water **8**(3) (2016). https://doi.org/10.3390/w8030084
4. Lake Model FLake Homepage. http://www.flake.igb-berlin.de/Lake17/
5. Stepanenko, V.M.: Numerical modeling of thermal regime in shallow lakes. Computational technologies, vol. 10 (Special issue. Proceedings of the International Conference and the School of Young Scientists "Computational and informational technologies for environmental sciences" (CITES 2005) Tomsk, 13–23 March 2005), part 1, pp. 100–106 (2005)
6. Lykosov, V.N.: O probleme zamykaniya modeley turbulentnogo pogranichnogo sloya c pomoshchyu uravneniy dlya kineticheskoy energii i skorosti ee dissipatsii. Izv. AN SSSR. Fizika atmosphery i okeana, vol. 28, pp. 694–704 (1992)
7. Chebotarev, A.E.: Obshchaya gidrologiya. Leningrad (1975)
8. Ivanov, A.V., Troitskaya, YuI, Papko, V.V., Sergeev, D.A., Baydakov, G.A., Vdovin, M.I., Kazakov, V.I., Kandaurov, A.A., Afanasieva, I.M., Donskova, O.A., Shuvalova, N.M.: Stratification as a factor of influence on water quality of a plane reservoir. Privolzhsky Nauchnyy J. **2**, 149–156 (2015)
9. ESRL: NCEP/NCAR Reanalysis 1: Summary. https://www.esrl.noaa.gov/psd/data/gridded/data.ncep.reanalysis.html
10. Beljaars, A.C.M., Holtslag, A.A.M.: Flux parameterization over land surfaces for atmospheric models. J. Appl. Meteor. **30**, 327–341 (1991)
11. Kuznetsova, A.M., Baydakov, G.A., Papko, V.V., Kandaurov, A.A., Vdovin, M.I., Sergeev, D.A., Troitskaya, Yu.I.: Adjusting of wind input source term in WAVEWATCH III model for the middle-sized water body on the basis of the field experiment. Hindawi Publishing Corporation, Advances in Meteorology, vol. 13 (2016). https://doi.org/10.1155/2016/8539127

Modeling Geomechanical Processes in Oil and Gas Reservoirs at the True Triaxial Loading Apparatus

V. I. Karev(✉), D. M. Klimov, and Yu. F. Kovalenko

Ishlinsky Institute for Problems in Mechanics of Russian Academy of Sciences, Moscow, Russia
wikarev@ipmnet.ru

Abstract. The paper presents the results of the investigation of the strain-strength properties of rocks (dolomites) raised from a depth of more than 6 km of the exploration well of the Kainsayskaya Field. The experiments were carried out on a unique experimental setup – the Triaxial Independent Load Test System of IPMech RAS on cubic specimens with an edge of 40 mm. Two triaxial tests were performed on two specimens showing the presence of strong strength anisotropy of the rock. A physical simulation of the pressure reduction process in deep wells near an open borehole and the tip of a perforation hole was performed on three specimens. The conducted studies have revealed a rather low strength of the studied rocks, despite the great depths of their lying under the conditions of high rock pressure. They have shown that the beginning of the rock destruction depends vastly on the type of stress state arising in the formation. Carrying out physical modeling of geomechanical processes in oil and gas reservoirs using true triaxial loading is of great importance from the point of view of justifying methods of influence on deep-lying strata in order to increase the productivity of wells, as well as reduce risks of well destruction during their drilling and operation.

Keywords: Deep deposits · Rock · Borehole · Test apparatus
Deformations · Stresses · Strength · Elastic modules

1 Introduction

In the conditions of depletion of large and accessible oil and gas fields [1], it is becoming increasingly urgent to search for new sources of hydrocarbons and develop technologies for their extraction, since adequate replacement of hydrocarbons is not expected in the coming decades. Resources of the Arctic basin, unconventional sources of hard-to-recover hydrocarbon raw materials with large explored reserves (shale oil, shale gas), oil and gas from deep horizons, gas hydrates, etc., can be considered as such.

One of the most promising directions is the development of deep horizons of oil and gas bearing sedimentary basins lying beneath currently exploited formations, that is, deeper than 5.5 km. Of course, the extraction of deep hydrocarbons is associated with difficulties, but their usefulness is determined by a number of factors and, above all, by the established infrastructure of the developed oil and gas provinces [2].

V. Karev et al. (Eds.): PMMEEP 2017, SPRINGERGEOL, pp. 336–349, 2018.
https://doi.org/10.1007/978-3-319-77788-7_35

A number of problems arises when hydrocarbons are extracted at great depths. In particularly, there are very acute geomechanical problems associated with the stability of deep borehole during drilling and operation of wells, as well as the specific features of the deformation, fracture and filtration processes in deep reservoir. Under these conditions, the rocks in the vicinity of the borehole may be in the limiting stress state and even pass to the critical deformation. Complications during drilling of oil and gas wells associated with loss of the borehole stability are usually accompanied by a large expenditure of funds for the elimination of their consequences, and therefore the forecasting and prevention of this type complications play an important role in reducing the cost of the field development. In conditions of large depths, the value of this factor multiplies many times. In addition, the destruction of well bores is one of the main factors limiting the maximum production rates of wells.

To solve these problems, it is necessary to create adequate mechanical-mathematical models describing the processes of deformation and destruction of reservoir rocks at great depths taking into account the specific features of reservoir systems in conditions of high rock and reservoir pressures. A characteristic feature of rocks is the essential dependence of their strain-strength characteristics on the type and level of stresses created in them. At great depths under very high pressures, the processes of deformation and fracture of rocks that occur during the construction of wells and various technological operations may differ significantly from those at depths of 2–3 km. An experimental equipment that allows in the laboratory not only to determine the strain and strength characteristics of rocks in conditions of great depths, but also to model real deformation and filtration processes occurring in deep-lying layers is needed. Such a possibility is provided by the unique Triaxial Independent Load Test System (TILTS) created at the Institute of Problems of Mechanics of the Russian Academy of Sciences [3]. The apparatus allows studying the strain, strength and filtration properties of rocks by testing of a cubic specimens of rock with an edge of 40 or 50 mm (Fig. 1).

Fig. 1. Triaxial Independent Load Test System (TILTS).

2 Experimental Apparatus

An original kinematic scheme is used in the design of the loading unit of the TILTS. It allows the pressure plates to approach in three directions without creating obstacles to each other, and to load the specimen independently along each of the three axes. This makes it possible to determine the elastic and viscoplastic characteristics of rocks (including anisotropic) under the conditions of true triaxial stress states, and also to recreate in the course of the experiments any stress states that arise in rock massifs during conducting technological operations and to study the deformation processes and filtration properties of rocks.

TILTS belongs to the class of electrohydraulic testing machines; it is equipped with an automated control system with feedback. TILTS allows to control the loading of the specimen both by force and by movements. This makes it possible to control the deformation of the specimen in the plastic region, without leading it to failure.

3 Object of Study

Experiments were carried out at the TILTS to determine the strain and strength properties of specimens of reservoir rocks selected from the deep-lying horizons of the Kainsayskaya Field (Orenburg).

The cubic specimens of rock with a 40 mm edge were made for tests. The specimens were cut out of core 1 m long, selected from a depth of 6114.2–6115.2 m. The rock according to the lithological description of the core is a fine-grained cover-nodular-porous strong massive fractured dolomite.

4 Triaxial Tests of Specimens

So-called triaxial tests of the specimens were carried out at the TILTS to determine the parameters of the bulk strength of rocks – the coefficient of adhesion and the angle of internal friction [4]. The program of triaxial testing of rock specimens is aimed at estimating the stress required to fracture a rock specimen, depending on the value of the all-round specimen compression, Fig. 2, where s_1, s_2, s_3 are the stresses applied during the experiment to the faces of the specimen. In order to find the necessary parameters, it is required to determine the value of the fracture stress for at least three values of the all-round compression.

The difficulty lies in the fact that, due to the shortage of core material taken from great depths and the substantial heterogeneity of rock formations, it is desirable to carry out experiments for all three volumetric compression levels on the same specimen. In this case, each loading cycle must be brought to a level when the specimen begins to deform plastically, but will still remain intact.

A constant control over the state of the specimen is necessary for this. For this purpose, deformation of the specimen is measured along each of three axes in the course of the experiment and a diagram of the specimen deformation along the loading axis $\sigma - \varepsilon$ is displayed on the computer screen in the online mode, Fig. 3.

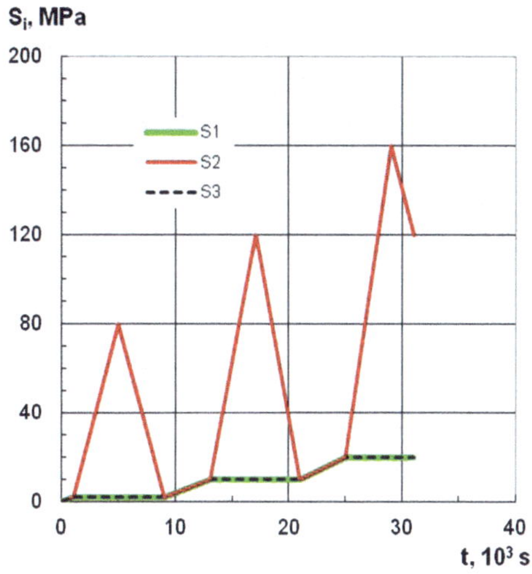

Fig. 2. The program of specimen triaxial test.

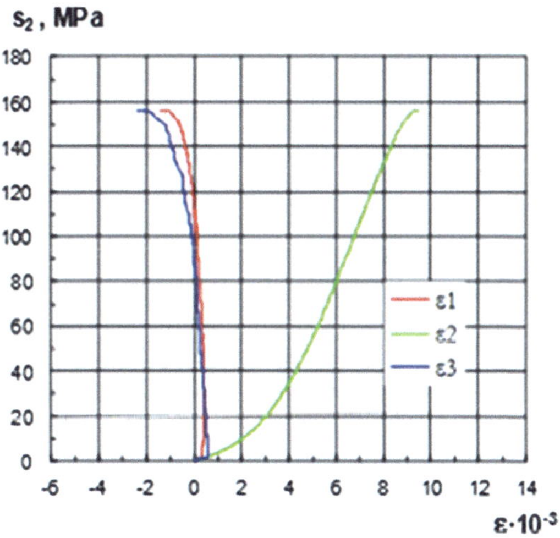

Fig. 3. The strain curves of specimen 1 during the test cycle 1.

In addition, the current tangent module $\Delta\sigma/\Delta\varepsilon$ is calculated for each point of this curve by linear approximation to 21 experimental points (10 points to the point under consideration and 10 points after it), Fig. 4. The tangent module is also displayed on the screen. The loading in the each of the two first cycles ends when the tangent modulus at the plastic deformation section decreases to 25–30% of the module value for the linear portion of the curve $\sigma - \varepsilon$.

In addition, the specimen loading must be carried out at control on displacement, not force. Otherwise, it is practically impossible to stop the loading of the specimen in the plastic region in time, without leading it to destruction.

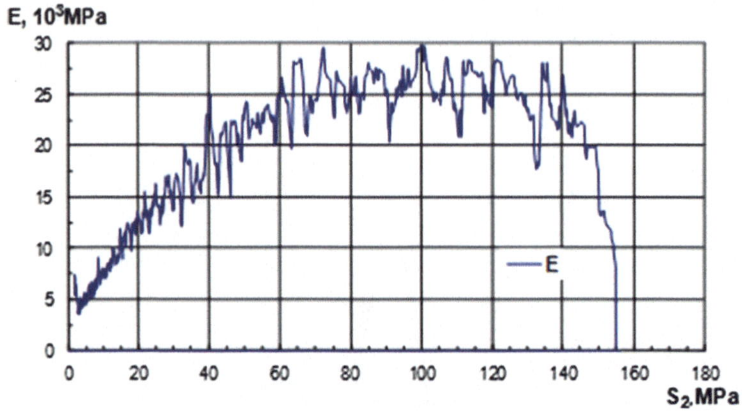

Fig. 4. The tangent modulus change of specimen 1 during the test cycle 1.

An important distinctive feature of triaxial tests of cubic samples at the TILTS in comparison with the triaxial tests of cylindrical specimens at installations based on the Karman principle is the possibility of registration of the deformation of specimens in three directions. It is needed in determining the strain-strength properties of anisotropic rocks.

In order to determine the influence of the bedding on the strain and strength properties of the rock under study, two specimens cut from one piece of core were tested according to the triaxial test program: specimen 1 and specimen 2. In the test of specimen 1, the load during each cycle was carried out orthogonally to bedding planes, and along bedding planes when specimen 2 was testing. During each loading cycle strains were measured along each of three axes of the specimen.

As an example, the strain curves got in the first test cycle of specimen 1 along three axes are plotted on Fig. 3. Here the sample was previously uniformly compressed along all three axes to a stress of 2 MPa, followed by additional compression along axis 2 with constant stresses along axes 1 and 3. Figure 4 shows the variation of the tangent module during the experiment.

The Young's modulus and the Poisson's ratio for each cycle as well as the strength constants such as the adhesion coefficient and the angle of internal friction were calculated on the base on the results of the triaxial tests for specimens 1 and 2. These values for specimen 1 were measured in the direction perpendicular the bedding, and for specimen 2 along the bedding. The results are given in Tables 1 and 2.

Here E is the Young's modulus of the rock in each of the cycles, v is the Poisson's ratio in each of the cycles, K is the adhesion coefficient, ρ is the angle of internal friction, S_2^* is the stress of the transition to plastic deformation. Young's moduli and Poisson's ratios were determined on rectilinear sections of the deformation curves at each loading cycle, and the adhesion coefficients and the internal friction angles were determined by using the Mohr circles constructed on the results of the triaxial tests.

Table 1. Elastic and strength constants in the direction perpendicular to the bedding (specimen 1).

Cycle no.	Uniform compression, MPa	$E * 10^{-4}$, MPa	ν	S_2^* MPa	K, MPa	ρ degrees
1	2	2,83	0,17	156	32,5	41
2	5	3,26	0,20	165		
3	15	3,34	0,20	220		

Table 2. Elastic and strength constants in the direction along to the bedding (specimen 2).

Cycle no.	Uniform compression, MPa	$E * 10^{-4}$, MPa	ν	S_2^* MPa	K, MPa	ρ degrees
1	2	2,30	0,18	83	19	38
2	10	2,94	0,21	126		
3	20	3,33	0,20	164		

Tables 1 and 2 show the rock is anisotropic in strength properties. Its strength constants under loading along the plane of the bedding are much lower than under loading in the orthogonal direction.

As for the Young's modulus, the picture here is more ambiguous. It can be seen from Tables 1 and 2 that the Young's modulus along the bedding plane during the first test cycle with a slight volumetric compression is significantly lower, than one in the orthogonal direction. But with the increase in the value of the preliminary volumetric compression this difference decreases (the cycle 2), and it disappears on the third cycle.

5 Study of Elastic-Plastic Properties of Rocks from Deep Horizons

In order to study the elastic-plastic properties of the rocks under investigation, a methodology for physical modeling of geomechanical processes was developed at the IPMech RAS. On its basis the studies were made on the rocks of tens of oil and gas deposits of different lithologic composition from a depth of several hundred meters to 6500 m.

According to the method, the rocks under investigation are tested on three basic loading programs creating in the specimens three types of stress state: generalized stretching, generalized shift and generalized compression. These stress states are characterized by the corresponding values of the Lode–Nadai parameter: " + 1", "0", "−1" [5].

These stress states are directly related to the situations occurring in the formations in the vicinity of oil and gas wells. The generalized shift is realized in the vicinity of an uncased wellbore, loaded with internal pressure on and rock pressure at infinity. The generalized stretching is realized in the vicinity of a spherical cavity loaded with internal pressure and external pressure at infinity, for example, in the vicinity of the tip of the perforation hole. The calculation of loading programs and the results of testing three specimens on each of the programs are presented. The fact that the rocks under study are permeable and effective stresses act on the soil skeleton is taken into account [6].

5.1 Testing the Specimen According to the Program "Generalized Stretching"

As indicated above, such a loading program corresponds to a stress state arising near a spherical cavity loaded with internal pressure and external pressure at infinity, for example, in the vicinity of the tip of the perforation hole.

According to the well-known solution from the theory of elasticity, the stresses acting in the soil skeleton in the vicinity of a spherical cavity and of the radius R_w (a tip of a perforation hole) are [5]

$$\begin{aligned} \tilde{\sigma}_r &= -(q+p_w)(R_w/r)^3 + q + p(r) \\ \tilde{\sigma}_\theta &= 1/2(q+p_w)(R_w/r)^3 + q + p(r) \\ \tilde{\sigma}_\varphi &= 1/2(q+p_w)(R_w/r)^3 + q + p(r) \end{aligned} \tag{1}$$

Here p_w is pressure in the well, $p(r)$ is the reservoir pressure at a distance r from the perforation hole.

On the surface of the tip of the perforation hole, i.e. when $r = R_w$, (1) gives

$$\begin{aligned} \tilde{\sigma}_r &= 0 \\ \tilde{\sigma}_\theta &= 3/2(q+p_w) \\ \tilde{\sigma}_\varphi &= 3/2(q+p_w) \end{aligned} \tag{2}$$

Specimen 3 was tested on the "generalized stretching" program. Figure 5 shows a loading program corresponding to a reduction in the pressure in the well p_w. The stresses S_1, S_2, S_3 refer to the axes of the loading unit of the TILTS and correspond to effective stresses $\tilde{\sigma}_\varphi, \tilde{\sigma}_\theta, \tilde{\sigma}_r$ acting near the hole.

The loading of specimen 3 consisted of three stages.

Stage 1. The specimen was squeezed uniformly from all sides to the stress equal to the difference between the value of the rock pressure q at a depth of $h = 6115$ m and the value of the reservoir pressure, which was assumed to be approximately equal to the hydrostatic pressure, i.e. $p_0 = 60$ MPa. At an average density of overlying rocks 2.3 g/cm^3 it equals to 80 MPa. This point corresponds to the stresses acting in the soil skeleton prior to drilling the well.

Stage 2. The two stress components (S_1 и S_2) corresponding to the circumferential stresses $\tilde{\sigma}_\varphi, \tilde{\sigma}_\theta$ increased equally, and the third stress S_3 corresponding to the radial stress decreased. The average stress $S = (S_1 + S_2 + S_3)/3$ was kept constant. The end of the second stage corresponds to the stress state around the perforation hole before the well is put into operation. At this point of the loading program the stresses are given by (2).

Stage 3. Two components of the stress continued to grow, and the third one remained virtually zero. In the stage 3 a change in the stresses acting near the perforation hole at pressure decreasing in the well was simulated.

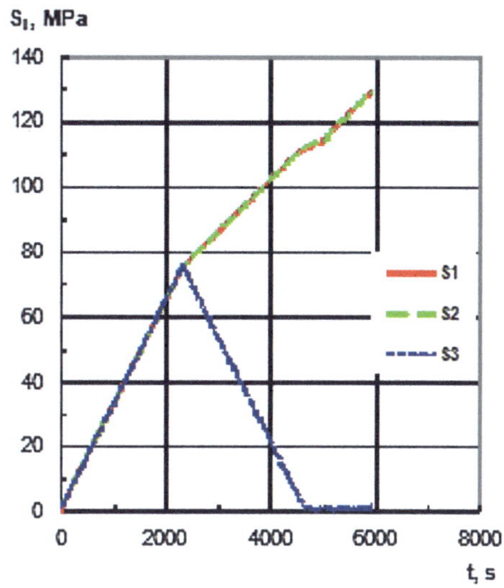

Fig. 5. Load program of specimen 3 "Generalized stretching".

In the course of the experiment specimen 3 was oriented in the loading unit of the TILTS in such a way that in stage 2 the specimen was unloaded in the bedding plane.

Figure 6 shows the strain curves of specimen 3 during the experiment. It depicts the components of the strain of the specimen along three axes $\varepsilon_1, \varepsilon_2, \varepsilon_3$ that are related to the axes of the loading unit of the TILTS in dependence on the loading parameter – the monotonically increasing stress component S_2.

It can be seen from Fig. 6 that under stress $S_2 = 110\,M\Pi a$, which according to (2) corresponds to the pressure at the bottom of the well $p_w = 67\,M\Pi a$, the specimen began to plastically deform and it was destroyed at $S_2 = 140$ MPa. This corresponds to a downhole pressure of 47 MPa, i.e. the destruction of the specimen occurred at a pressure drawdown of 13 MPa.

5.2 Testing the Specimen Using the Program "Generalized Shift"

Such a loading program corresponds to a stress state arising in the vicinity of an uncased vertical well bore loaded with internal pressure and rock pressure at infinity.

For permeable rocks the stresses acting in the soil skeleton in the vicinity of an open hole of radius R_w according to the well-known Lame's task solution are [5]

$$\begin{aligned}\tilde{\sigma}_r &= -(q+p_w)(R_w/r)^2+q+p(r)\\ \tilde{\sigma}_\theta &= (q+p_w)(R_w/r)^2+q+p(r)\\ \tilde{\sigma}_z &= q+p(r)\end{aligned} \quad (3)$$

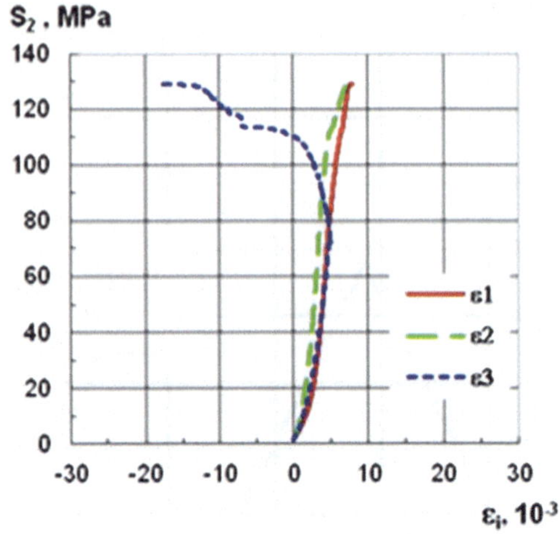

Fig. 6. The strain curves of specimen 3 during the experiment.

On the wellbore wall, i.e. at $r = R_w$

$$\begin{aligned} \tilde{\sigma}_r &= 0 \\ \tilde{\sigma}_\theta &= 2(q + p_w) \\ \tilde{\sigma}_z &= q + p_w \end{aligned} \tag{4}$$

Under the "generalized shift" program, specimen 4 was tested. In Fig. 7 its load program was shown which corresponds to a reduction in pressure in the well. The stresses S_3, S_2, S_1 shown on it refer to the axes of the loading unit of the TILTS and correspond to the effective stresses $\tilde{\sigma}_z, \tilde{\sigma}_\theta, \tilde{\sigma}_r$.

The loading of specimen 4 consisted of three stages.

Stage 1. The specimen was squeezed uniformly along the three axes to a stress equal to the difference between the value of the rock pressure q at the depth H and the value of the formation pressure p_0. End point of the stage corresponds to the stresses acting in the soil skeleton before drilling the well (80 MPa).

Stage 2. In the next stage one stress component (S_2) continued to grow, the second stress (S_3) remained constant, and the third one (S_1) decreased, and the loading was changed in such a way that the average stress $S = (S_1 + S_2 + S_3)/3$ throughout the stage 2 remained constant.

The end point of the stage corresponds to the state when the well is drilled and filled with process water.

At this point according to (4): $S_2 = 2(q + p_0)$, $S_3 = (q + p_0)$, $S_1 = 0$.

Stage 3. At the third stage, the process of pressure reduction in the well is modeled. In this case, as it can be seen from the relations (4), the ring and vertical stresses grew, but the vertical stresses increased approximately two times slower.

The third stage was the last and lasted until the specimen was destroyed.

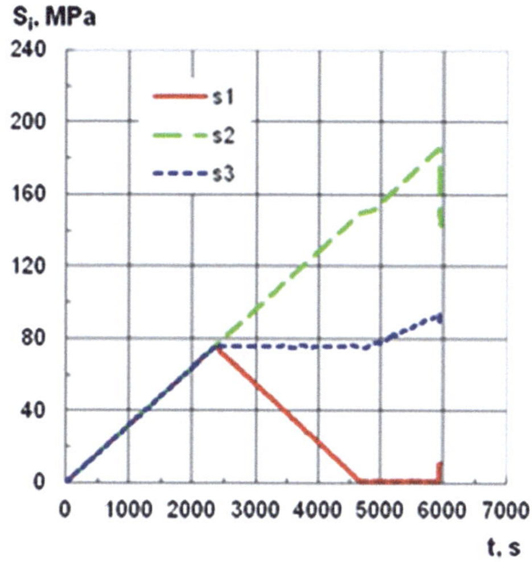

Fig. 7. The loading program of specimen 4 “Generalized shift”.

During the experiment, the strains of the specimen were measured in three directions.

In the course of the experiment, specimen 4 was oriented in the loading unit of the TILTS in such a way that in the stage 2 the specimen was loaded by the stress S_2 acting long the axis 2 of the apparatus in the bedding plane and its unloading was conducted along the axis of the specimen perpendicular to the bedding.

Figure 8 shows the strain curves of specimen 4 during the experiment. It depicts the strain components of the specimen along three axes that are related to the axes of the loading unit of the TILTS in dependence on the loading parameter S_2 - the monotonically increasing stress component.

Figure 8 shows that at the stress $S_2 = 180$ MPa, which, according to (4), corresponds to the pressure at the bottom of the well $p_w = 50$ MPa, the specimen began to deform plastically and disintegrated at $S_2 = 200$ MPa. This corresponds to the pressure at the well bottom 40 MPa, i.e. the destruction of the specimen occurred at a pressure drawdown of 20 MPa.

5.3 Testing the Specimen Using the Program “Generalized Compression”

Under the program “Generalized compression” specimen 5 was tested. The loading program of specimen 5 is shown in Fig. 9. It consisted of two stages.

Stage 1. The specimen was squeezed uniformly from all directions to a stress equal to the difference between the value of the rock pressure q at the depth h and the value of the reservoir pressure p_0. The end point of the stage corresponds to the stresses acting in the soil skeleton in an intact reservoir (80 MPa).

Stage 2. The two components of the stresses (S_1 and S_2) decreased identically to almost zero, and the third stress S_3 grew. The average stress $S = (S_1 + S_2 + S_3)/3$ was kept constant.

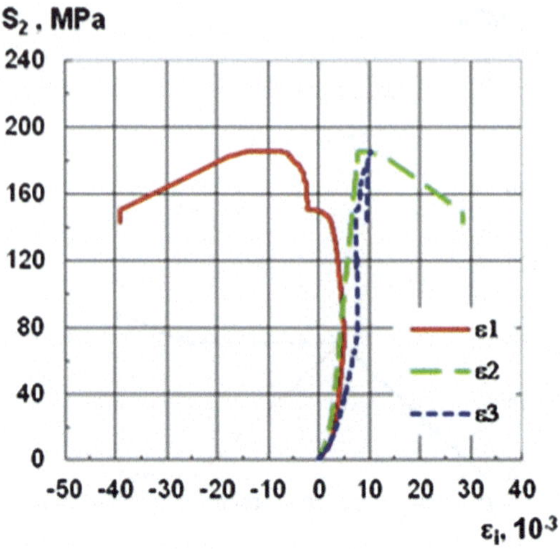

Fig. 8. The strain curves of specimen 4 in the course of the experiment.

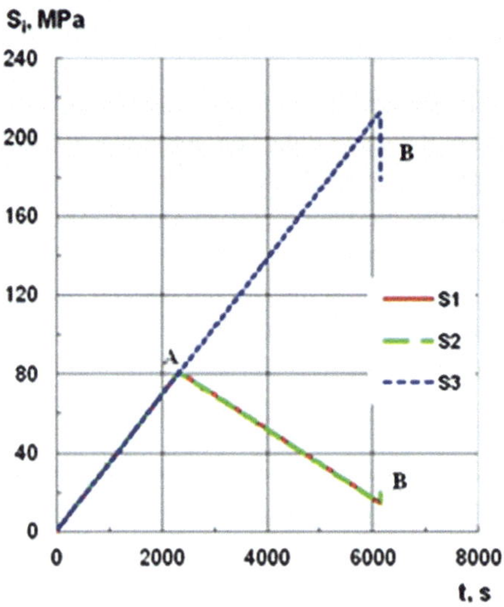

Fig. 9. Load program of specimen 5 "Generalized compression".

In the course of the test the specimen 5 was oriented in the loading unit in such a way that in stage 2 it was loaded along the axis 3 of the apparatus by stress S_3 perpendicularly to the bedding. Accordingly, unloading of the specimen was carried out along two axes 1 and 2 in the bedding plane.

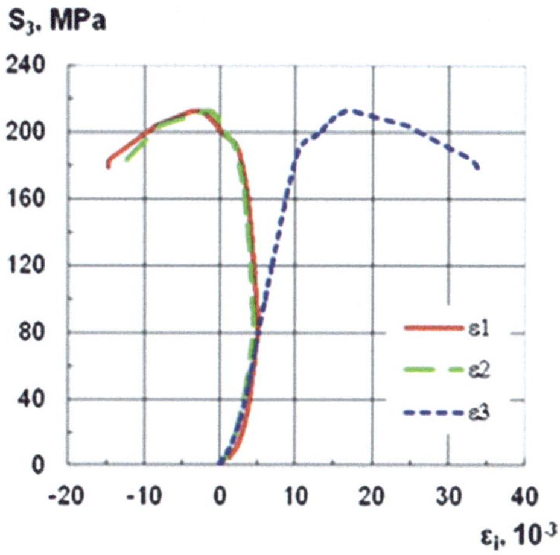

Fig. 10. The strain curves of specimen 5 during the experiment.

In Fig. 10 the strain curves of specimen 5 are shown during the test. The figure depicts the strain components of the specimen along three axes $\varepsilon_1, \varepsilon_2, \varepsilon_3$ related to the axes of the TILTS loading unit in the dependence on the loading parameter S_2 the monotonically increasing stress component.

6 Discussion of the Results

On the Triaxial Independent Load Test System, 5 cubic rock specimens (dolomite) from the exploration well of the Kainsayskaya area from the depth of more than 6 km were tested. The purpose of the research was to study the deformation and strength properties of rocks lying at great depths under the conditions of high rock pressure. The loading programs simulating the stresses that actually appeared in oil and gas reservoirs in the vicinity of oil and gas wells were implemented on three specimens.

As a result, a number of interesting and important results were obtained:

- Triaxial tests of the specimens 1 and 2 revealed a significant strength anisotropy of the rocks under study - their strength constants under loading in the bedding planes were significantly lower than in the orthogonal direction. The presence of such anisotropy can have a significant effect on the stability of well bores, especially directed obliquely, during their drilling and operation [6, 7];
- The simulation of real stress states at the depth of 6 km in the vicinity of the borehole (the specimens 3 and 4) at the TILST showed that the rocks under investigation, despite the fact that they were under conditions of compression by high rock pressure, nevertheless have strength, comparable to the strength of rocks from much lower depths [8];

– The simulation of the process of pressure reduction in the vicinity of the open hole in the vertical well (the specimen 4) and near the tip of the perforation hole (the specimen 3) at the TILST showed that the beginning of the rock failure essentially depends on the type of stress state arising in the formation.

Thus, the destruction of the specimen 4 at the simulation of an open hole in a vertical well occurred at stresses corresponding to the bottomhole pressure of 40 MPa, i.e. at the drawdown pressure of 20 MPa, and when the stress state near the tip of the perforation hole was simulated (the specimen 3) destruction occurred at the stresses corresponding to the bottomhole pressure 47 MPa, i.e. at the drawdown pressure of 13 MPa.

A similar picture was observed at the tests of rock specimens from other deposits, including deep-lying ones. In particular, two sandstone specimens cut from one piece of core sampled from a depth of 6 km of the Astrakhan gas condensate field were tested on the programs 'a well' and 'a sphere' [9]. The situation in the vicinity of the open borehole of the vertical well was modeled on the first specimen, and the stresses near the tip of the perforation hole were simulated on the second specimen. The specimen tested on program 'a well' was deformed almost elastically up to the maximum possible pressure drawdown, and the specimen tested on program 'a sphere' was intensively plastically deformed and destroyed.

7 Results

The obtained results prove the importance of experimental study of deformation-strength properties of rocks from deep horizons at facilities that allow physical modeling of real triaxial stress states arising in oil and gas reservoirs during drilling and operation of wells.

Carrying out of such researches allows to define strength parameters of productive strata and enclosing rocks, to estimate the risks connected with the danger of loss of stability of well bores at technological operations in deep wells, to develop recommendations on the necessary well design, and also to substantiate methods of impact on deep-lying strata in order to increase the productivity of wells.

This is important for choosing the right technical and technological solutions to optimize the strategy and tactics of developing deep-lying deposits.

Acknowledgement. The work was done under financial support of Russian Science Foundation, project No. 16-11-10325.

References

1. Varshavskaya, I.E., Volozh, Yu.A., Dmitrievskii, A.N., Leonov, Yu.G., Miletenko, N.V., Fedonkin, M.A.: A new concept of developing hydrocarbon resources. Herald Russian Acad. Sci. **82**(1), 17–26 (2012)
2. Volkova, A.Yu., Gouzenkov, I.K., Karev, V.I.: The development of deep-seated deposits as an opportunity to expand the resource base of hydrocarbons. Process. GeoMedia **1**, 21–31 (2015)

3. Karev, V.I., Kovalenko, Yu.F.: Well stimulation on the basis of preliminary triaxial tests of reservoir rock. In: Proceedings of EUROCK (The ISRM International Symposium), Rock Mechanics for Resources, Energy and Environment, pp. 935–940. CRC Press/Balkema, Leiden (2013)
4. Coulomb, C.A.: Essai sur une application des règles des maximis et minimis à quelques problèmes de statique relatifs, à l'architecture. Mem. Acad. Roy. Div. Sav. **7**, 343–387 (1776)
5. Timoshenko, S.P., Goodier, J.: Theory of Elasticity. Nauka Publishing, Moscow (1979)
6. Klimov, D.M., Karev, V.I., Kovalenko, Y.F., Ustinov, K.B.: Mechanical-mathematical and experimental modeling of well stability in anisotropic media. Mech. Solids **48**(4), 357–363 (2013)
7. Karev, V.I., Klimov, D.M., Kovalenko, Yu.F, Ustinov, K.B.: Fracture model of anisotropic rocks under complex loading. Phys. Mesomech. **19**(6), 34–40 (2016)
8. Klimov, D.M., Ter-Sarkisov, R.M., Chigai, S.E., Kovalenko, Yu.F, Ryzhov, A.E.: Determination of the strength characteristics of the rocks of the Shtokman gas condensate field and assessment of the risks of sand removal during its development. Gazovaya Promyshlennost **11**, 57–60 (2010)
9. Klimov, D.M., Karev, V.I., Kovalenko, Yu.F, Sidorin, Yu.V: Deformation strength and filtration properties of rocks from deep horizons under the conditions of true triaxial loading. Process. GeoMedia **4**(9), 327–332 (2016)

Modeling of Deformation and Filtration Processes Near Wells with Emphasis of their Coupling and Effects Caused by Anisotropy

V. I. Karev(✉), D. M. Klimov, Yu. F. Kovalenko, and K. B. Ustinov

Ishlinsky Institute for Problems in Mechanics of Russian Academy of Sciences, Moscow, Russia
wikarev@ipmnet.ru

Abstract. The approach to modeling geomechanical processes in the well vicinity including mathematical modelling of deformation, fracture and filtration as well as experimental determining the parameters involved, under conditions, corresponding to the real in situ ones is presented. The approach involves three stages: (i) choosing the mechanical model and its adopting to the considered problem; (ii) determining the model parameters by using the direct experiments; (iii) mathematical modeling of deformation, fracture and filtration processes in question.

The important mechanical model feature is that it accounts for anisotropy of mechanical and filtration properties and dependence of yield transition on volumetric stresses and pore pressure. Another important peculiarity consists in using the experimentally determined dependences of permeability on stress-strain state.

The results of the experimental determination of the model parameters for two lithotypes of Kirinsky field and one lithotype of Filanovsky field using the Triaxial Independent Loading Test System (TILTS) are given. Numerical simulation for the used model for the cases of uncased and perforated bottomhole is presented. The stress concentrations and production rate are calculated. The results of the work carried out demonstrate the capability of the approach to solve geomechanical problems in order to optimize technological processes.

Keywords: Rocks · Stress-strain state · Effective stress · Pore pressure Permeability strength anisotropy · Non-associative plasticity

1 Introduction

Among the problems related to processing of wells (both producers and injectors) there are two groups determined essentially by deformational processes in near well zones. These undesirable phenomena are: loss of well stability and drop of permeability. To prevent or minimize the damage caused by these processes one needs better understanding of their basis. Therefore, modeling of deformation and filtration processes became nowadays an important part of design of oil and gas deposit processing.

V. Karev et al. (Eds.): PMMEEP 2017, SPRINGERGEOL, pp. 350–359, 2018.
https://doi.org/10.1007/978-3-319-77788-7_36

A number of specialized programs, often referred to as simulators, have been created and used by producing and service companies for this purpose. That does not mean, however, that all the problems are solved. In particular such a property as strength anisotropy although has been investigated (e.g. [1]), but, has not been addressed in the existing simulators. Another important, but not yet comprehensively studied factor is the influence of stress-strain state and its history on permeability. This property may be determined only experimentally for particular rocks. Many scientific centers around the world are involved in the research aiming at filling the gaps in the existing models [2].

Both experimental and numerical modeling have been carried out in Laboratory of Geomechanics of Institute for Problems in Mechanics of RAS aiming at development of the complex approach to solving geomechanical problems. The approach consists in the following main stages illustrated on Fig. 1.

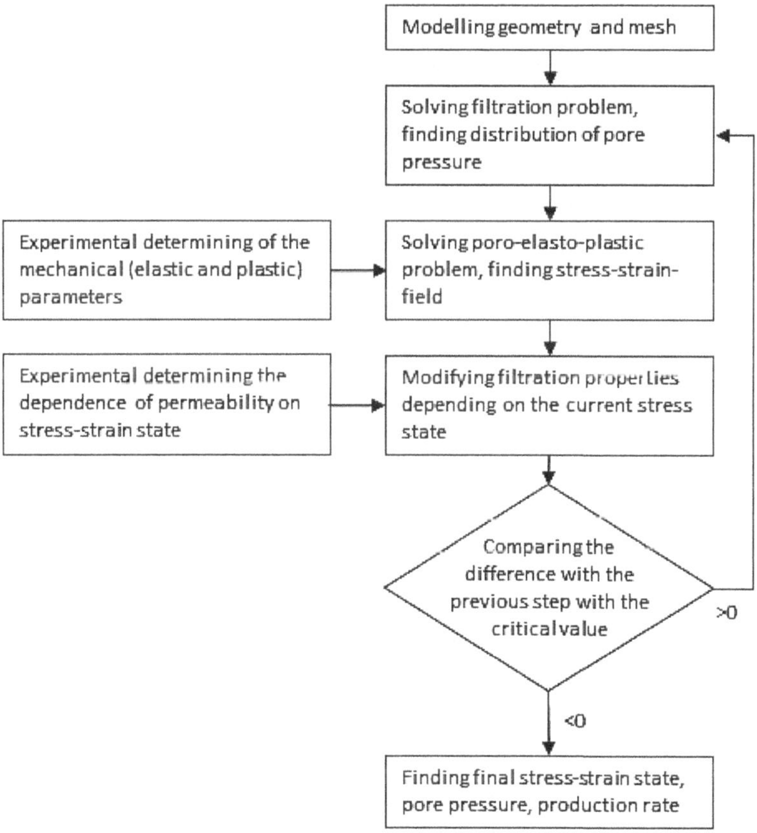

Fig. 1. Block-scheme of the used algorithm.

The approach has been applied for the number of oil and gas deposits [3–6]. In the present paper its application is demonstrated for collectors of Kirinskoe and Filanovskoe deposits.

2 Models of Mechanical Behavior

In order to yield reliable results the model need to be able to describe the key peculiarities of mechanical behavior, among which the following need to be stressed:

- influence of pore pressure on deformation;
- nonlinear stress-strain relation, appearance of inelastic strains, dependence the yield criteria not only on shear, but also on normal stresses, as well as on pore pressure;
- possible presence of inelastic volumetric strains, and their non-obvious relation with the normal and shear stresses;
- possible elastic, strength and filtration anisotropy;
- dependence of permeability on the stress-strain state and its history.

All of the mentioned features are described by the existed models, and some of them are included in the program codes of both wide-range packages and the specialized simulators, the problem of the complex description of all of the mentioned key features when modeling mechanical and filtration processes within the hydro-carbonate reservoirs still exists.

The base used model of deformation and filtration includes the following system of equations.

- Equations of filtration flow, which in case of stationary have the form

$$q_i = -\kappa_{ij}\, p_{i,j} \quad q_{i,i} = 0 \tag{1}$$

where q_i is the vector of fluid flow; p is the pore pressure; κ_{ij} is permeability tensor, that may depend generally on pore pressure, coordinates, stress state and other factors; indexes after the comma stand hereafter for derivative over the corresponding coordinate.

- Experimentally determined dependence of permeability on the history of stress-strain state for each type of rocks; in case of pronounced filtration anisotropy all components of permeability tensor should be determined, or, at least, permeability in the direction of interest

$$\kappa_{ij} = \kappa_{ij}(\sigma) \tag{2}$$

  where σ is a parameter characterizing stress state.
- The set of equations of poro-elasto-plasticity, in which the elastic part is described by the traditional system of equations by Biot [7], and inelastic part is described by a variant of theory of plastic flow for anisotropic media [8, 9] with accounting for the volumetric stress similar to [10] and non-associative flow rule of the type of [11, 12].

The outlined scheme of poro-elasto-plasticity is expanded as follows. The kinematics is described by the total strain tensor ε_{ij}^{T}, which is decomposed to elastic, ε_{ij}^{E}, and plastic, ε_{ij}^{P}, parts

$$\varepsilon_{ij}^{T} = \varepsilon_{ij}^{E} + \varepsilon_{ij}^{P} = \frac{1}{2}\left(u_{i,j} + u_{j,i}\right) \tag{3}$$

where u_i is the displacement vector. The statics is described by equation of equilibrium

$$\sigma_{ij,i} = 0 \tag{4}$$

and equation

$$s_{ij} = \sigma_{ij} + \alpha_P p \delta_{ij} \tag{5}$$

relating the total σ_{ij} and effective (acting on solid skeleton) s_{ij} stress tensors, and pore pressure p. $0 \leq \alpha_P \leq 1$ is Bio's coefficient, characterizing the influence of pore pressure on the stress state and depending on porosity structure; for well permeable collectors α_P approaches unity from below [13].

The elastic strain tensor related to the stress tensor in accordance with the Hooke's law

$$s_{ij} = \Lambda_{ijkl}\varepsilon_{kl}^{E} \tag{6}$$

Here Λ_{ijkl} are components of elasticity tensor (in case of elastic isotropy $\Lambda_{ijkl} = \lambda\delta_{ij}\delta_{kl} + 2\mu\delta_{ik}\delta_{jl}$, $\lambda = \frac{\nu E}{(1+\nu)(1-2\nu)}$, $\mu = \frac{E}{2(1+\nu)}$ are Lame constants, E, ν are Young's moduli and Poisson's ratios).

According to the concept of plastic flow theory, the increments (also called "rates") of plastic strains $d\varepsilon_{ij}^{p}$ during active loading are calculated as values proportional (with coefficient $d\lambda$) to partial derivatives of plastic potential Q

$$d\varepsilon_{ij}^{P} = d\lambda\frac{\partial Q}{\partial \sigma_{ij}} \tag{7}$$

Besides, it is supposed that during active loading a particular combination of components of stress tensor satisfied an equation, called yield surface, F. In the classical theory of plasticity the yield surface and plastic potential coincide, $Q = F$. Meanwhile, using this equality for non-metal plasticity leads to a strong contradiction between the observed and predicted behavior: in particular, the predicted volumetric inelastic deformations are usually much greater than the observed ones. That leads to putting forward laws with $Q \neq F$, called non-associative laws, to describe deformation of rocks and soils [15, 11]. In order to describe adequately the role of inelastic volumetric change the concept of dilatancy was suggested [16, 17], according to which the change in volumetric deformation are related to the shear strain intensity, rather than to the volumetric stress. That leads to violating the associativity of the volumetric strain, while the "deviatoric associativity" remains. In accordance with this concept, yield function, F, and plastic potential, Q, are chosen of the same form, with the differences in factors at the linear terms of stresses

$$\begin{Bmatrix} F \\ Q \end{Bmatrix} = \left(G^0_{(23)}(s_{22}-s_{33})^2 + G^0_{(13)}(s_{11}-s_{33})^2 + G^0_{(12)}(s_{11}-s_{22})^2 + 2L^0_{(23)}s^2_{23} + 2L^0_{(13)}s^2_{31} + 2L^0_{(12)}s^2_{12}\right)^{1/2} + \left(B^m_{(1)}s_{11} + B^m_{(2)}s_{22} + B^m_{(3)}s_{33}\right) - A \quad (8)$$

Here $G^0_{(ij)}$, $L^0_{(ij)}$, $B^m_{(ij)}$ are material constants;. $m = 0$ for the yield function, $m = 1$ for the plastic potential; the dimensionless parameter A, may be considered as a characteristic of hardening. Yield function, F, and plastic potential, Q, (7) are written in the axes of the material anisotropy.

It should be noted, that maintaining the same coefficients at quadratic terms of stresses $G^0_{(ij)}$, $L^0_{(ij)}$ in (7), preserves the deviatoric associativity.

Isotropy imposes the restrictions:

$$G^0_{(13)} = G^0_{(23)} = G^0_{(12)} = G,\ L^0_{(13)} = L^0_{(23)} = L^0_{(12)} = 3G,\ B^m_{(1)} = B^m_{(2} = B^m_{(3)} = B/3, \quad (9)$$

that leads to reducing the criterion to the Druckep-Prager criterion [10] for effective stresses.

$$\begin{aligned} &F = \sigma_i + \alpha \cdot \sigma_m - \tau_s = 0 \\ &\sigma_m = \tfrac{1}{3}s_{kk}, \quad \sigma_\mathrm{i} = \sqrt{\tfrac{3}{2}\left(s_{jk} - \tfrac{1}{3}s_{ii}\delta_{jk}\right)\left(s_{jk} - \tfrac{1}{3}s_{ii}\delta_{jk}\right)} \end{aligned} \quad (10)$$

Here α, τ_s are the constants of Drucker-Prager criterion, similar to the angle of internal friction and cohesion of More-Coulomn criterion.

3 Experimental Determining of the Model Parameters

Experiments are carried out using Triaxial Independent Loading Test System (TILTS) of IPMech RAS [14]. It allows loading of cubic specimens either 40 × 40 × 40 mm or 50 × 50 × 50 mm in three perpendicular directions independently with measuring stresses and strains in all directions, and simultaneous measurement of permeability in one direction. Loading system of this kind allows to reconstruct the stress states sufficiently realistically, and to determine materials properties in the conditions as close to the reality as possible.

Experiments are carried out according to various loading programs; in most cases, two types of programs are used that correspond to the generalized shear and axial compression with lateral constant stresses [5, 6]. By analyzing the obtained dependencies the following characteristics are determined:

- the set of elastic constants, namely Young's modulus and Poisson's ratio for isotropy, which was the case for the studied types of rocks (in case of transversal anisotropy five elastic constants are to be measured);
- the set of constants, characterizing transition to plasticity, or more generally to non-elastic behavior: Drucker-Prager constants in case of isotropy, or critical stresses, corresponding to loading in various direction with different lateral pressure, allowing to reconstruct the model of anisotropic plasticity for the investigated material;
- dependence of permeability on the history of stress-strain state (usually the dependence of permeability on shear stress intensity), i.e. specifying function determined by (2).

In the present work the results of modeling for two lithotypes of Kirinsky and one lithotype of Filanovsky fields are presented. Rocks of Kirinsky field showed isotropy of elastic, strength and filtration properties. Rocks of Filanovsky fields revealed pronounced strength transverse anisotropy related to their layered texture, however neither noticeable elastic, nor filtration anisotropy were observed.

Mechanical parameters of both lithotypes of Kirinsky field was obtained as follows: $E = 6 \cdot 10^3$ MPa; $v = 0.25$; $E_p = 1.37 \cdot 10^3$ MPa; $\tau_s = 18$ MPa; $\alpha = 1.28$. Here E_p is the modulus of plasticity, i.e. the ratio of the stress increments to increment of the total displacement after reaching yield point measured in compression with the lateral pressure. Relation of E_p with the parameter $d\lambda$ for the considered type of models is described in [6].

Mechanical parameters of Filanovsky field are $E = 7.78 \cdot 10^3$ MPa; $v = 0.15$; $E_p = 0.78 \cdot 10^3$ MPa. The parameters of the yield criteria [8] were determined from the whole set of experiments of both types (the property corresponded to transverse isotropy [8]) by means of the least square method $G^0_{(12)} = 2.7 \cdot 10^{-3}$, $G^0_{(13)} = G^0_{(23)} = 1.27 \cdot 10^{-3}$, $B^0_{(1)} = B^0_{(2)} = 1.95 \cdot 10^{-2}$, $B^0_{(3)} = 3.28 \cdot 10^{-2}$, $L^0_{(13)} = 1.0 \cdot 10^{-2}$. The last parameter was not possible to determine from the available experiments, its value was chosen by analogue with samples of a similar lithotype [5].

Parameters of plastic potential (7) diverged from the corresponding parameters of the yield criteria are chosen in accordance to the condition of vanishing dilatancy (that was observed)

$$B^1_{(i)} = B^0_{(i)} - \frac{1}{3}\sum_{j=1}^{3} B^0_{(j)}, \tag{11}$$

The corresponding values are: $B^1_{(1)} = -4.44 \cdot 10^{-3}$, $B^1_{(3)} = 8.88 \cdot 10^{-3}$.

The obtained dependencies of permeability on stress intensity are shown in Fig. 1 together with the approximations

$$k_1/k_{10} = \begin{cases} 1 - 0.0079\tau - 7 \cdot 10^{-5}\tau^2 & \tau < 63 \text{ MPa} \\ 0.05 & \tau \geq 63 \text{ MPa} \end{cases}$$
$$k_2/k_{20} = \begin{cases} 1 - 0.0062\tau & \tau < 55.4 \text{ MPa} \\ 0.0526\tau - 2.26 & \tau \geq 55.4 \text{ MPa} \end{cases} \quad (12)$$
$$k_3/k_{30} = \begin{cases} 1 - 0.0176\tau & \tau < 54 \text{ MPa} \\ 0.05 & \tau \geq 54 \text{ MPa} \end{cases}$$

for the soils of the first and second types of Kirinskoe, and Filanovskoe fields, respectively,

The dependencies prove that the deformation processes may lead to both closing of filtration channels, and hence to decrease in permeability, and to appearance and growth of new systems of cracks, leading to increase in permeability [18].

4 Numerical Modeling

Numerical modeling consists in solving the coupled problem of poro-elasto-plasticity and filtration using the chosen model and the experimentally determined parameters. For each considered geometry the mathematical modelling is carried out according to the steps stated on Fig. 1.

The problem was solved in 3-D for various geometries of the well face (open face, various types of perforation, in particular six perforation holes, aligned along the well length (three on each side), Fig. 2. For calculations the following parameters are used: well radius, R = 0.1 m; perforation hole length L = 0.1 m; hole radius r = 0.01 m. Meshes with 27 761 nodes and 136 421 elements corresponding (due to symmetry) to the 1/8 of the initial areas were used. The mechanical properties are listed in Sect. 3.

The boundary conditions have been chosen corresponding to the stress state of the field under consideration. For Kirinsky field: normal stress and pore pressure at the outer boundary are 63.8 MPa and 27.7 MPa, respectively. For Filanovsky field: normal stress and pore pressure at the outer boundary are 31 MPa and 13 MPa, respectively. For the other boundaries the conditions coincides. Normal stress and pore pressure at the well and cut surfaces are zero. For the normal to the well surfaces, zero displacements and zero fluid flow are prescribed.

Production rates, normalized to the rates of the "ideal" well (for which the permeability is constant and equal to the initial permeability) without perforation cuts are presented in the Table 1. Distributions of the intensity of stresses, plastic strains and pore pressure for configuration with the perforation holes for the conditions of Filanovsky field are presented on Fig. 3.

5 Summary

The approach to modeling geomechanical processes in the well vicinity is presented. The approach includes mathematical modelling of deformation, fracture and filtration as well as experimental determining the parameters involved, under conditions,

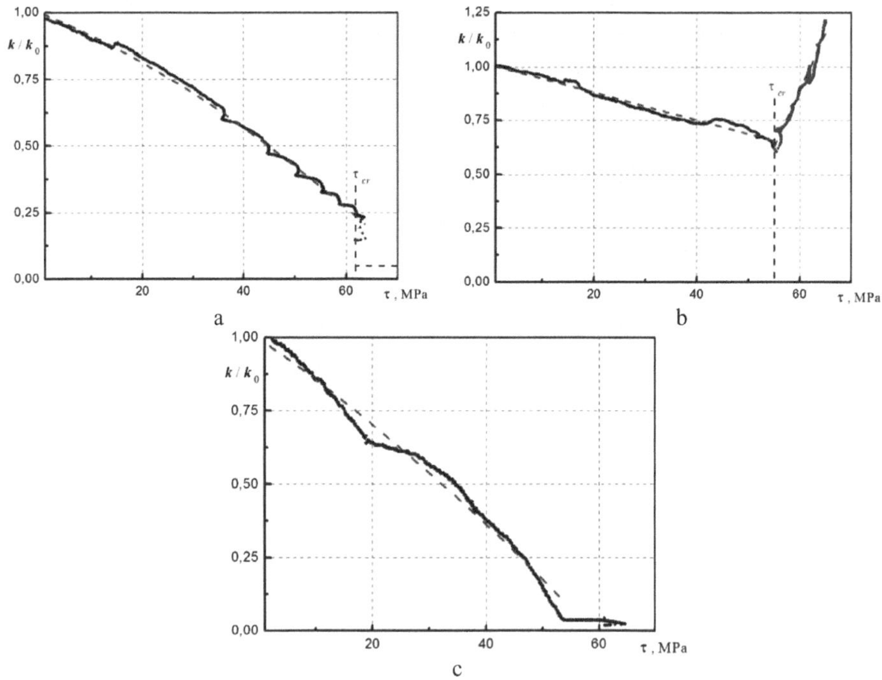

Fig. 2. Dependence of permeability on stress intensity (a) first lithotype of Kirinsky field; (b) second lithotype of Kirinsky field;. (c) Filanovsky field.

Table 1. Production rates, normalized to the rates of the 'ideal' well.

Geometry	Field	No dependence	Elasticity	Plasticity
Open wellbore	Kirinsky 1-st	1	0.82	0.91
Wellbore with cuts	Kirinsky 1-st	1.05	0.92	0.95
Open wellbore	Kirinsky 2-nd	1	0.96	0.94
Wellbore with cuts	Kirinsky 2-nd	1.05	1.20	1.15
Open wellbore	Filanovsky	1	0.77	0.79
Wellbore with cuts	Filanovsky	1.05	0.81	0.82

corresponding to the real in situ ones. The approach involves the following stages: (i) choosing the appropriate mechanical model and its adopting to the problem under consideration; (ii) determining the model parameters using the direct testing with the true triaxial loading system; (iii) mathematical modeling of deformation, fracture and filtration processes for the particular well configurations.

All stages of modeling are presented for three types of rocks two lithotypes of Kirinsky field and one lithotype of Filanovsky field. The results demonstrate the capability of the approach to solve problems appearing in oil and gas fields processing.

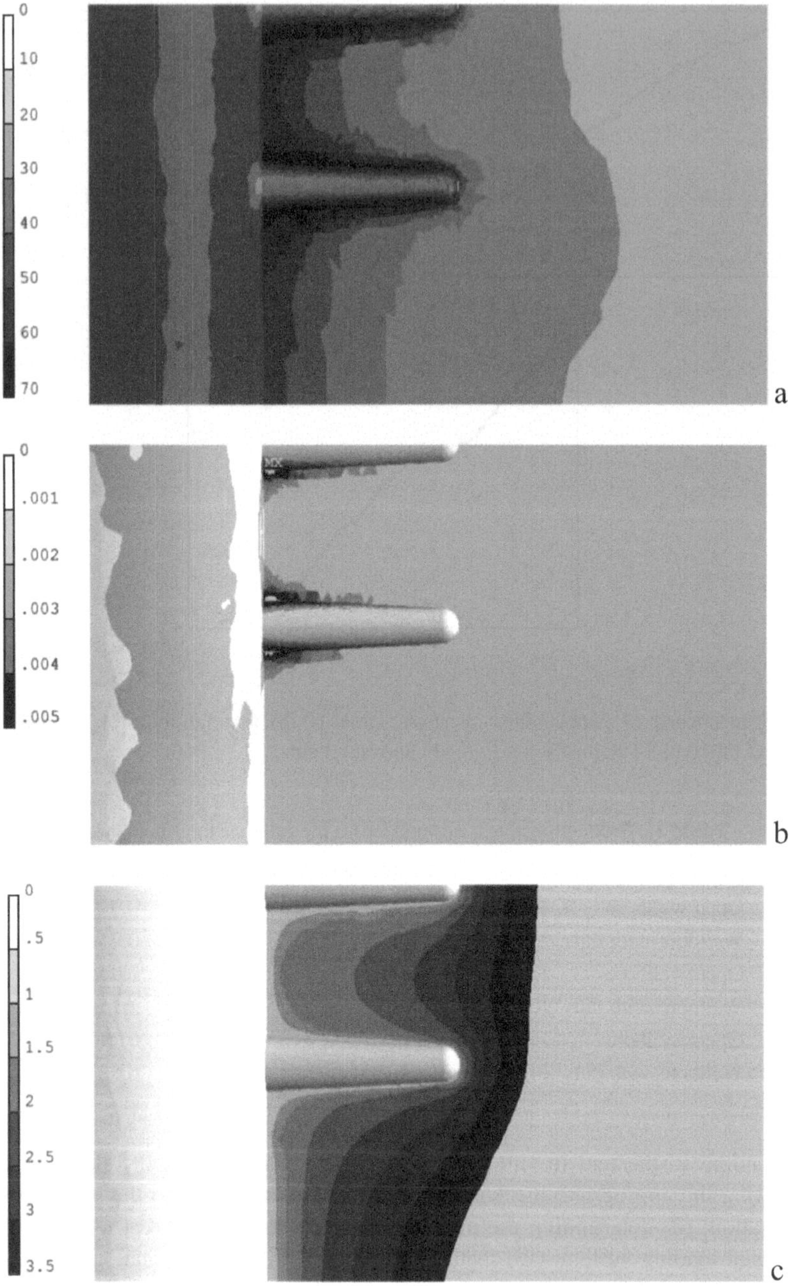

Fig. 3. Distributions of the intensity of stresses (a), intensity of plastic strains (b) and pore pressure (c) for the conditions of Filanovsky field.

Acknowledgement. The work was done under financial support of Russian Science Foundation, project No. 16-11-10325.

References

1. Singh, M., Samadhiya, N.K., Kumar, A., Kumar, V., Singh, B.: A nonlinear criterion for triaxial strength of inherently anisotropic rocks. Rock Mech. and Rock Eng. **48**(4), 1387–1405 (2015)
2. Le Calvez, J., Malpani, R., Xu, J., Stokes, J., Williams, M.: Hydraulic Fracturing Insights from Microseismic Monitoring. Oilfield Rev. **28**(2), 16–33 (2016)
3. Karev, V.I., Kovalenko, Y.F., Zhuravlev, A.B., Ustinov, K.B.: Model of filtration in a well with allowance for the dependence of permeability on stresses. Processes in GeoMedia **4**(4), 35–44 (2015)
4. Karev, V.I., Klimov, D.M., Kovalenko, Y.F., Ustinov, K.B.: Fracture of Sedimentary Rocks under a Complex Triaxial Stress State. Mech. Solids **51**(5), 522–526 (2016)
5. Karev, V.I., Klimov, D.M., Kovalenko, Y.F., Ustinov, K.B.: Destruction model of anisotropic rocks under complex loading. Phys. Mesomech. **19**(6), 34–40 (2016)
6. Karev, V.I., Kovalenko, Y.F., Ustinov, K.B.: Deformation and cracking of rocks in the vicinity of the horizontal well. Fiziko-tekhnicheskie problemy razrabotki poleznykh iskopaemykh (FTPRPI) **3**, 12–22 (2017)
7. Biot, M.A.: Le problème de la consolidation des matières argileuses sous une charge. Ann. Soc. Sc. de Brux. Ser. B **55**, 110–113 (1935)
8. Hill, R.: A theory of the yielding and plastic flow of anisotropic metals. Proc. Roy. Soc. London A. **193**, 281–297 (1948)
9. Caddel, R.M., Raghava, E.S., Atkins, A.G.: A yield criterion for anisotropic and pressure dependent solids such as oriented polymers. J. Mater. Sci. **8**, 1641–1646 (1973)
10. Drucker, D.C., Prager, W.: Soil mechanics and plastic analysis for limit design. Quart. Appl. Math. **10**(2), 157–165 (1952)
11. Nikolaevskii, V.N.: Geomehanika i flyuidodinamika. Nedra, Moscow (1996)
12. Ustinov, K.B.: On application of models of plastic flow to description of inelastic behavior of anisotropic rocks. Processes in GeoMedia **3**(7), 278–287 (2016)
13. Khristianovich, S.A., Zheltov, Yu. P.: Hydrofracturing of Oil Reservoir. Izv. AN SSSR. OTN. 53–41 (1955)
14. Karev, V.I., Kovalenko, Y.F.: Triaxial loading system as a tool for solving geotechnical problems of oil and gas production, pp. 301–310. True Triaxial Testing of Rocks. CRC Press/Balkema, Leiden (2013)
15. Nikolaevskii, V.N.: On interconnection of volume and shear strains and on shock waves in soft soils. Proc. USSR Acad. Sci. (Doklady) **177**, 542–545 (1967)
16. Reynolds, O.: On the dilatancy of media composed of rigid particles in contact, with experimental illustrations. Philos. Mag. Ser. **5–20**(127), 469–481 (1885)
17. Mead, W.J.: The geologic rôle of dilatancy. J. Geol., 33 **5**(6), 85–98 (1925)
18. Klimov, D.M., Karev, V.I., Kovalenko, Y.F.: Experimental study of the influence of a triaxial stress state with unequal components on rock permeability. Mech. Solids **50**(6), 633–640 (2015)

Effect of a Tidal Wave Caused by Large Gliding Satellite on Formation of 220 km Seismic Boundary and Split of the Mantle into Blocks

S. Kasyanov[1(✉)] and V. Samsonov[2]

[1] Zubov State Oceanographic Institute, Moscow, Russia
skas53@yandex.ru
[2] Institute of Mechanics, Lomonosov Moscow State University, Moscow, Russia

Abstract. The movement of a large satellite around a model planet, which is a thick spherical layer of a heavy, perfect liquid with a solid nucleus, is being discussed. The tidal disturbance caused by the satellite is propagated by long gravity waves. When the satellite is gliding along the layer surface the energy and orbital period decrease, and the tidal height and period of free gravity waves supporting the tide growth increase. To maintain the tide the orbital period should not be less than the period of free long gravity waves. The value of static tide is assessed for the case when orbital period and free gravity waves expending over the surface of spherical liquid layer periods coincide. When this condition is met gliding becomes impossible and the satellite fully submerges. With regard to the Earth, movement of the satellite projection is supersonic for the material composing the mantle above a certain horizon, and subsonic for underlaying horizons. The pressure of the tide bulge has effect on the entire lower layer and on a narrow front area of shock wave in the upper layer. Therefore, a considerable horizontal stress occurs at the abovementioned horizon. When the stress exceeds the shear strength limit, the upper layer of the mantle starts to move at a finite velocity in relation to the lower mantle. A melt forms in the friction layer, so the velocity of longitudinal seismic waves decreases. The boundary of supersonic and subsonic velocities gradually lowers increasing the melt layer thickness and forming an abrupt jump in seismic wave velocities. After the boundary reaches the horizon at 220 km, the satellite submerges into the mantle. This is the reason why at present moment, after the melt solidification there is a low velocity zone of seismic waves in the upper mantle and a 220 km seismic boundary.

Keywords: Asteroid · Satellite · Spherical layer · Ideal fluid · Gliding Long gravity waves · Earth's mantle · Tide · Seismic waves velocity Melt · Seismic boundary · Low velocity zone of seismic waves

V. Karev et al. (Eds.): PMMEEP 2017, SPRINGERGEOL, pp. 360–370, 2018.
https://doi.org/10.1007/978-3-319-77788-7_37

1 Introduction

In this article we study the satellite movement with radius of approximately 700–900 km around a certain model planet, represented as a spherical layer of heavy ideal fluid with outer radius of R_e approximately 6,378 km and a solid core with radius of approximately 1,100 km [1]. The satellite, which moves at the orbit, causes a tidal perturbation in the spherical layer of the fluid, which is propagated by long gravity waves.

2 Condition of Possibility of Long Gliding of the Satellite

As it is known [1] in the inertial coordinates system the orbiting period T_a of a satellite along an undisturbed Keplerian orbit is as follows:

$T_a = 2\pi(a^3/\mu)^{0.5}$, $a=0.5(\mathrm{r}_\alpha + \mathrm{r}_\pi)$, where: $\mu=f(M+M_a)$, $f = 6.67 \cdot 10^{-11} m^3/(kg \cdot \mathrm{s}^2)$, M_a - asteroid weight, M - planet weight, r_α - distance to apogee,r_π - distance to perigee, a - semi-major axis of asteroid orbit ellipsoid. Let $h_\alpha = r_\alpha - R_e$ and $h_\pi = r_\pi - R_e$ be heights of apogee and perigee, R_e- the planet radius. The orbiting period of the satellite with fixed perigee height is minimal in a circular orbit.

The orbiting satellite causes the tide in stationary liquid sphere in the area of maximum proximity to the orbit perigee. Let's estimate the value of static tide at the line connecting the planet center of mass and satellite center of mass in the orbit perigee when the satellite is moving in equatorial plane.

On this line the planet's gravity potential is equal to:

$V(x) = f \cdot M/x$, where x- distance from the top of tide bulge to the center of planet. Tide-producing potential equals to:

$\delta W = fM_a(1/\mathrm{r}' - 1/\mathrm{r}_\pi - x/\mathrm{r}_\pi^2)$, where r'- length of radius-vector from a point on the surface of tide bulge into the satellite center of mass [2].

In the top point of the tide bulge on the line connecting the planet center of mass and satellite, the summary potential of gravity forces equals to:

$$W = f \cdot M/x + fM_a(1/\mathrm{r}' - 1/\mathrm{r}_\pi - x/\mathrm{r}_\pi^2).$$

Let's find the position of disturbed surface of gravity forces potential level. Assume that the constant of gravity forces potential is equal to potential of undisturbed surface of free planet surface level without disturbing body. So we get a proportion determination of distance x from the top of tide bulge to the planet center:

$$W = f \cdot M/x + fM_a(1/\mathrm{r}' - 1/\mathrm{r}_\pi - x/\mathrm{r}_\pi^2) = W_0 = f \cdot M/R_e.$$

On the line connecting the planet center of mass and satellite center of mass, $r' = r_\pi - x$. So the equation for determination of tide height is like the following:

$$1/x + (M_a/M)(1/(r_\pi - x) - 1/r_\pi - x/r_\pi^2) = 1/R_e.$$

With this equation we determine the value x, which is the radius of maximum tidal wave in relation to the center of the planet.

The satellite gliding along the planet surface can continue for a long time with its small submerges into lithosphere and mantle, which happens at small angles of attack. The lifting force while gliding prevents the body from prompt submergence. So the tide bulge maintains the body gliding along the planet surface. With an increase of the angle of attack, as it was seen in the numerical integration of the satellite movement equations with account of environment resistance [1], there happens an abrupt submergence of the satellite into the spherical liquid layer of the planet.

A considerable tide bulge will grow only when the appearing disturbance will be distributed by long gravity waves.

The velocity of long gravity waves in the spherical layer in question laying from the planet surface to the solid core can be estimated as long gravity waves velocity in equatorial channel of given depth with vertical sides and without friction, which is equal to: $u_w = \sqrt{gH}$, where $g = fM/x^2 - fM_a/(r_\pi - x)^2$- is a module of gravity acceleration at the wave surface, $H = x - R_c$- thickness of layer of the environment where the tidal wave is distributing, R_c-radius of planet solid core. In the liquid layer in question the velocity of free long gravity waves decreases with the growth of wave height as the gravity acceleration at the wave surface decreases more quickly that the wave height increases. The wave velocity at the top of wave is less than the velocities of waves in lower points of the surface. So the wave maximum in the tide bulge is shifted back; the leading edge of wave becomes smoother, and the tailing edge becomes sharper. Therefore, the long gravity waves in the liquid layer under consideration fall frontwards, but not backwards.

The dependency of velocity of free long gravity waves with height, equal to the height of static tide caused by gravity force of the satellite, on the height of satellite orbit in perigee point, is shown in Fig. 1. The period of long gravity waves, distributing along the surface of this layer is equal to $T_w = 2\pi R_e/\sqrt{gH}$. The tidal disturbance can develop only in case if the satellite orbiting period is less than the period of long gravity waves circle around the sphere surface. With the planet rotation at an angular velocity ω_0 the period T_s between the satellite crossings of the fixed point on the planet surface at the equator is equal to $T_s = 2\pi R_e/(2\pi R_e/T_a + \omega_0 R_e) = 2\pi/(2\pi/T_a + \omega_0)$.

Let the orbit of the satellite be such that in perigee point it glides along the liquid surface. If the period of the satellite orbiting is less than the period of gravity waves ($T_s < T_w$), the satellite will catch the sharp trailing edge of the wave and will submerge into the tide bulge at a high angle of attack. If the satellite orbiting period is more than the period of gravity waves ($T_s \geq T_w$), the tidal wave will pursue the satellite with its

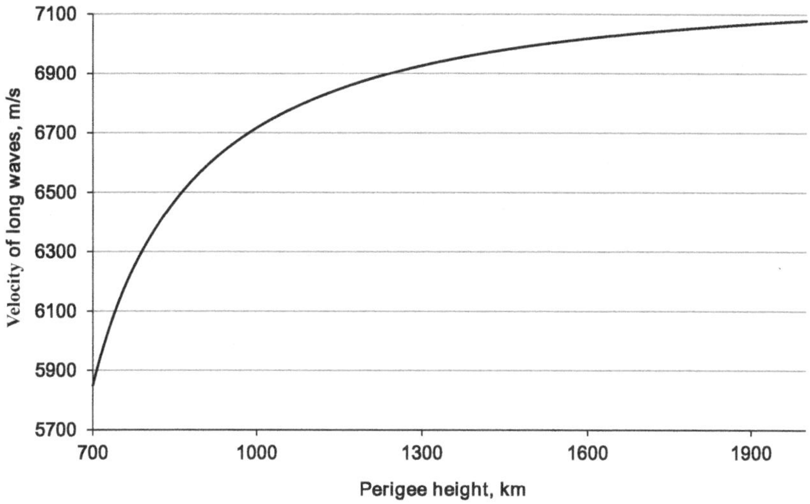

Fig. 1. The dependency of velocity of free long gravity waves with height, equal to the height of static tide caused by gravity force of the satellite, on the height of satellite orbit in perigee point.

smooth edge and will keep the satellite moving. The angle of attack will be low, and instead of submergence of the satellite into the tidal wave the satellite will rebound. While gliding along the spherical layer surface the satellite energy should be lost, the orbit in perigee point should be lower, orbiting period should reduce, and the height of tide bulge should increase. So, in accordance with Fig. 1, the velocity of long gravity waves should decrease, and their period should extend.

Thus, with satellite slowdown and lowering of its orbit satellite orbiting period decreases, and, by contrary, the period of long gravity waves, maintaining the tide bulge, increases.

Movement of the satellite is possible until the period of long gravity waves becomes less that the satellite orbiting period. Cease of satellite gliding and final submergence happens when the two abovementioned periods are equal. Let's estimate the apogee distance r_α and perigee distance r_π, at which this happens, where the equality is true, $T_s = T_w$. At the result we get a curve on the plane of orbital parameters, shown in Fig. 2 in the coordinates (h_π, h_α). In the area of parameters above this curve the satellite orbiting period is longer than the period of long waves circling around the planet and the satellite gliding is possible. In the curve the equality is true, $T_s = T_W$.

Along the curve $T_s = T_w$ in the parameters plane h_π, h_α the satellite velocity in the perigee point on the orbit there is a decreasing function of perigee height h_π, shown in Fig. 3. The value of static tide h_p along the curve $T_s = T_w$ on the parameters plane h_π, h_α is a function of orbit perigee height h_π (see Fig. 4).

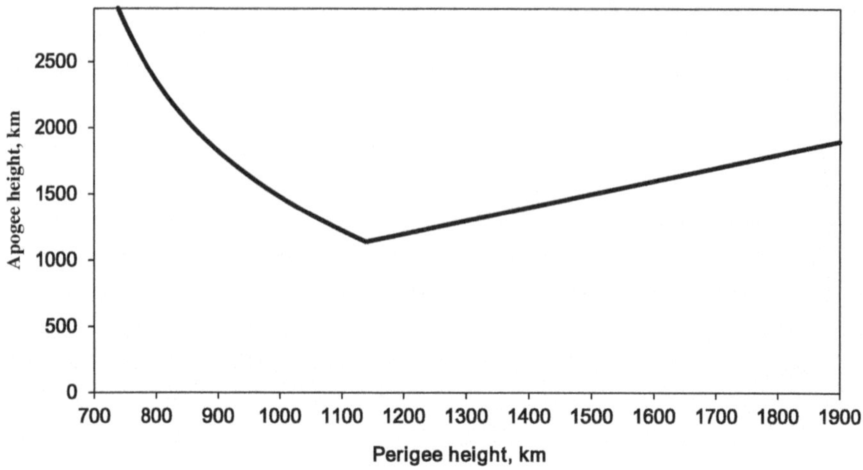

Fig. 2. The curve $T_s = T_w$ in the plane of orbital parameters h_π, h_α in the domain above which the period of satellite orbiting is longer than the period of long waves circling around the planes and the gliding is possible.

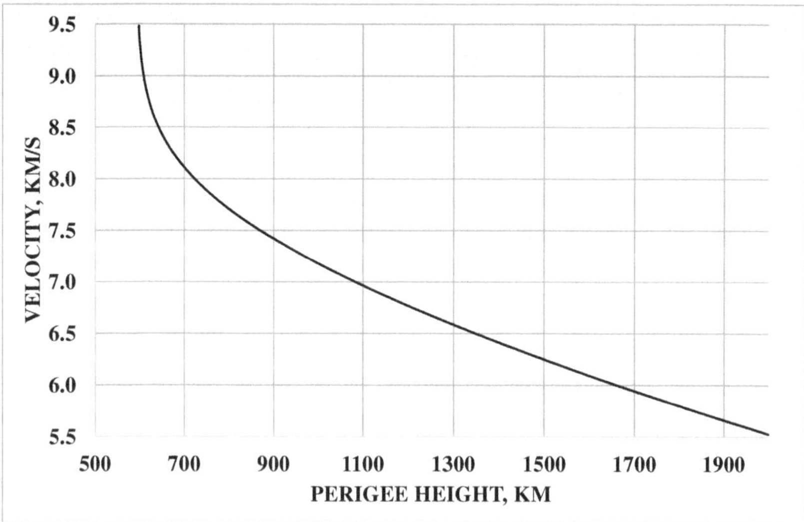

Fig. 3. Velocity of satellite projection movement to the day surface in the orbit perigee as the function of perigee height h_π.

So, along the curve $T_s = T_w$ the height of static tide (see Fig. 4) and perigee velocity (see Fig. 3) increase with the satellite slowdown and decrease of orbit perigee height.

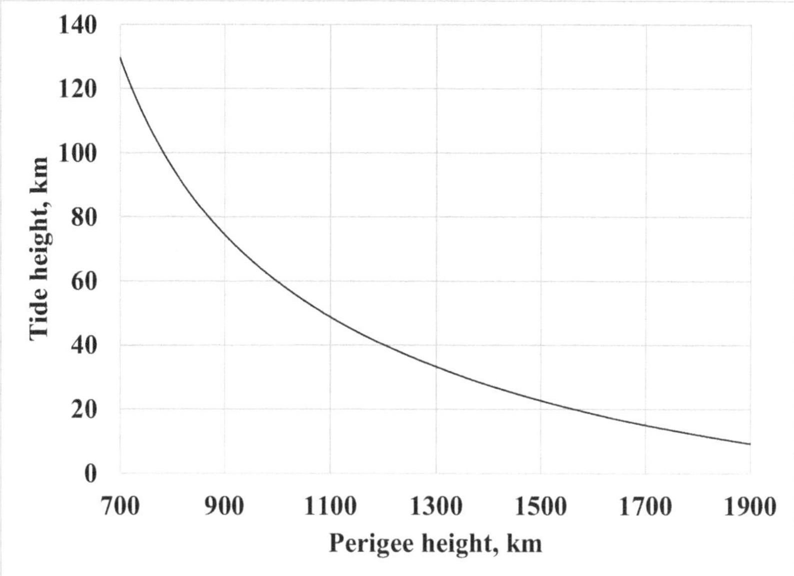

Fig. 4. Dependency of static tide height h_p on orbit perigee height h_π.

3 The Effect of the Tidal Wave on the Upper Mantle of the Earth

Now we turn to comparison with conditions on the Earth. Let's compare the velocities of long gravity waves propagation in spheric layer of ideal liquid with the speed of sound, which is the velocity of longitudinal seismic waves V_P in the mantle of the Earth and speed of satellite projection movement at different horizons. Figure 5 shows the change of longitudinal seismic waves velocity V_P with change of depth in the model PREM [3]. Satellite movement in a certain horizon in the planet mantle is supersonic for some horizons above the certain horizon $z_{M=1}$, and subsonic for some horizons below a certain horizon. So the pressure of the tide bulge will be transferred to the entire lower layer and a narrow area of shock wave front in the upper layer. As a result, on a certain surface $z = z_{M=1}$ (in other words, surface M = 1), where the velocity of satellite movement becomes equal to V_P, a considerable horizontal stress occurs. If the horizontal stress on this surface overcomes the shear strength of rocks composing the mantle, the entire upper layer, locates above the horizon $z_{M=1}$ will shift at a finite velocity in relation to the lower layers. Figure 6 shows the dependency of the M = 1 surface $z_{M=1}$ depth in the mantle on the satellite perigee height, which was built using the velocities V_P of PREM model.

Figure 7 shows the dependency of shear stress occurring at the surface M = 1 in the mantle on the satellite perigee height assuming that the layer with length along the route of the satellite coincides with the distance from the tide maximum to minimum, which is a quarter of the planet circle at the depth $z_{M=1}$.

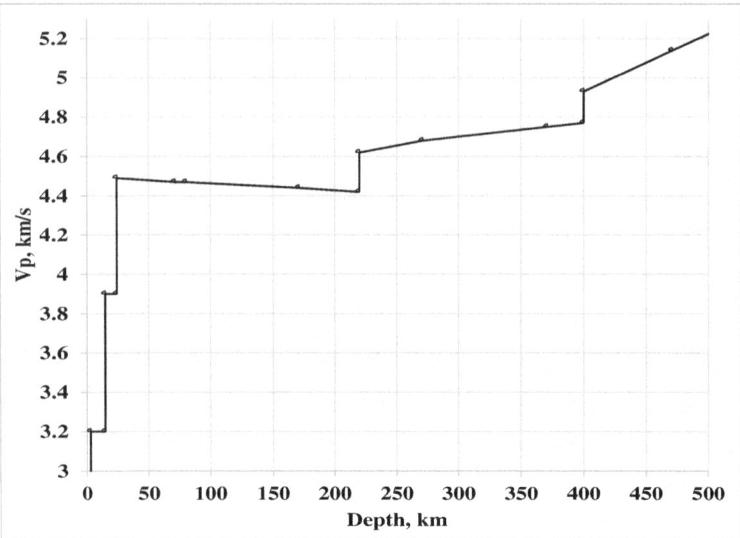

Fig. 5. Velocity of longitudinal seismic waves in isotropic model PREM [3].

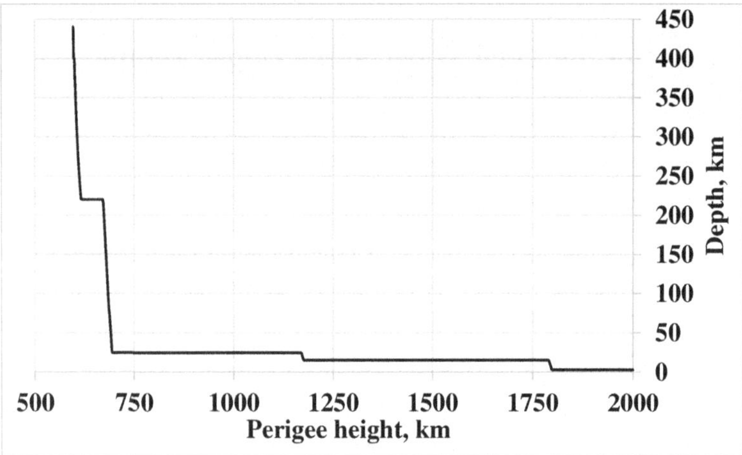

Fig. 6. Dependency of M = 1 surface depth in the mantle on the satellite perigee depth.

As the maximum compression strength usual for olivine a value of 220 MPa is taken, which corresponds to the range of values from 0.28*220 MPa to 0.35*220 MPa that is from 61.6 MPa to 77 MPa for maximum shear strength.

Figures 6 and 7 show that with the satellite slowdown the M = 1 surface depth $z_{M=1}$ increases, the shear stress on the surface increases and reaches the maximum shear strength of 77 MPa at the depth of approximately 170 km, and then even exceeds this value. That is why the upper layer will start to shift along the lower layer at approximately 170 km horizon.

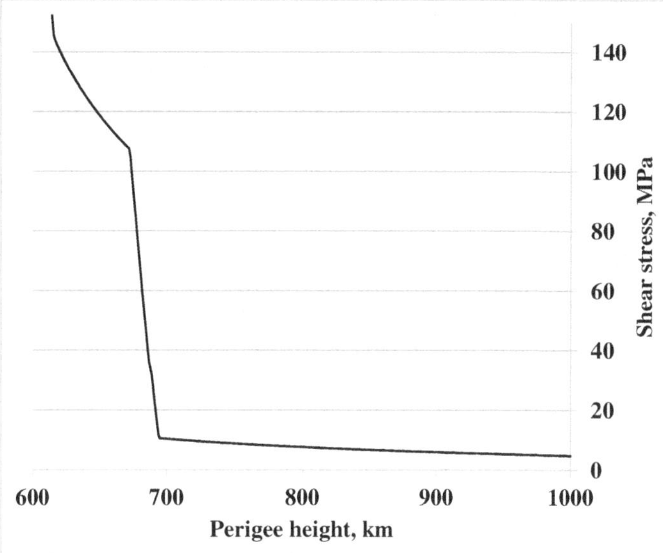

Fig. 7. Dependence of shear stress at M = 1 surface on the satellite perigee height.

4 Formation of the Horizontal Low Velocity Zone of Seismic Waves and Seismic Boundary of 220 km

Due to a high pressure during shifting a thin layer of mantle material will start to heat in the friction area, where melting occurs and upper layer of the mantle starts to slide along the thin melt. The value V_P decreases. Therefore with every circle of the satellite orbit the M = 1 surface gradually lowers and leads to an abrupt jump of velocity V_P, which gradually goes down. Finally at the height of satellite orbit perigee of approximately 618 km, which coincides with $z_{M=1} = 220\,\text{km}$, gliding ceases and the satellite promptly submerges.

Couette flow appears in the melt layer formed due to the friction. The heat produced due to viscous dissipation goes for melting of the mantle material, so the melt layer thickness grows and the melt temperature remains close to the melting points. This explains the well-known fact that at the up-to-date moment (after solidification of the melted layers in the upper mantle above the horizon of 220 km), their temperature is close to melting point.

Therefore, as the effect of tidal wave caused by gliding satellite in the temperature and density fields there remains a trait, proving the existence of low velocity zone of seismic waves in the mantle and of a jump, corresponding to the seismic boundary of 220 km.

Figures 6 and 7 show that the revealed gradual deepening of M = 1 surface and formation of seismic boundary of 220 km will occur even with less values of shear stress. That is why the conclusions made are still valid despite the shear stress for olivine depends on temperature and its' value in mantle of the Earth should be less that values measured at the room temperature.

Let us note that according to the data [4] (see Fig. 8) at the depth of approximately 127 km there is a local minimum of V_P. According to Fig. 7 around this depth the shear

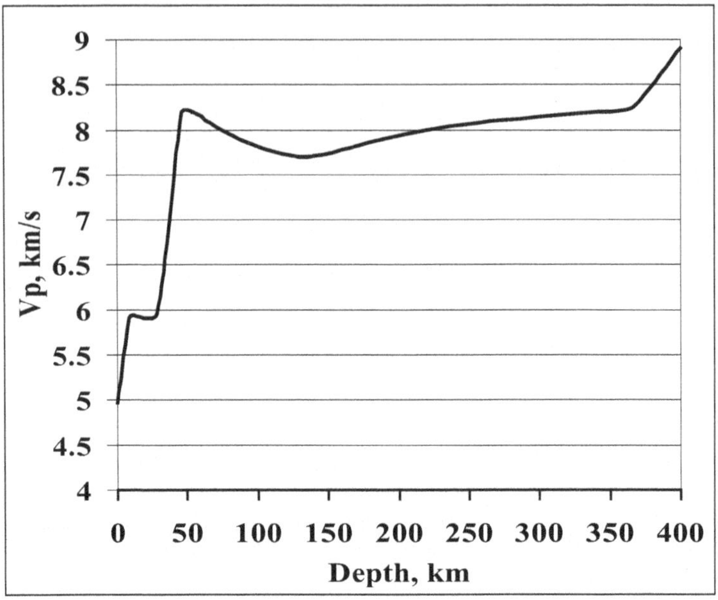

Fig. 8. Velocity of longitudinal seismic waves V_P according to [4].

stress values reach 62 MPa, which is close to the minimum of range used for measurement of character value of shear stress of 61.6 MPa.

So the existence of local minimum velocities of seismic waves V_P near horizon of 127 km (see Fig. 8) is also explained by shift of layers located above the horizon of 127 km in relation to the lower layer; by heating of a thin layer, where the sliding occurs, and gradual lowering of this layer at each next circle of the satellite orbit. At the same time the commencement of the upper layer sliding in the horizon of 127 km, when the melt, decreasing the friction, has not appeared yet, was accompanied by inconsiderable energy release. This is why a thin layer with the highest temperature and the lowest velocities of longitudinal seismic waves remained near the horizon of 127 km, shown in Fig. 8.

5 Formation of Inhomogeneities with Vertical Boundaries in the Mantle

Pressure of the ide bulge can also lead to vertical shear of the mantle layer with height of 2,800 km from the liquid core to the day surface.

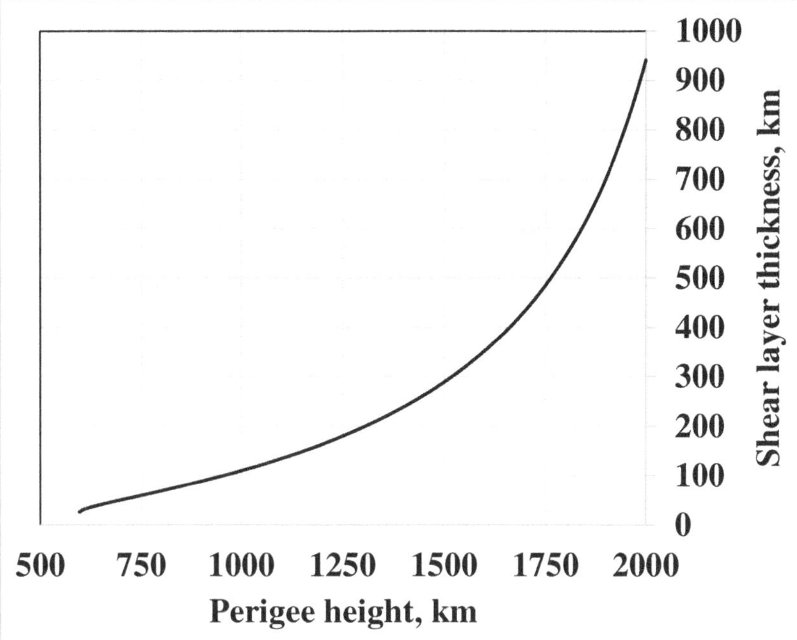

Fig. 9. Thickness of vertical shear layer.

The width of the moving layer (see Fig. 9) is determined by the tide height h_p and decreases with the reduction of orbit perigee height.

In particular at the height of perigee 618 km, corresponding to the position of M = 1 surface at the seismic boundary of 220 km ($z_{M=1} = 220\,\text{km}$), the width of the vertical shear layer is maximum and is equal to 34.3 km. With further lowering of the orbit a full submergence of the satellite happens. When the layers of the mantle move vertically, due to a high pressure the thin friction layer is heated and forms a thin melt film. As the result of this process, small inhomogeneities with vertically oriented boundaries are formed in the mantle.

6 Conclusion

Therefore, under the multiple impact of tidal wave caused by a gliding satellite the mantle was divided into blocks at the boundaries of which there were layers of thin melt substance which later where solidified.

The result of this is formation of low velocity zone of seismic waves in the mantle and formation of the seismic boundary at the depth 220 km (Lehmann discontinuity).

The effect of tidal wave caused by gliding satellite could lead to modification of initially existed profile of seismic wave velocities in the upper mantle, monotonically grown with the depth, into modern profile with local minimum of velocities in asthenosphere and jump of seismic waves velocities at the seismic boundary of 220 km.

Acknowledgement. The work was done with the support of Grant RFBR No. 16-01-00466.

References

1. Kasyanov, S.Y.: Model of the orbit of a large asteroid moving down and into the lithosphere and mantle of the earth. Phys. Issues Environ. (Environ. Phys.) **18**, 151–164 (2012)
2. Panteleyev, V.L.: Earth and Planet Physics. A Course of Lectures. Lomonosov Moscow State University, Moscow (2001)
3. Dziewonski, A.M., Anderson, D.L.: Preliminary reference Earth model. Phys. Earth Plan. Int. **25**, 297–356 (1981)
4. Dobretsov, N.L., Kirdyashkin, A.G., Kirdyashkin, A.A.: Deep-Level Geodynamics, Novosibirsk, 409 p. (2001)

Influence of Baroclinicity on Sea Level Oscillations in the Baltic Sea

Evgeny Zakharchuk[1,2(✉)], Natalia Tikhonova[1,2], Anatoly Gusev[4,5], and Nikolay Diansky[3,4]

[1] Saint Petersburg State University, Universitetskaya Emb. 13B, 199034 Saint-Petersburg, Russia
eazakharchuk@yandex.ru
[2] Saint-Petersburg Branch of N.N.Zubov State Oceanographic Institute, Beringa Street 38, 199397 Saint-Petersburg, Russia
[3] N.N. Zubov State Oceanographic Institute, Kropotkinsky Lane 6, 119034 Moscow, Russia
[4] Institute of Numerical Mathematics of the Russian Academy of Sciences, Gubkina Street 8, 119333 Moscow, Russia
[5] P.P. Shirshov Institute of Oceanology of the Russian Academy of Sciences, Nakhimovsky Avenue 36, 117997 Moscow, Russia

Abstract. On the basis of numerical experiments with the ocean model INMOM adapted for the Baltic Sea conditions, the influence of baroclinic processes on sea level oscillations is investigated. It is shown that baroclinic perturbations make a significant contribution to the total Baltic Sea level oscillations. Baroclinic effects have the dominate impact on formation of the mean sea level. The spectral analysis testifies the most considerable contribution of baroclinic fluctuations in the ranges of seasonal and mesoscale variability. The highest amplitudes of sea level baroclinic perturbations are noted in eastern part of the Gulf of Finland where they reach +30 cm, as well as in Bay of Bothnia and Gulf of Riga (+20 : 25 cm). The greatest intensity of the baroclinic sea level oscillations is noted during the autumn and winter period in the local regions of open Baltic, the Bay of Bothnia, eastern part of the Gulf of Finland, Gulf of Riga, as well as the Kattegat and the Danish Straits.

Keywords: Baltic sea · Sea level · Baroclinic effects

1 Introduction

Baroclinic sea level oscillations are generated and evolve in a stratified medium only. Their spatio-temporal variability is associated largely with variability of water density formed by buoyancy (heat and salt) fluxes. Vertical stratification and horizontal heterogeneity of the density fields in seas and oceans lead to generation of baroclinic modes in different types of wave motion [1, 14, 16]. Comparison of low-frequency level oscillation characteristics in the open Baltic Sea and Gulf of Finland, revealed by spectral analysis of tide-gauge and satellite altimetry data, with theoretical dispersion relations for different types of low-frequency waves showed that in the synoptic range

V. Karev et al. (Eds.): PMMEEP 2017, SPRINGERGEOL, pp. 371–380, 2018.
https://doi.org/10.1007/978-3-319-77788-7_38

of scales they can be identified as either barotropic or baroclinic topographic Rossby waves [19, 20]. Still unanswered is a question about relative contribution of wave motion barotropic and/or baroclinic components to the total sea level oscillations at the synoptic scale.

As regards the mesoscale range (periods from hours to days), the ADCP measurements show that during propagation of the low-frequency waves forming dangerous level rises in eastern part of the Gulf of Finland, mesoscale fluctuations of currents with a period of about 1 day do contain a well pronounced baroclinic component [21], while it is not clear how such a component relates to the sea level oscillations.

Yet poorly studied are the issues of estimating the baroclinic fluctuation magnitude, influence of baroclinicity on the mean Baltic Sea level, the distribution of baroclinic effects across individual spectral frequency bands, as well of the respective dispersion spatio-temporal patterns.

In this article, by comparing the results of two numerical experiments with a hydrodynamic model and their statistical analysis, the contribution of baroclinic component to the spatio-temporal variability of the Baltic Sea level oscillations is investigated.

2 Description of the Model and Numerical Experiments

In order to describe the baroclinic effects in the Baltic Sea level variability formation, the two numerical experiments on reproducing water dynamics were performed. The first experiment, aimed to estimate sea level fluctuations in the baroclinic sea, used the 3D baroclinic nonlinear hydrodynamic model. The latter is a version of the oceanic and marine circulation σ-model developed in the INM RAS and internationally referred to as INMOM (Institute of Numerical Mathematics Ocean Model) [6].

The INMOM is based on the complete set of nonlinear ocean hydrothermodynamics primitive equations in spherical coordinates with the hydrostatic and Boussinesq approximations. The dimensionless variable $\sigma = (z - \zeta)/(H - \zeta)$ is used as a vertical coordinate, where $\zeta = \zeta(\lambda, \phi, t)$ is the sea level deviation from the undisturbed surface; $H = H(\lambda, \phi)$ is bottom topography depth. Model variables are horizontal components of circulation velocity, potential temperature, salinity and sea level deviation from the undisturbed state. To compute ocean density, the state equation is used taking into account sea water compressibility, which is specially designed for numerical models [4].

The INMOM includes the sea ice dynamics and thermodynamics model consisting of 3 modules. The thermodynamics module [18] describes the freezing of ice, the snow fallout and its further transformation into ice, as well as melting of snow and ice due to thermal processes. The ice dynamics module computed ice drift velocity, which evolves due to wind, oceanic surface currents, Earth diurnal rotation, the sea level slope and internal ice sheet interaction described by elastic-viscous-plastic rheology [8]. The ice transport module describes evolution of the ice and snow due to drift and uses a monotonic transport scheme [3] to ensure non-negativity of ice-snow concentration and mass.

The main feature of the INMOM, which distinguishes it from other known ocean models such as MOM (Modular Ocean Model) [13], INMIO (Institute of Numerical Mathematics and Institute of Oceanology ocean model) [9, 10], POM (Princeton Ocean Model) [2] and ROMS (Regional Ocean Modelling System) [15] consists in using the method of splitting into physical processes and spatial coordinates by its numerical implementation.

For the Baltic Sea, the INMOM was implemented on the area with grid step of 2 miles. The extension of the area is from 9.375 °E to 30.375 °E on longitude and from 53.625 °N to 65.9375 °N on latitude. In vertical, it contains 25 unevenly distributed σ-levels.

The original data representing the Baltic Sea topography with resolution $2' \times 1'$ (http://nest.su.se) were smoothed to eliminate their local features, and interpolated to the model grid.

To set the initial state and open lateral boundary conditions we used hydrographical data collected for the project Copernicus [5]. They include monthly mean fields of hydrophysical characteristics of the Baltic Sea (water temperature, salinity, horizontal velocity) with resolution 5 m on depth and horizontal resolution 5.6 km for 1990–2009, obtained by assimilating the ship and satellite data in the 3D baroclinic hydrodynamic ocean model Hiromb-BOOS-Model (HBM-V1).

On the solid lateral boundary, heat and salt fluxes are taken to be zero, while for velocities no normal flow and free slip conditions are set.

At the open boundary in the Kattegat passage, along 57°44'N in buffer zone of 14.8 km width, were set climatic monthly mean values of temperature and salinity, while the sea level were taken from the tide-gauge hourly measurements at stations Aarhus (56°09′N, 10°13′E) and Sjaellands Odde (55°58′30″N, 11°22′20″E) (http://boos.org), located in the southern Kattegat, at the entrance to the Danish Straits. The sea level data were interpolated to the grid points inside the buffer zone along the whole open boundary. Free flow condition for velocity was applied there. Prescribing density (in terms of temperature and salinity) and sea level on the open boundary implies "spontaneous" implementation of a combined condition for setting both the gradient velocities and, like an overlay, a longwave component of the water dynamics induced by the sea level spatio-temporal variability. Meanwhile, the available data do not allow one to accurately estimate the absolute sea level slope in the Kattegat. Thus, the mean sea level values for the years 2009–2010 at stations Aarhus and Sjaellands Odde were subtracted, and therefore the mean water exchange through the Kattegat, caused by the sea level slope, becomes zero. Hence, the model takes into account only water exchange anomalies from the mean, which, in other case, should be close to zero.

Changes of the seawater salinity by freshwaters from 29 rivers of the Baltic basin were estimated using data on the monthly mean discharge of these rivers.

The sea surface boundary conditions in the INMOM atmosphere module were set using the following meteocharacteristics taken from the array CFSR (NOAA) with spatial resolution of 0.5°: air temperature and humidity (2 m above sea surface), sea level pressure, wind speed (10 m above sea surface), downwelling shortwave and longwave radiation, and precipitation.

Numerical simulations of the Baltic Sea circulation were performed for the period from January 1, 2009 to December 31, 2010 with time step 2.5 min. Sea level data output was done once per 6 h of model time.

The assessment of the Baltic Sea level oscillations performed with the 3D baroclinic model INMOM in the first numerical experiment contain the total contribution of both barotropic and baroclinic components to the sea level change. To highlight the baroclinic component, we need to subtract its barotropic part from the total one obtained with the full 3D model. To this end, we performed one more numerical experiment to assess the Baltic Sea barotropic sea level oscillations. The run was made using the 3D barotropic model, which is the same model INMOM having a constant water density everywhere and every time during the entire interval of integration. Boundary conditions were the same as in the first numerical experiment, but with account of barotropic run.

Finally, sea level baroclinic component was obtained by subtracting the 2nd barotropic experiment's results from corresponding total ones obtained in the 1st experiment.

3 Comparative Statistical Analysis of the Results of Two Numerical Experiments

Table 1 summarizes the results of comparison between the sea level oscillations obtained in the two experiments and ones measured at several tide-gauge coastal stations in the Baltic Sea. It is clearly seen that the sea level oscillations simulated in the two experiments are rather well agreed with their instrumental measured data: the correlation coefficients between them vary from 0.67 to 0.88, although in some barotropic simulations there is a slight decrease of correlation by 0.02 ÷ 0.04.

Figure 1 illustrates the averaged for 2009–2010, simulated Baltic Sea level fields, centered with the mean value, to display: (1a) joint contribution from its baroclinic and barotropic components, and (1b,c) individual contributions from these components. The centering procedure is necessary because there is no accurate sea level estimate possible if there is no account of exact water balance for the entire Baltic Sea area with influence of the river runoff, the water exchange across the open boundary, precipitation and evaporation. Even contemporary datasets do not allow one to take into account all this factors. In addition, the method used in this work to allow for river runoff in terms of their influence on the salinity regime, the above mentioned account of water exchange with the North Sea through the Kattegat, the accuracy of precipitation and evaporation setting with rather deficient and coarse meteorological data prevents from accurate determination of the Baltic Sea water balance.

Moreover, the available meteorological data make it impossible to ensure the Baltic Sea heat balance, while the imbalance can lead to errors in estimating the mean sea level.

However, it is just the average water level of the Baltic Sea to which the above remarks concern. Thus, it must be subtracted from the results of calculation.

In the average field of the total sea level elevation (Fig. 1a) there is clearly expressed sea level slope, well-known from the results of the analysis of longterm

Table 1. Coefficients of the maximum correlation between data series of the sea level oscillations measured on coastal tide-gauge stations in the Baltic Sea and the ones simulated in the scope of the two numerical experiments.

Tide-gauge stations	Baroclinic model	Barotropic model
Warnemünde	0.76	0.73
Kangzholmsfort	0.71	0.71
Klaipeda	0.81	0.77
Daugava	0.86	0.86
Helsinki	0.85	0.85
Kronstadt	0.81	0.81
Spikarna	0.80	0.80
Kemi	0.88	0.88
Degerby	0.77	0.77
Calix	0.87	0.87
Kaskinen	0.84	0.82
Colca	0.84	0.84
Sillamae	0.88	0.88
Tyne	0.73	0.73

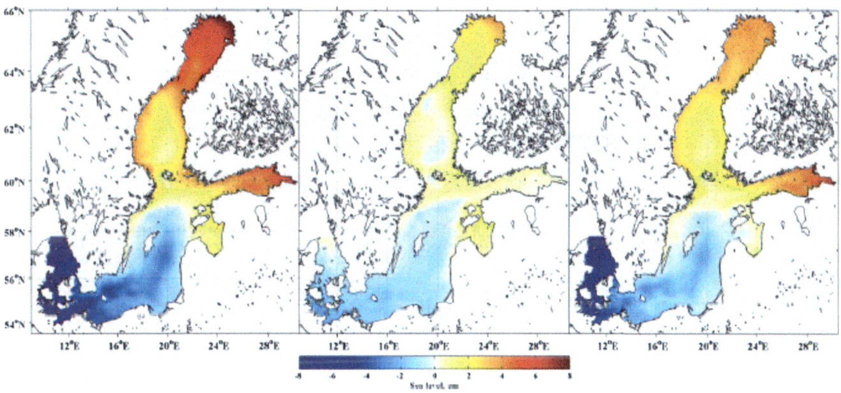

Fig. 1. Averaged for the years 2009–2010, centered Baltic sea level fields from numerical experiments with INMOM: (a) total contribution of baroclinic and barotropic components; (b) only barotropic component, and (c) only baroclinic component.

tide-gauge observations of sea level [7, 11, 17]: the sea level gets higher from −8.6 cm in the Kattegat up to +8.4 cm towards the head of the Gulf of Bothnia or the Gulf of Finland. Figure 1c shows that the sea level slope is mostly formed by baroclinic component, with the effect of sea level rise by 14.3 cm if moving from the Danish Straits toward the Gulf of Finland or the Gulf of Bothnia. Barotropic component contribution to the mean field of the Baltic Sea level is not significant, at most from −2.8 to +4.2 cm.

Figure 2 shows maximum sea level in baroclinic simulation. Its estimates become greater from the southwest to the north-northeast of the sea. The largest amplitudes of baroclinic sea level disturbance are observed in the east of the Gulf of Finland where they reach 30 cm. In the central and western Gulf of Finland, as well in the Gulf of Bothnia and the Gulf of Riga the largest baroclinicity effects reach up to 20 ÷ 25 cm; in the open Baltic Sea 10 ÷ 20 cm; in the Kattegat and the Danish Straits 5 ÷ 12 cm.

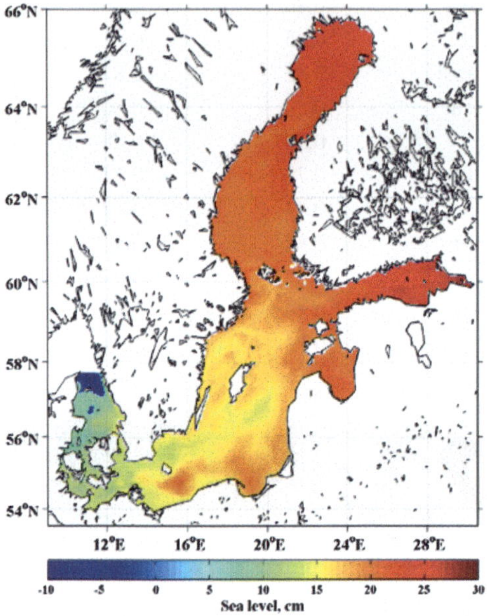

Fig. 2. Values of the maximum sea level elevation in baroclinic fluctuations (cm).

For investigation of the seasonal variability of baroclinic dynamics, their variance was calculated for the four seasons (Fig. 3). A noticeable increase of the baroclinic variance is observed for the winter and autumn in the north of the Gulf of Bothnia, in the east of the Gulf of Finland, locally in the open Baltic Sea, in the Gulf of Riga, as well in the Kattegat and the Danish Straits.

To answer the question about a frequency range in which baroclinic effects on the sea level oscillations are especially strong, we estimated for various places of the sea, and compared baroclinic spectra based on the results of the two numerical experiments – see Fig. 3. In every of these places, in the period range from several days to 2 weeks, there is no noticeable differences in spectral densities of either total or barotropic sea level oscillations, which evidences their barotropic nature in the considered range of time scales. In all the cases, significant differences were observed in spectral estimates for the seasonal range of oscillations. In addition, significant differences are also present in the mesoscale range (from 12 h to several days) in the Gulf of Bothnia and open Baltic Sea. In the Gulf of Finland and in the southwestern sea, baroclinic effect on the sea level oscillations is weak (see Fig. 3b and d).

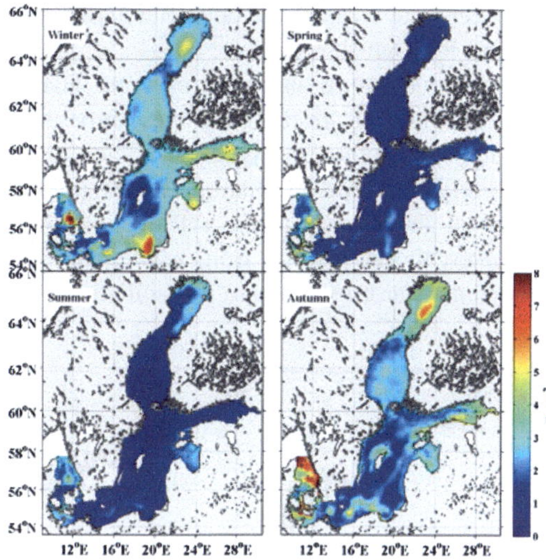

Fig. 3. Variance of baroclinic sea level fluctuations in the Baltic Sea, estimated for 4 seasons: (a) December, 2009 – February, 2010; (b) March – May, 2010; (c) June – August, 2010; (d) September – November, 2010.

4 Conclusions

The comparative analysis of the statistical characteristics of the sea level oscillations obtained in the two numerical experiments shows one a significant contribution of baroclinic effects to formation of the Baltic Sea level. Comparison of the mean fields of the total and barotropic sea level oscillations illustrates that quasi-stationary response of the Baltic Sea level to the impact of external and internal forces is predominantly baroclinic. It is the baroclinic component that has a predominant influence on the total mean sea level slope formation, from -8.6 cm in the southwest sea to +8.4 cm in the Gulfs of Bothnia and of Finland. The reason is the considerable variability of the Baltic Sea density formed due to specific features of the water balance: in the Gulf of Bothnia and the Gulf of Finland, the water density is low because of freshening by terrestrial runoff, but the southwestern sea has much denser waters as it is often influenced by salty waters inflow from the North Sea. Influence of the barotropic component on the formation of the mean Baltic sea level slope is much less: the barotropically induced sea level is in a range −2.8 to +4.2 cm.

Distribution of the maximum instantaneous baroclinic sea level values is non-uniform: the largest amplitudes, up to 30 cm, are in the eastern Gulf of Finland, in the central and western Gulf of Finland, as well in the Gulf of Bothnia and in the Gulf of Riga, the largest baroclinic sea level values reach 20 ÷ 25 cm; in the open Baltic 10 ÷ 20 cm; in the Kattegat and in the Danish Straits ±5 ÷ 12 cm.

The non-stationarity of baroclinic sea level dispersion illustrated at Fig. 3, demonstrates that the noticeable increase of the baroclinic sea level dispersion is observed during the autumn-winter time in the north of the Gulf of Bothnia, in the east of the Gulf of Finland, in separate areas of the open Baltic Sea, in the Gulf of Riga, as well as in the Kattegat and the Danish Straits.

Comparison of spectra for the total and barotropic sea level oscillations (Fig. 4) reveals an appreciable difference in spectral densities throughout the seasonal and mesoscale range of variability. In other words, these ranges contain the largest baroclinic contribution to the total sea level oscillation. The smallest differences between spectral densities of the total and the barotropic sea level oscillations are observed in the synoptic range of timescale, at periods from several days to 2 weeks. This is indicative for a predominantly barotropic nature of oscillations in this range.

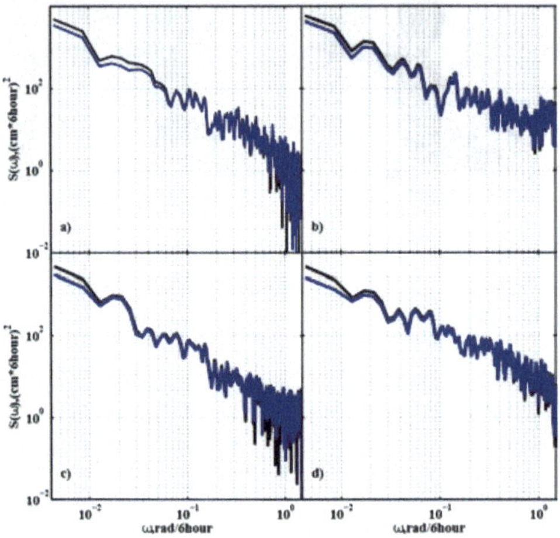

Fig. 4. Comparison of the spectra for simulated baroclinic (black) and barotropic (grey) sea level oscillations in: (a) the Gulf of Bothnia, (b) the Gulf of Finland, (c) the open Central Baltic Sea, and (d) the southwestern Baltic Sea.

In [12], the spectra of tide-gauge measured sea level data were compared with the ones obtained with the barotropic model (not allowing for the water exchange with the North Sea). The compared spectra do not coincide considerably with each other in the seasonal and mesoscale ranges of sea level oscillations. To account for this for the seasonal scale, the author refers to the neglect of the water exchange with the North Sea, and for the mesoscale range to inexact reproduction by the model of reanalysis of meteorological fields of short-period fluctuations of wind and air pressure [14]. Our results of comparative statistical analysis of the sea level oscillations estimated with numerical simulations indicate that the spectral differences, noted in [12], concerned with neglecting the impact of baroclinic effects when modelling the Baltic Sea level oscillations.

Thus, to sum up the results of our study, it can be said that baroclinic perturbations contribute significantly to the overall Baltic Sea level oscillations. The most noticeable baroclinic impact is in the seasonal and mesoscale ranges of variability, as well as in the formation of the mean sea level.

The research was supported by the Russian Foundation for Basic Research (grants no. 16-05-00534, no. 18-05-01107) and the Saint-Petersburg State University (agreement no. 18.37.140.2014).

References

1. Belonenko, T., Zakharchuk, E., Fuks, V.: Gradient-vorticity waves in the ocean. Saint-Petersburg State University (2004)
2. Blumberg, A., Mellor, G.: A Description of a Three-Dimensional Coastal Ocean Circulation Model, pp. 1–16. American Geophysical Union (2013)
3. Briegleb, B., Bitz, C., Hunke, E., Lipscomb, W., Holland, M., Schramm, J., Moritz, R., et al.: Scientific description of the sea ice component in the community climate system model, version three (2004)
4. Brydon, D., Sun, S., Bleck, R.: A new approximation of the equation of state for seawater, suitable for numerical ocean models. J. Geophys. Res. Oceans **104**(C1), 1537–1540 (1999)
5. Copernicus, Marine Environment Monitoring Service. http://marine.copernicus.eu. Accessed 30 Nov 2017
6. Diansky, N.: Modelling of the ocean circulation and investigation of its responce to the short-range and long-range atmospheric forcing. Physmatlit, Moscow (2013)
7. Hünicke, B., Zorita, E., Soomere, T., Madsen, K., Johansson, M., Suursaar, Ü.: Recent change – sea level and wind waves. In: T.B.I.A. Team (ed.) Second Assessment of Climate Change for the Baltic Sea Basin, pp. 155–185. Springer, Cham (2015)
8. Hunke, E., Dukowicz, J.: An elastic-viscous-plastic model for sea ice dynamics. J. Phys. Oceanogr. **27**(9), 1849–1867 (1997)
9. Ibrayev, R.: Model of enclosed and semi-enclosed sea hydrodynamics. Russ. J. Numer. Anal. Math. Modell. **16**(4), 291–304 (2001)
10. Ibrayev, R., Khabeev, R., Ushakov, K.: Eddy-resolving 1/10° model of the world ocean. Izv. Atmos. Oceanic Phys. **48**(1), 37–46 (2012)
11. Leppäranta, M., Myrberg, K.: Physical Oceanography of the Baltic Sea. Springer, Heidelberg (2009)
12. Medvedev, I.: Spectrum of the Baltic Sea level variability in the range of transitions from hours to days. Ph.D. thesis, IO RAS (2014)
13. Pacanowski, R., Griffies, S.: Mom 3.0 manual. GFDL Ocean Group Technical Report 4, 680 (1999)
14. Pedlosky, J.: Geophysical Fluid Dynamics. Springer, Heidelberg (1982)
15. Shchepetkin, A., McWilliams, J.: Accurate boussinesq oceanic modeling with a practical "stiffened" equation of state. Ocean Model. **38**(1), 41–70 (2011)
16. Tareev, B.: Dynamics of baroclnic perturbations in the ocean. Moscow State University (1974)
17. Terziev, F., Rozhkov, V., Smirnova, A. (eds.): Project "Seas of USSR". Hydrometeorology and hydrochemistry of USSR's seas, vol. III, Baltic Sea, Issue I. Hydrometeorological conditions. Hydrometeoizdat, Saint-Petersburg (1992)
18. Yakovlev, N.: Reproduction of the large-scale state of water and sea ice in the Arctic Ocean in 1948–2002: part I. numerical model. Izv. Atmos. Oceanic Phys. **45**(3), 357–371 (2009)
19. Zakharchuk, E. (ed.): Dynamics of the Baltic Sea waters in synoptic range of spatio-temporal scales. Hydrometeoizdat, Saint-Petersburg (2007)

20. Zakharchuk, E., Klevantsov, Y., Tikhonova, N.A.: Space-time structure and identification of synoptic disturbances of the Baltic Sea level from satellite altimetric measurements. Russ. Meteorol. Hydrol. **31**(5), 53–60 (2006)
21. Zakharchuk, E., Sukhachev, V.: On the problem of the Neva river flood waves identification. Russ. Meteorol. Hydrol. **38**(3), 185–190 (2013)

Author Index

V. Karev et al. (Eds.): PMMEEP 2017, SPRINGERGEOL, pp. 381–382, 2018.
https://doi.org/10.1007/978-3-319-77788-7

The manufacturer's authorised representative in the EU is Springer Nature Customer Service Centre GmbH, Europaplatz 3, 69115 Heidelberg, Germany. If you have any concerns regarding our products, please contact ProductSafety@springernature.com

Printed and bound by CPI Group (UK) Ltd, Croydon, CR0 4YY
05/10/2026
02228139-0016